Ethical Issues in AI for Bioinformatics and Chemoinformatics

This unique volume presents AI in relation to ethical points of view in handling big data sets. Issues such as algorithmic biases, discrimination for specific patterns, and privacy breaches may sometimes be skewed to affect research results so that certain fields appear more appealing to funding agencies. The discussion on the ethics of AI is highly complex due to the involvement of many international stakeholders such as the UN, OECD, parliaments, industry groups, professional bodies, and individual companies. The issue of reliability is addressed, including the emergence of synthetic life, 5G networks, intermingling of human artificial intelligence, nano-robots, and cyber security tools.

Features

- Discusses artificial intelligence and ethics, the challenges and opportunities
- Presents the issue of reliability in the emergence of synthetic life, 5G networks, intermingling of human artificial intelligence, nano-robots, and cyber security tools
- Examines ethical responsibility and reasoning for using AI in big data
- Addresses practicing medicine and ethical issues when applying artificial intelligence

Ethical Issues in AI for Bioinformatics and Chemoinformatics

Edited By
Yashwant Pathak, Surovi Saikia, Sarvadaman Pathak,
Jayvadan Patel and Jigna Prajapati

CRC Press
Taylor & Francis Group
Boca Raton London New York

CRC Press is an imprint of the
Taylor & Francis Group, an **informa** business

First edition published 2024
by CRC Press
2385 Executive Center Drive, Suite 320, Boca Raton FL 33431

and by CRC Press
4 Park Square, Milton Park, Abingdon, Oxon, OX14 4RN

© 2024 selection and editorial matter, **Yashwant Pathak, Surovi Saikia, Sarvadaman Pathak, Jayvadan Patel and Jigna Prajapati**; individual chapters, the contributors

CRC Press is an imprint of Taylor & Francis Group, LLC

Library of Congress Cataloging-in-Publication Data
Names: Pathak, Yashwant, editor. | Saikia, Surovi, editor. | Pathak,
Sarvadaman, editor. | Patel, Jayvadan K., editor. | Prajapati, Jigna,
editor.
Title: Ethical issues in AI for bioinformatics and chemoinformatics /
edited by Yashwant Pathak, Professor and Associate Dean for Faculty
Affairs, College of Pharmacy, University of South Florida, Surovi
Saikia, Translation Research Laboratory, Department of Biotechnology,
Bharathiar University, Coimbatore, India, Sarvadaman Pathak, Chief
Medical Officer, HIPAA compliance officer, Universiti LLC, Tampa, USA,
Jayvadan Patel, Formulation Scientist at Aavis Pharmaceuticals, USA and
Professor Emeritus, Faculty of Pharmacy, Sankalchand University, India,
Jigna Prajapati, Ganpat University, Mehsana, Gujarat, India.
Description: First edition. | Boca Raton : CRC Press, 2024. | Series: Drugs
and the pharmaceutical sciences | Includes bibliographical references
and index. | Summary: "This unique volume presents AI in relation to
ethical points of view in handling big data sets. Issues such as
algorithmic biases, discrimination for specific patterns and privacy
breaches may sometimes be skewed to affect research results so that
certain fields to appear more appealing to funding agencies. The
discussion on the ethics of AI is highly complex due to the involvement
of many international stakeholders such as the UN, OECD, parliaments,
industry groups, professional bodies, and individual companies. The
issue of reliability is addressed including the emergence of synthetic
life, 5G networks, intermingling of human artificial intelligence,
nano-robots and cyber security tools"-- Provided by publisher.
Identifiers: LCCN 2023025067 (print) | LCCN 2023025068 (ebook) | ISBN
9781032396583 (hb) | ISBN 9781032405827 (pb) | ISBN 9781003353751 (eb)
Subjects: LCSH: Artificial intelligence--Medical applications--Moral and
ethical aspects. | Cheminformatics--Moral and ethical aspects. |
Bioinformatics--Moral and ethical aspects. | Artificial
intelligence--Moral and ethical aspects.
Classification: LCC R859.7.A78 E84 2024 (print) | LCC R859.7.A78 (ebook)
| DDC 610.285/63--dc23/eng/20230828
LC record available at https://lccn.loc.gov/2023025067
LC ebook record available at https://lccn.loc.gov/2023025068

ISBN: 978-1-032-3-9658-3 (HB)
ISBN: 978-1-032-4-0582-7 (PB)
ISBN: 978-1-003-3-5375-1 (EB)

DOI: 10.1201/9781003353751

Typeset in Times
by Deanta Global Publishing Services, Chennai, India

Contents

Preface

The book deals with the "not so good" part of AI, relating to ethical point of view in handling big data sets. Issues such as algorithmic biases, discrimination for specific patterns, privacy breaches, etc. may sometimes go beyond immediate effects on research to skew certain research fields to appear more appealing for funding agencies. The discussion on the ethics of AI is highly complex due to the involvement of many international stakeholders such as the UN, OECD, parliaments, industry groups, professional bodies, individual companies, etc.

The chapters describe how ethical issues are involved in handling big data and what it costs for being responsible with reasoning. The ethical reasons that exist especially cannot be solved in an isolated forum but will emerge from the existing way of procedure only if all the "not so good" are not made to "appear" good and are presented as they are, such as in the context of Machine Learning, prediction of organic synthesis etc., where it solely depends on the kind of data fed to the train the model.

The issue of reliability also needs to be addressed, which always takes a backseat in the context of AI. The emergence of synthetic life, 5G networks, intermingling of human artificial intelligence, nano-robots, cyber security tools, etc. are very fascinating and at the same time can be quite hazardous. To be aware of such hazardous breaches and how they may unknowingly harm humanity, we included some chapters relating to the enforcement of international law in this regard so that the readers are aware of what they can do and should not do. This book examines the scientific viewpoint on AI and the interplay of law, how carrying out such research may potentially breach ethical ground unintentionally, so we expect this book will be very helpful for younger research community who are on the brink of dealing with AI in every aspect of research in the near future.

MATLAB® is a registered trademark of The MathWorks, Inc. For product information, please contact:

The MathWorks, Inc.
3 Apple Hill Drive
Natick, MA 01760-2098 USA
Tel: 508 647 7000
Fax: 508-647-7001
E-mail: info@mathworks.com
Web: www.mathworks.com

Editor Biographies

Dr. Yashwant Pathak has over 9 years of versatile administrative experience in an academic setting as Dean (and over 17 years as faculty and as a researcher in higher education after his Ph.D) and has worked in student admission, academic affairs, research, graduate programs, Chair of the International Working group for USF Health (consisting of USF Medical, Nursing, Public Health, Pharmacy and Physiotherapy colleges), Director of Nano Center and now holds the position of Associate Dean for Faculty Affairs. Dr. Yash Pathak is an internationally recognized scholar, researcher, and educator in the areas of health care education, nanotechnology, drug delivery systems, and nutraceuticals. Dr Pathak has edited several books in the field of nanotechnology, drug delivery systems, antibody mediated drug delivery systems, and in the field of nutraceuticals, artificial neural networks, and conflict resolution and cultural studies, with over 300 research publications, reviews, chapters, abstracts, and articles in educational research. He has authored many popular articles published in newspapers and magazines. He fluently speaks many languages and was trained and taught in English all through his career.

Dr. Surovi Saikia works as a Dr. DS Kothari Postdoctoral Fellow at the Department of Biotechnology, Bharathiar University, Coimbatore. She completed her Ph.D (Bioinformatics) from Dibrugarh University, Assam in 2020 and UG and PG from Tinsukia college and Dibrugarh University. She has 8 years of research experience with 18 publications in peer reviewed national and international journals. Currently, she is working on algorithmic based computational work.

Research Experience

- Dr. D.S Kothari Postdoctoral Fellow at Bharathiar University, Coimbatore, Tamil Nadu from 23/11/2021 to present.
- Senior Research Fellow-NFST Fellowship for PhD at CSIR-NEIST, Natural Product Chemistry Group, Jorhat, Assam from 01/06/2018 to 01/10/2020.
- Junior Research Fellow-NFST Fellowship for PhD at CSIR-NEIST, Natural Product Chemistry Group Jorhat, Assam from 01/06/2016 to 31/05/2018.
- Project Fellow- at CSIR-NEIST, Natural Product Chemistry Group, Jorhat, Assam from 08/04/2013 to 31/05/2016.
- DBT-studentship at CSIR-NEIST, Biotechnology Division from 02/07/2012 to 02/01/2013.

Dr Sarvadaman Pathak: 2016–2017 Global Clinical Scholars Research and Training Program. Harvard Medical School, Boston MA. A master's level program which, provides clinicians and clinician-scientists advanced training in the methods and conduct of clinical research. The program also covers data analysis techniques using STATA, and access to datasets from various NIH trials.

2013–2014 Post-Doctoral Fellowship, in Traditional Chinese Medicine with a focus on Integration of Eastern and Western Medicine, Dalian Medical University, Dalian, P.R. China.

2007–2011 Doctor of Medicine (M.D.), Summa Cum Laude, Avalon University
School of Medicine, Bonaire, Netherlands Antilles (Caribbean Island)

2004–2007 Pre-Medicine/Pre-Pharmacy Track, University of Houston, Houston, TX.

Professional experience includes

2021–Present Chief Medical Officer, HIPAA compliance officer

Universiti LLC – A cloud-based healthcare provider focusing on behavioral and mental healthcare delivery using AR/VR and AI technology.

Dr. Jayvadan Patel is a Formulation Scientist at Aavis Pharmaceuticals, USA and Professor Emeritus, Faculty of Pharmacy, Sankalchand University, India. He has more than 26 years of research, academic, and industry experience. Dr. Jayvadan has published 270 research papers, presented 52 papers, both in India and abroad, and holds one patent jointly, and three are filed. He has co-authored 20 books and contributed 116 chapters to books published in the USA. His papers have been cited more than 5200 times, with more than 27 papers getting more than 50 citations each. He has a 35 h-index and 123 i10-index for his credit. Department of Science and Technology, Ministry of Science and Technology, Government of India has published International Comparative Research Base (2009–2014) and Dr. Patel has been put in an honors list of 4th Rank of the Top 10 authors in the subject area Pharmacology, Toxicology, and Pharmaceutics. His 16 projects, worth more than Rs. 2.2 Crore, are funded by government agencies and industries. He has guided 106 M. Pharm students and mentored 46 Ph.D. scholars and is presently guiding 8 more for their doctoral theses. He is already decorated with a dozen awards and prizes, both national and international. He is the recipient of the very prestigious "AICTE-Visvesvaraya Best Teachers Award-2020" by the All India Council for Technical Education, Government of India, and the APTI Young Pharmacy Teacher Award (2014) by the Association of Pharmaceutical Teachers of India. He is a member of many professional bodies/universities/councils and boards.

Dr. Jigna Prajapati has rich experience of 17 years in academia, research, and the IT industry, holding doctorate in Computer Science. She claims her name on more than thirty international and national publications. She has secured many best paper awards in international and national conferences in the research category. Besides assistant professor, she is Motivator, Online and Offline Trainer and counsellor, leading a team which involves work of infrastructure/academic planning and development, network design and management, examination, key administration and research. She is actively working in field of artificial intelligence and Machine Learning focusing on healthcare. Her expert domains are as artificial intelligence, machine learning, software engineering, and emerging technologies. She works on research projects that explore new techniques and methods in their area of Machine Learning and collaborates with other researchers both within their institution and around the world.

She has played an important role in educating the next generation of computer scientists. She has also mentored 5000+ students who are working on various projects, helping them to develop their skills and explore new areas of computer science. She has delivered 11 invited session/hands-on in various conferences, seminars, and workshops.

Area Of Interest: Software Fault Tolerance, Data mining, Artificial Intelligence, ML classification, AI in Healthcare

List of Contributors

Firdush Ahmed
i/c Salakati Police Station, Kathalguri No.2,
    Kokrajhar, Assam- 783376, Assam

Vishnu Prabhu Athilingam
Translational Research Laboratory
Department of Biotechnology
Bharathiar University
Tamil Nadu, India

Priyanka Banerjee
Adamas University
West Bengal, India

Allyson Lim D
Taneja College of Pharmacy
University of South Florida
Tampa, Florida

Pinalkumar Engineer
Sardar Vallabhbhai National Institute of
    Technology
Surat, Gujrat, India

Partha Pratim Kalita
Faculty of Science
Assam down town University
Assam, India

Adithya Kaushal
University of Vermont
Burlington, Vermont

Divya Kaushal
Columbia University
New York, New York

Satyen M. Parikh
Ganpat University
Mehsana, India

Meghna B. Patel
Faculty of Computer Application
Ganpat University
Gujarat, India

Ronak B. Patel
SRIMCA
Uka Tarsadia University
Maliba Campus
Gujarat, India

Jagruti Patel
Faculty of Computer Application
Ganpat University
Gujarat, India

Devangi Patel
Faculty of Computer Application
Ganpat University
Gujarat, India

Savan Patel
Faculty of Computer Application
Ganpat University
Gujarat, India

Roshani J Patel
Faculty of Computer Application
Ganpat University
Gujarat, India

Rina K Patel
Faculty of Computer Application
Ganpat University
Gujarat, India

Param Patel
College of Pharmacy
University of South Florida
Tampa, Florida

Yashwant Pathak
College of Pharmacy
University of South Florida
Tampa, Florida

Jigna Prajapati
Ganpat University
Gujarat, India

Prajesh Prajapati
School of Pharmacy
National Forensic Sciences University
Gujarat, India

Deep Prajapati
Thadomal Shahani Engineering College
Mumbai, India

Bhupendra Prajapati
Shree S.K. Patel College of Pharmaceutical
    Education and Research
Ganpat University
Gujarat, India

Sapna Rathod
Shree S.K. Patel College of Pharmaceutical
    Education and Research
Ganpat University
Gujarat, India

Surovi Saikia
Translational Research Laboratory
Department of Biotechnology
Bharathiar University
Tamil Nadu, India

Ankur Pan Saikia
Faculty of Engineering and Technology
Assam down town University
Assam, India

Manash Pratim Sarma
Faculty of Science
Assam down town University
Assam, India

Saptarshi Sanyal
Adamas University
West Bengal, India

Amartya Sen
Adamas University
West Bengal, India

Rakesh Sharma
Massachusetts General Hospital
Harvard Medical School
Boston, Massachusetts
and
IITR-GMC Academy
Government Medical College
Sagharanpur, India

Divya Sheth
College of Pharmacy
University of South Florida
Tampa, Florida

Radhika Sreedharan
Department of Computer Science and
    Engineering
Presidency University
Bangalore, India

1 Artificial Intelligence and Ethics
Challenges and Opportunities

Meghna B. Patel, Ronak B. Patel,
Satyen M. Parikh, and Jigna Prajapati

1.1 INTRODUCTION OF AI ETHICS

Artificial intelligence (AI) has advanced at a previously unheard-of rate in just the last ten years. Crop yields have already increased, business productivity has increased, access to financing has improved, and disease detection has become quicker and more accurate (1).

More significantly, AI becomes smarter the more we use it and the more data we produce. And as these systems improve in capability, our world gets richer and more productive (2).

By 2030, it might add 14% to global GDP and more than $15 trillion to the global economy. According to a study on the influence of artificial intelligence (AI) on the Sustainable Development Goals (SDGs) published in *Nature*, 134, or 79%, of all SDG targets may be enabled by AI (3).

However, just as AI has the ability to enhance the lives of billions of people, it may also duplicate, exacerbate, and create new issues.

Ethics includes moral principles that help us discern between right and wrong. The production and outcomes of artificial intelligence are governed by a set of principles known as AI ethics. Recency and confirmation bias are just two examples of the many cognitive biases that affect us and are reflected in our actions and, therefore, in our data. Since data serves as the basis for all machine learning algorithms, we required to keep this in mind while we design experiments and algorithms. These human biases can be amplified and scaled by artificial intelligence at a previously unheard-of rate.

As a direct consequence of the emergence of big data, businesses are putting in much more effort to automate processes and implement data-driven decision-making throughout their entire operations. Even though the goal is frequently, if not always, to improve commercial outcomes, businesses are seeing unexpected implications from some of their AI applications. This is particularly the case as a result of poor upfront study design and biased datasets (4).

In response to ethical questions concerning AI, new regulations have emerged, primarily from research and data science groups, as instances of unfair outcomes have been exposed. Leading AI enterprises have a stake in influencing these principles because some of the consequences of disregarding ethical norms in their products have already started to impact those companies. A lack of caution in this area could expose one to threats to their reputation, the law, and regulations, as well as result in costly penalties. As with all technological breakthroughs, innovation often outpaces state regulation in new, emergent fields. As the government sector develops the requisite expertise, we might expect further AI regulations that firms must abide by in order to protect civil liberties and human rights (5).

A broad range of ethical AI norms that take into account issues of safety, security, human welfare, and the environment are together referred to as "AI Ethics." Definitions change.

DOI: 10.1201/9781003353751-1

1.2   WHY IS IT IMPORTANT?

The use of excellent AI Ethics is not only appropriate behaviour, as demonstrated by AI experience, but is also necessary for AI to deliver good commercial value. Product failures, legal issues, brand harm, and other business risks associated to ethics are only a few examples.

Countries, including cities and counties, are approaching AI Ethics in their unique ways. As a result, it follows that businesses must be aware of national and even local laws governing artificial intelligence. The "Right to Explanation" clause in the EU General Data Protection Rules and the pertinent sections of the California Consumer Privacy Act are two examples of well-known AI-related regulations. Cities in the US are deciding how to use algorithms, particularly those used in law enforcement, at a more local level (6). The impending AI Act in the European Union is one of the most significant legislatives (7).

1.2.1   INITIATIVES IN AI

AI Ethics becomes a crucial component of AI literacy for most people as people from all walks of life begin to see AI in their daily lives and at their places of employment (8). In addition to Responsible AI practices, AI Ethics is now taught in high school and middle school as well as in professional business courses. One might think "Why is ethics in artificial intelligence being brought up all of a sudden?" Machine learning (ML) that is based on neural networks is seeing rapid expansion, for the awareness of AI Ethics will spread as regulations like the AI Act become more common for the following three reasons: A large increase in the amount of the data sets that are popular, a major breakthrough in the machine learning algorithms, and an increase in the human ability to create them (9). Each of these three developments involves a greater concentration of authority, and as the proverb says, "With tremendous power comes big responsibility." This is certainly the case with each of these three developments (10).

1.3   CHALLENGES AND OPPORTUNITIES OF AI

Here, are some of the biggest ethical challenges of artificial intelligence (11, 12)

1.3.1   TECHNICAL SECURITY

Safety is an additional key ethical problem with machine learning, particularly in systems that interact with the actual world, such as those that handle crucial healthcare or autonomous vehicles (13). Even if safety is not now a primary concern, it will become more prominent in public discourse once technologies that are enabled by machine learning begin to interact with people on a more widespread basis (14).

As an example, a number of individuals have been killed in incidents using semi- autonomous cars because the cars were unable to make safe judgments when confronted with potentially hazardous circumstances. It is possible for a manufacturer's legal obligation to be reduced via the establishment of extremely precise contracts that restrict liability. However, from a moral sense, not only does responsibility stay with the manufacturer, but the contract itself may be considered as an immoral ploy to avoid lawful accountability. This is because the contract allows the manufacturer to avoid being held legally accountable.

The question of how a well operating technology might be utilised for good or for evil is distinct from the issue of technological safety and failure. Even though it's only a function-related question, it serves as the cornerstone for the rest of the examination.

1.3.2   PRIVACY

How to resolve privacy concerns (and get permission) with the usage of data has been a long-standing ethical issue with artificial intelligence (15). We need data in order to train AIs, but the question is where that data comes from and how it should be used. We don't always have this, but sometimes we imagine that all of the data is originating from fully mature adults who are able to decide on their own how to utilise it. This is not always the case, but it is something that we occasionally assume.

For example, Barbie now has a chatty doll that is equipped with artificial intelligence. What exactly does this mean when it comes to ethics? An algorithm is collecting information on the ways in which your kid interacts with this toy while they play with it. What purpose does this data serve, and where will it ultimately be stored?

As per a number of headlines over the last few months, there are a lot of companies out there that collect information and then sell it to other companies. Which rules and regulations regulate the collection of this kind of data, and which ones may be necessary to ensure the privacy of users' personal information?

1.3.3 Data Storage and Protection of Data

In order to properly train algorithms, the bulk of artificial intelligence development services need access to data on a massive scale. The production of large volumes of data, although increasing the number of commercial opportunities available, also creates challenges over storage capacity and data integrity. As more data is generated and as more people have access to it, there is a greater possibility that some of it may find its way into the hands of a person operating on the dark web. The fact that this data is created by millions of users all over the globe has resulted in challenges with data storage and security that have extended to a global level. Enterprises have a responsibility to ensure that this is the case, since the use of the best possible data management environment for sensitive data and training algorithms for AI applications is necessary (16, 17).

1.3.4 Beneficial Use and Capacity for Good Optimal Application and Goodwill Capability

As was the case with all earlier forms of technology, the primary objective of artificial intelligence is to assist people in leading longer, more affluent, and more fulfilled lives. Since this is a positive development, we have reason to rejoice and express our gratitude for the benefits that artificial intelligence (AI) brings to our lives, insofar as it assists people in the aforementioned ways (18).

More intelligence is likely to lead to improvements in almost every area of human life, such as data analytics, energy efficiency, biomedical research, communication, space exploration, archaeology, legal services, farming, education environmental protection, space exploration, resource management, finance, medical diagnostics, waste management, transportation and so on.

Some modern agricultural machinery is equipped with computer systems that can detect weeds and apply very low dosages of pesticide to them. This is but one illustration of how AI might be helpful. This preserves not just the environment by lowering the total quantity of pesticides that are applied to crops, but also the health of humans by minimising the amount of these chemicals to which they are exposed.

1.3.5 Malicious Use and Capacity for Evil

When used as intended, a technology that works perfectly, like a nuclear weapon, can cause a lot of harm. It is inevitable that evil will be done with artificial intelligence, just as it has been done with human intellect.

For instance, AI-enabled surveillance is already rather popular, and it is used not just in settings where it is suitable (like airport surveillance cameras), but also in situations where it may be problematic (such as items with always-on microphones in our homes) (e.g., products which help authoritarian regimes identify and oppress their citizens) (19). Other unfavourable instances include AI-assisted computer hacking and lethal autonomous weapons systems (LAWS), which are often referred to as "killer robots." Concerns such as those shown in the films *2001: A Space Odyssey*, *Wargames*, and *Terminator* are also among those that people have (20, 21).

Even though films and weapons techniques may look like egregious examples about how AI could help evil, we should keep in mind that competition and war have always been the major causes of technological advancement, and that militaries and corporations are currently working on these technologies. Additionally, we should keep in mind that competition and war have always been the primary drivers of technological progress. The lessons of history also teach us that the most terrible events are not necessarily caused intentionally. As an example, the First World War was an unintended occurrence, and during the Cold War, there were a number of near misses with nuclear weapons. Even if you don't want to utilise your destructive ability, just having it puts you in danger of some kind of catastrophic event happening. Because of this, the most sensible course of action would be to prohibit, restrict access to, or renounce the use of some forms of technology.

1.3.6 BIASES

Our AI algorithms need data to learn, and we need to do everything we can to get rid of any bias in that data (22).

AI systems are only as good as the data they are trained on. Good services for building artificial intelligence depend on having good data. If a company doesn't have good data, it will have trouble putting AI to use because of biases. When machine learning algorithms create outcomes that are based on discriminating assumptions formed during the process of machine learning or biases in the training data, this is an example of an anomaly known as a bias (23). Bad data often goes hand in hand with all kinds of prejudices, including racist, sexist, communal, and ethnic biases.

Such biases need to go away. Real change could come from either training AI systems with unbiased data or making algorithms that are easy to understand and read. Also, various companies that make AI put a lot of money into building control frameworks and techniques to make AI algorithms more trustworthy and open, and to find bias in them (24).

In the ImageNet database, for instance, there is a significant majority of white faces relative to the number of non-white faces. When we train our AI algorithms to recognise facial traits using a database that doesn't contain the proper balance of faces, this results in a built-in bias that can have a significant effect. As a result, the algorithm won't perform as well when applied to non-white faces, and this can have a negative impact.

Rather than shrugging our shoulders and presuming that we are teaching our AI to accurately represent our culture, I believe that it is imperative that we eliminate as much bias as possible while we are training it. The first step in completing this process is coming to terms with the reality that our AI solutions may include some kind of bias (25).

1.3.7 AI'S EFFECT ON JOBS

Although the loss of jobs is the fundamental concern of the general public about artificial intelligence, this concern should probably be reframed. A change in the market's demand for certain employment positions is triggered by the introduction of every disruptive new technology (26). For instance, if we take a look at the automotive industry, we can see that some manufacturers, such as General Motors, are shifting their focus to the development of electric cars in order to support various environmentally conscious efforts. Even though the energy industry is still in operation, it is transitioning from a fuel economy to an electric economy as the primary source of energy. Similar considerations apply to artificial intelligence, which will transfer the demand for labor to other sectors. Due to the daily growth and change in data, there will need to be personnel to assist in managing these systems. The industries most likely to be impacted by changes in job demand, such as customer service, will nevertheless require resources to handle more complicated issues. Helping people move to these new areas of market need will be a key component of artificial intelligence and its impact on the job market (27).

1.3.8 Increasing Socio-Economic Discrimination

The first set of challenges that arise as a direct consequence of living in a digital era is the economics. The discussion of employment and unemployment is perhaps the one that receives the greatest attention among them. It has been known for a very long time that technologies connected to AI have the potential to bring in a new age of automation, which would result in the elimination of employment (28). In point of fact, Norbert Wiener predicted that the employment situation would become so dire if computers began to compete with people for jobs: "It is perfectly clear that this will produce an unemployment situation, in comparison with which the current recession and even the depression of the thirties will seem like a pleasant joke."

It is believed that artificial intelligence would have a negative impact on employment, despite the fact that this pessimistic forecast has not (yet) come true. For the first time, artificial intelligence (AI) is perceived as a danger, which is distinct from earlier anxieties over information and communications technology (ICT) or other automation technologies. This is due to the fact that the occupations that seem to be at risk are higher-paying ones. It's possible that artificial intelligence may make it more difficult for middle-class professionals to earn a livelihood (29). The loss of a job presents challenges not just in terms of financial stability but also in terms of one's relationships with others and their mental health (30). The topic of what the actual consequences of AI on the labour market will be is one that requires empirical investigation, at least in part. Occupations may not be eliminated entirely, but they could transform, and new jobs might be created, which would raise new concerns about the distribution and fairness of employment opportunities (31).

Some people have suggested a universal basic income (UBI) as a solution, but this would require major changes to the economies of each country. There may be various approaches to resolving this issue, but fundamental changes in both society and government are required for any of them to work. Because, in the end, this is a political issue and not a technical one, the answer, just like the solutions to many of the other difficulties outlined in this article, has to be dealt with on the political level.

1.3.9 Effects on the Environment

There are moments when we fail to take into account how AI will impact the surrounding environment. We are operating under the assumption that after we have completed the process of training an algorithm using data on a cloud computer, we will utilise that data to fuel the recommendation engines on our website. However, the data centres that support our cloud architecture use a significant amount of power (32).

For example, training in artificial intelligence can produce 17 times the amount of carbon emissions that the typical American produces in the course of about a year. Because of the significant amount of energy that is required, the cost of training machine learning models may easily approach tens of millions of dollars or even more. If this energy originates from fossil fuels, it would, without a doubt, have a big negative impact on climate change, in addition to being damaging in other regions along the hydrocarbon supply chain. This is something that should go without saying.

Machine learning may not only assist in finding answers to problems in resource management, environmental research, biodiversity, and other related fields, but it also has the potential to dramatically improve the efficiency of electricity distribution and consumption. Artificial intelligence (AI) is a technology that is, in some very basic ways, focused on efficiency. One of the ways that AI's talents might be directed is toward energy efficiency.

It would seem that AI may have a net positive influence on the environment; however, this would only be the case if it were really employed for that goal, and not merely to use energy for other reasons (33).

1.3.10 ETHICS AUTOMATION

One of the benefits of artificial intelligence is that it can automate decision-making, therefore reliev-ing the stress placed on people and maybe speeding up some decision- making processes dra-matically (34). Nevertheless, if these choices are beneficial for society, then society will profit from them; but, if they are terrible for society, then society will suffer. Because of this, the automation of decision-making processes will bring up substantial challenges for society.

As AI agents are given greater freedom to make decisions on their own, they will need to have some kind of ethical standards and guidelines built into them. Ethical decision- making may require as little as following a programme to fairly distribute a benefit, where the decision is made by humans and executed by algorithms, but it also may require much more detailed ethical analysis, even though we humans would prefer that it did not have to involve such analysis. This is owing to the fact that AI will be able to do tasks far quicker than humans can, thereby taking humans "out of the loop" of control in some scenarios where human error is present. Cyberattacks and high-frequency trading are two instances of this phenomenon that are currently taking place (both of which present various ethical difficulties that are often ignored), and the problem is only going to become worse as AI plays a more important part in society.

Because artificial intelligence has the potential to be so strong, the ethical standards that we establish for it need to be quite high (35, 36).

1.3.11 MORAL DEBILITATION AND DESKILLING

As our use of artificial intelligence increases, we are increasingly expecting machines to make cru-cial decisions (37). If we hand over the ability to make choices to robots, we will find that our own decision- making skills suffer as a result. An international treaty, for instance, has a provision that requires the utilization of autonomous drones in their operations. If a drone has the capability of firing a rocket that has the potential to kill someone, then the decision-making process must include the involvement of a human decision-maker (38). We have been able to get over some of the biggest control difficulties with AI so far with the help of a patchwork of rules and standards comparable to this.

The autopilot can handle every aspect of flying an aircraft, from take-off to landing, but then airline pilots purposefully opt to manually operate the aircraft during critical times (such as takeoff and landing) in order to keep their skills sharp.

The problem is that artificial intelligences need to be able to make snap decisions more and more of the time. For instance, in high-frequency trades, over 90 percent of all financial dealings are now managed by algorithms; hence, there is no prospect of placing a human person in charge of the judgments.

It is important for us to keep in mind the possibility that humans may become less skilled at these tasks since one of the uses of AI will be to either assist or replace people in some sorts of decision-making (e.g., spelling, driving, stock-trading, etc.). If artificial intelligence (AI) begins making deci-sions for us in terms of ethics and politics, then humans will get poorer at those things. This is the most severe version of the possibility. When our ability to make decisions and exercise power are at their peak and most important, we run the danger of restricting or stifling our moral growth.

The same is true for automobiles that drive themselves. It is vital that the AI be in command of the situation since they need to respond swiftly if a child runs out into the road. This brings up some important ethical questions in relation to AI and control.

As a direct consequence of this, the study of ethical issues as well as training in ethics is now more important than it has ever been. It is important for us to do research into the ways in which artificial intelligence might improve our ethical education and training. We will never again allow oneself to become ethically incompetent and incapable; otherwise, when our technology finally presents us with challenging decisions and problems to solve—decisions and challenges that maybe our ancestors might have been able to solve—future people would not be able to do so (39, 40). This is why we must never allow ourselves to become ethically incompetent and incapable.

1.3.12 ARTIFICIAL GENERAL INTELLIGENCE AND SUPERINTELLIGENCE

If or when AI reaches human levels of intelligence, it will be an Artificial General Intelligence (AGI), and the only other intellect of its kind to exist on Earth at the human level (41). If or when AI reaches human levels of intelligence, it will be able to perform everything that humans are capable of doing as well as the average human.

If and when artificial general intelligence (AGI) exceeds human intellect, it will evolve into a superintelligence, which will be an entity that is potentially substantially more intelligent and competent than humans are. Myths, religions, and storytelling have been the only places where humans have ever made a reference to superintelligence (41).

According to the techno-optimistic interpretation of artificial general intelligence (AGI), there will come a time when AI will be sufficiently developed to begin improving itself, which will eventually lead to the singularity (42). The singularity is an explosion of intelligence that is caused by a positive feedback loop of AI upon itself. As a direct consequence of this, superintelligence will be created. The implication is that artificial general intelligence will ultimately achieve consciousness and self-awareness in addition to outperforming humans in the vast majority of all cognitive pursuits. The people taking part in this discussion have divergent opinions on the probable sequence of events. It is possible that the superintelligent AGI may either be beneficial and increase the quality of life for humans, that it would see humans as competitors and eradicate us, or that it will reside in another sphere of awareness and completely ignore mankind (43).

It's important to note that AI technology is advancing really quickly. Governments and multinational companies are vying for control of AI's potential. Furthermore, there is no reason why AI development should end at AGI. Scalable and quick AI. In contrast to a human brain, AI will perform tasks faster and faster as we provide it with more technology.

1.3.13 THE RELIANCE ON AI

Technology has become essential to people's everyday lives. Since the beginning of time, we have depended heavily on various forms of technology, to the point where it is almost impossible to separate ourselves from other species. However, what was formerly limited to rocks, wood, and fur clothing has grown considerably more intricate and delicate. Being without power or phone service can be a severe issue on a psychological or even physical level (if there is an emergency). In addition, there is no other dependency that can be compared to the reliance on one's intelligence.

Intelligence reliance may be thought of as a kind of dependency comparable to that which a child feels toward an adult. Children typically depend on adults to make decisions for them, and since ageing brings on cognitive decline in certain individuals, older people also regularly rely on those who are younger in age to make decisions for them. Imagine for a second if middle-aged people who provide care for children and the elderly depend on artificial intelligence to guide them. Only "grownups" created by artificial intelligence would be left, with no "adults" created from humans. Our AI caregivers would have pampered us and catered to our every want as if we were helpless toddlers.

Naturally, this raises the issue of what a human species that has been infantilized would do in the event that our AI parents encountered any kind of difficulty. We run the danger of becoming like orphans if we get dependent on that AI because if we don't have it, we won't be able to care for either ourselves or our modern civilization. This "lostness" already happens, for example, when the navigation programmes on people's smartphones malfunction (or the power just runs out).

The trend toward technological dependence has already begun. How can we get ready right away to prevent the risks posed by dependence on AI for intelligence in particular?

1.3.14 AI-Driven Addiction

Smartphone app makers have turned the study of addiction into a science, and AI- powered video games and apps have the potential to be just as addictive as illegal drugs. Artificial intelligence has the potential to exploit a wide variety of human frailties and desires, including, but not limited to, purpose-seeking, gambling, avarice, libido, and aggressiveness (37).

Aside from using its power over us to manipulate and control us, addiction also prevents us from accomplishing other, more important duties, such as those in the areas of economics, education, and social interaction. It enslaves us and takes up time that should be spent engaging in things that are more useful. What hope do we have of evading the AI's control when it is always gathering more information on us and making greater efforts to keep us clicking and scrolling? Or, to be more exact, the grasps of the app developers who construct these AIs with the intention of ensnaring us, since it is not the AIs themselves who make the decision to exploit humans in this manner.

Any time I discuss this subject with a group of students. I find that they are all "addicted" to one or more apps. Although the students identify it that way, it may not be a clinical addiction, and they are aware that they are being taken advantage of and hurt. App developers need to cease doing this since AI shouldn't be built with the goal of taking advantage of flaws in human psychology.

1.3.15 Being Alone and Isolated

The problem of loneliness is becoming more widespread in modern society. According to recent study there are 200,000 elderly citizens in the UK who haven't talked to a friend or family in more than a month because they haven't had the opportunity to do so (44). This is a dreadful scenario due to the fact that being lonely might potentially lead to one's death (45). It is detrimental to the interpersonal ties that serve as the fundamental basis for society and a catastrophe for public health. It is easy to forget that things could be different, and in fact, they were quite different only a few decades ago. This is especially true in light of the fact that technology has been linked to a number of negative social and psychological trends, such as feelings of loneliness and isolation, hopelessness, tension, and anxiety.

Even while "social" media, cellphones, and artificial intelligence can seem as if they might be beneficial, the reality is that they are major contributors to loneliness since they need people to connect with screens rather than one another. What really helps is having strong interpersonal links, yet these are the same things that are being replaced by addictive (and often AI-powered) technology.

Giving up technological devices and working on creating meaningful relationships with people in real life are both effective ways to overcome loneliness. In other words, use caution.

At the societal level, it may be quite challenging to buck the tendencies we have up till now. This may not be an easy task. But since a better, more compassionate society is conceivable, we should resist. Technology doesn't have to make the world less intimate and compassionate; if we wanted it to, it could really have the opposite effect.

1.3.16 Impacts on Human Spirit

The aforementioned research topics will all have an impact on how people view themselves, interact with one another, and go about their daily lives. However, there is also a more existential query. If intelligence plays a role in humanity's identity and purpose, are we relegating ourselves to a lower status among our own creations if we externalise it and enhance it beyond human intelligence?

The development of artificial intelligence has prompted the exploration of a more profound question, one that penetrates deeper than the realms of philosophy, spirituality, and religion, all the way into the centre of the idea of what it is to be human (46). What will happens to the human spirit if or when our own innovations begin to exceed us in all that we do? Will the meaning of human existence lose its significance? Will we come to a new understanding of who we are, one that is not limited to our intelligence?

Perhaps handing over intelligence to computers will let us realise that intellect is not as crucial to our identity as we might think it is. If, on the other hand, we locate our humanity not in our heads but in our hearts, there is a good chance that we will come to the conclusion that compassion, love, caring and kindness are, in the end, what characterise us as human beings and what contribute to the value of life. AI may be able to assist us in realising this dream of a more compassionate world by alleviating some of the monotony of existence.

1.4 CONCLUSION

Artificial intelligence (AI) equips us with incredible new capabilities that enable us to help people and make the world a better place, and new technologies are continually being created for the sake of the good they can accomplish. But if we want to make the world a better place, we have to make the conscious decision to behave in an ethically responsible way. We can only hold out hope that advances in AI technology will enable us to build a better society by coordinating the efforts of a large number of individuals and institutions working together.

In this chapter, we have attempted to highlight some of the most major ethical dilemmas that are associated with AI; nevertheless, there are more. It's possible that we'll need a lot more time to discuss topics like AI-powered surveillance, the role of AI in disseminating false information and misinformation, the significance of AI in international and politics relations, the AI governance, and other related topics.

REFERENCES

1. Zhang B, Zhu J, Su H. Toward the third generation artificial intelligence. *Science China Information Sciences* 2023;66(2):1–19.
2. Huynh-The T, Pham Q-V, Pham X-Q, Nguyen TT, Han Z, Kim D-S. AoAI: Artificial intelligence for the metaverse: A survey. 2023;117:105581.
3. Zhou Z. *Life Cycle Optimization Analysis of Bridge Sustainable Development*: Universitat Politècnica de València; 2023.
4. Mayta-Tovalino F, Munive-Degregori A, Luza S, Cárdenas-Mariño FC, Guerrero ME, Barja-Ore J. Applications and perspectives of artificial intelligence, machine learning and "dentronics" in dentistry: A literature review. *Journal of International Society of Preventive & Community Dentistry* 2023;13(1):1–8.
5. Schuett J. Defining the scope of AI regulations. Innovation, technology 2023:1–23.
6. Brkan M. The regulation of data-driven political campaigns in the EU: from data protection to specialized regulation. *Yearbook of European Law* 2023.
7. Bode I, Huelss H. Constructing expertise: The front-and back-door regulation of AI's military applications in the European Union. *Journal of European Public Policy* 2023:1–25.
8. Catanzariti M. What role for ethics in the law of AI? In: *Artificial Intelligence, Social Harms and Human Rights*: Springer; 2023, pp. 141–159.
9. Księżak P, Wojtczak S. Artificial intelligence and legal subjectivity. In: *Toward a Conceptual Network for the Private Law of Artificial Intelligence*: Springer; 2023, pp. 13–35.
10. Peterson TL, Ferreira R, Vardi MY. Abstracted power and responsibility in computer science ethics education. *IEEE Transactions on Technology and Society* 2023;4(1):96–102.
11. Feuerriegel S, Dolata M, Schwabe G. Fair AI: Challenges and opportunities. *Business & Information Systems Engineering* 2020;62(4):379–384.
12. Al Ridhawi I, Otoum S, Aloqaily M, Boukerche AJIN. Generalizing AI: Challenges and opportunities for plug and play AI solutions. *IEEE Network* 2020;35(1):372–379.
13. Ali ANF, Sulaima MF, Razak IAWA, Kadir AFA, Mokhlis H. Artificial intelligence application in demand response: Advantages, issues, status, and challenges. IEEE Access 2023.

14. Chen J, Lai P, Chan A, Man V, Chan C-H. AI-assisted enhancement of student presentation skills: Challenges and opportunities. *Sustainability* 2023;15(1):196.

15. Chakrobartty S, El-Gayar OF. Fairness challenges in artificial intelligence. In: *Encyclopedia of Data Science and Machine Learning*: IGI Global; 2023, pp. 1685–1702.

16. Wu C, Xu H, Bai D, Chen X, Gao J. Jiang X. Public perceptions on the application of artificial intelligence in healthcare: A qualitative meta-synthesis. *Qualitative Research* 2023;13(1):e066322.

17. Meurisch C, Mühlhäuser M. Data protection in AI services: A survey. *ACM Computing Surveys* 2021;54(2):1–38.

18. Ali O, Abdelbaki W, Shrestha A, Elbasi E, Alryalat MAA, Dwivedi YK. A systematic literature review of artificial intelligence in the healthcare sector: Benefits, challenges, methodologies, and functionalities. *Journal of Innovation & Knowledge* 2023;8(1):100333.

19. Borda A, Molnar A, Neesham C, Kostkova P. Ethical issues in AI-enabled disease surveillance: Perspectives from global health. *Applied Sciences* 2022;12(8):3890.

20. Mittelsteadt M. Artificial intelligence: An introduction for policymakers. *Technology and Innovation* 2023.

21. Secondo Matteo IVJP. Stanley Kubrick's 2001: An existential Odyssey; 2023.

22. Roselli D, Matthews J, Talagala N, editors. Managing bias in AI. *Companion Proceedings of the 2019 World Wide Web Conference*; 2019.

23. Alon-Barkat S, Busuioc M. Human–AI interactions in public sector decision making:"automation bias" and "selective adherence" to algorithmic advice. *Journal of Public Administration Research and Theory* 2023;33(1):153–169.

24. Nelson GS. Bias in artificial intelligence. *North Carolina Medical Journal* 2019;80(4):220–222.

25. Mehrabi N, Morstatter F, Saxena N, Lerman K, Galstyan A. A survey on bias and fairness in machine learning. *ACM Computing Surveys* 2021;54(6):1–35.

26. Nguyen TM, Malik A. A two-wave cross-lagged study on AI service quality: The moderating effects of the job level and job role. *British Journal of Management* 2022;33(3):1221–1237.

27. Agrawal A, Gans JS. Goldfarb A. Artificial intelligence: The ambiguous labor market impact of automating prediction. *Journal of Economic Perspectives* 2019;33(2):31–50.

28. Giovanola B, Tiribelli S. Beyond bias and discrimination: Redefining the AI ethics principle of fairness in healthcare machine-learning algorithms. *AI & Society* 2022:1–15.

29. Alber EJ. Préparer la Retraite: New age-inscriptions in west African middle classes. *Anthropology & Aging* 2018;39(1):66–81.

30. Kaplan A, Haenlein M. Siri, Siri, in my hand: Who's the fairest in the land? On the interpretations, illustrations, and implications of artificial intelligence. *Business Horizons* 2019;62(1):15–25.

31. Lords HO. AI in the UK: Ready, willing and able? 2018;13;2021.

32. Leike J, Martic M, Krakovna V, Ortega PA, Everitt T, Lefrancq A, Orseau L, Legg S. AI safety gridworlds. arXiv preprint arXiv:1711.09883. 2017 Nov 27.

33. Makridakis SJF. The forthcoming artificial intelligence (AI) revolution: Its impact on society and firms. *Futures* 2017;90:46–60.

34. Ramaswamy S. Joshi H. Automation and ethics. 2009:809–833.

35. Bostrom N, Yudkowsky E. *The Ethics of Artificial Intelligence. Artificial Intelligence Safety and Security*: Chapman and Hall/CRC; 2018, pp. 57–69.

36. Ryan M. In AI we trust: Ethics, artificial intelligence, and reliability. *Science and Engineering Ethics* 2020;26(5):2749–2767.

37. Green BP. Artificial intelligence and ethics: Sixteen challenges and opportunities. *Markkula Center for Applied Ethics* 2020.

38. Wierzbicki AP, Wessels J. *The Modern Decision Maker*: Springer; 2000.

39. Heyd D. Procreation and value can ethics deal with futurity problems? *Philosophia* 1988;18(2–3):151–170.

40. Lunenburg FC, editor. The decision making process. *National Forum of Educational Administration & Supervision Journal*; 2010.

41. Wang P, Goertzel B, editors. Introduction: Aspects of artificial general intelligence. *Proceedings of the 2007 Conference on Advances in Artificial General Intelligence: Concepts, Architectures and Algorithms: Proceedings of the AGI Workshop*. 2006;2007.

42. Williams AE, editor. A model for artificial general intelligence. *Artificial General Intelligence: 13th International Conference, AGI 2020, St Petersburg, Russia, September 16–19, 2020, Proceedings*;13; 2020: Springer.

43. Everitt T, Lea G, Hutter M. AGI safety literature review. 2018.

44. McCabe M, You E, Tatangelo G. Hearing their voice: A systematic review of dementia family caregivers' needs. *Gerontologist* 2016;56(5):e70–e88.
45. Leigh-Hunt N, Bagguley D, Bash K, Turner V, Turnbull S, Valtorta N, Caan W. An overview of systematic reviews on the public health consequences of social isolation and loneliness. *Public Health* 2017;152:157–171.
46. Herzfeld NL. *In Our Image: Artificial Intelligence and the Human Spirit*: Fortress Press; 2002.

2 Basic Ethical Issues in Bioinformatics and Chemoinformatics

Meghna B. Patel, Jagruti Patel, Devangi Patel,
Prajesh Prajapati, and Jigna Prajapati

2.1 INTRODUCTION

Chemoinformatics techniques can be applied in a variety of different ways in the chemical and related sectors. Chemoinformatics gives us the chance to convert the data obtained through the integration of the two fields into knowledge, which can then be extended to make proper and better decisions in areas like drug discovery, comprehending chemical interactions, standardizing drug manufacturing protocols, etc. Chemoinformatics may assist us in doing virtual studies, which offer perceptions into how our body may actually react to a medicine. In addition, the procedure is very quick, saving time because it eliminates the need for practical labor, which is a bonus. Unfortunately, since everything in chemical informatics is speculative, we cannot be certain that the medication will behave as expected. Solutions will need bits from other disciplines, like chemoinformatics, bioinformatics, and others, as more studies on medication discovery are conducted in universities, institutions, and tiny businesses.

Around 40 years have passed since the creation of the first computer-based system. The term "chemoinformatics" was defined in 1998 by F.K. Brown. Chemoinformatics is a new subject that was created as a result of the chemical difficulties that exist today, including complicated linkages, plenty of data, and a lack of vital data. Chemoinformatics combines information technology and chemistry to provide quick analysis without real testing. The creation of models that connect chemical structure and different molecular characteristics is one of the crucial uses of chemical informatics. In the context of drug discovery and data organization, IT and chemistry have therefore been crucial. Nowadays, there are roughly 45 million chemical molecules that are understood. All of the information gathered in this way is kept in a database and is available to everyone. In a certain sense, this has the potential to usher in a revolution in wealth of knowledge accessible that could be utilized to comprehend the chemistry involved in drug discovery. Being at the intersection of chemical in chemistry, bio in biology, and software as a system in computer science [1], it is a significant field of science.

The vast topic of chemoinformatics encompasses design as a style, development, organization, maintenance, retrieval, analysis, dissemination, visualization, and application of chemical information. Obviously, every discipline of chemistry, other than drug design, requires the translation of data turned into information and of information converted into knowledge. The data gathered may be utilized for analyzing data in businesses including paper and pulp, dye, and associated industries in addition to drug development [2]. Using past knowledge and information that has been condensed into knowledge, three primary tasks—design of reaction/synthesis designs, structure elucidation, and structure–property / activity relationships—are addressed [3]. The amount of data that has to be processed is frequently rather substantial. Only electrical methods using computer power can handle this enormous volume of data. Chemoinformatics can be helpful in this way [4].

DOI: 10.1201/9781003353751-2

2.2 MAJOR ASPECTS OF CHEMOINFORMATICS

1. Information management and acquisition: Data collection techniques (primarily experimental); the development of databases for the storage and retrieval of data.
2. Model development, examination of the data and correlation are examples of information use.
3. Informational applications include virtual chemical library screening, system chemical biology networks, and the prediction of molecular features pertinent to pharmaceuticals.

2.2.1 CHEMOINFORMATICS' RELATIONS TO OTHER DISCIPLINES

Machine learning and chemoinformatics: Whilst chemoinformatics might be seen as a relatively particular field of machine learning's application, it is commonly employed for structural property modeling. Chemical objects' nature, the chemical universe's intricacy and the ability to consider additional information all contribute to the uniqueness of chemoinformatics. Instead of a straightforward constant-size vector of integers similar to how it is used in common applications of mathematical statistics and machine learning, the fundamental chemical entity is a graph (or hypergraph). This mandates the necessity of using graph theory, to construct innovative using machine learning, descriptors, and structured graph kernel algorithms adequate to coping with discrete structured data.

The following key difference is that the chemical data were obtained through an exploratory procedure in a vast chemical area rather than through well-planned sampling. No, they can't thus be regarded as an impartial, representative, and uniformly random selection from a clear distribution. As a branch of theoretical chemistry, chemoinformatics is therefore required to handle this challenge using unique methodologies, such as active learning, multiple methods for exploring the chemical universe and the "applicability domain" notion [6].

Physicochemical theory's linkages between various properties can also be used. For modelling the rate constants, the Arrhenius law, for instance, may be quite helpful. These links might be incorporated into the chemoinformatics workflow as external knowledge.

2.2.1.1 Chemoinformatics and Chemometrics

Chemometrics is "a chemical discipline that applies mathematics, statistics, and formal logic (a) to design and select optimal experimental procedures; (b) to provide the most pertinent chemical information by analyzing chemical data; and (c) to learn about chemical systems," according to Massmart [5]. Chemometrics doesn't often need knowledge of chemical structure; therefore, the only area where it and chemoinformatics intersect is with the use of machine learning methods. It is frequently used in domains that need a thorough analysis of multivariate data, such as spectra treatment, chemical engineering, analytical chemistry, and experiment design.

2.2.1.2 Chemoinformatics and Bioinformatics

Bioinformatics employs computational methods to research the composition and purpose of biomolecules, in contrast to chemoinformatics, which deals with "chemical size" molecules (proteins, nucleic acids). Here is a vast area that mostly involves modelling in 1D (sequence alignment) and 3D (force field and quantum mechanics computations). The second displays a biomolecule serving as series of characters (building blocks). Bioinformatics hardly ever employs models of the graph and fixed-size vectors utilized within chemoinformatics.

The creation of fingerprints or descriptions of protein-ligands based on three-dimensional data on protein-ligand complexes is another technique to merge bio- and cheminformatics approaches. Consequently, Tropsha et al. created CoLiBRI descriptions for a fictitious molecule made of interacting protein and ligand atoms. For each protein atom interacting with the ligand, Varnek et al. [8] have created "interaction fingerprints" that take into consideration eight different forms of interactions. Ionic (proteins with a positive charge) atoms protein donor atom, protein acceptor atom, H-bond (protein donor atom), H-bond (protein acceptor atom), hydrophobic, aromatic (face to face), aromatic (edge to face), ionic (negatively charged protein atoms), and metal complexation. The creation of pharmacophoric ligand models using an examination of 3D protein-ligand structures has been discussed by Langer et al. [7].

The creation of PLKs, or protein-ligand kernels, as byproducts of "chemical" ligand-ligand (LLK) and "biological" protein-protein kernels is a potential technique to explain ligand-receptor complexes (PPK). The feature spaces corresponding to LLK and PPK are combined to form the feature space for PLK. PLK models for machine learning are predicated based on the idea that similar ligands bind to similar proteins. These kernels make it possible to predict the possibility for different ligands to bind to different proteins as well as different proteins to bind to a certain ligand. Erhan et al. merged "chemical" kernels based on MOE descriptors with "biological" kernels based on protein-ligand "interaction fingerprints." Using the unique molecular descriptors, Faulonetal. calculated "chemical" and "biological" Tanimoto kernels. Jacob and Vert combined a Tanimoto kernel for the ligands and other types of kernels for the proteins. They compared EC counts or protein sequences particularly for PPK. Bajorath et al. used a linear kernel for the ligands and protein-protein kernels obtained from the sequence identity matrix.

2.3   APPLICATIONS OF CHEMOINFORMATICS

a. Chemistry Fields

Chemoinformatics has a wide variety of uses; its techniques can be beneficial in many areas of chemistry.

A list of chemical subfields and examples of common chemoinformatics applications follow. It should be stressed that this is by no means a complete list of all potential uses.

b. Information on Chemicals
- Flood management by storing and retrieving chemical structures and associated information.
- Internet-based data dissemination.
- Data and information cross-referencing.

c. Biological Chemistry
- Predicting the results of organic reactions
- Developing organic syntheses.

d. Drug Design
- Finding new lead structures.
- improving lead structures.
- Quantitative structure activity connections.
- Comparing chemical libraries.
- Planning structural diversity of chemical libraries.
- High-throughput data analysis, and docking a ligand onto a receptor.

e. ADME-Tox Property Modelling
- Xenobiotic metabolism prediction.
- Biochemical pathway analysis. Despite the variety of these topics and the variety of their applications, the subject of chemoinformatics is still in its infancy.
- Chemoinformatics techniques may still be used to a wide range of issues and difficulties.

In searching for new applications and creating new methodologies, there is a lot of room for innovation.

2.4   CHEMOINFORMATICS INSTRUCTION

In order for chemists to understand where chemoinformatics could be useful to them and where they should best consult chemoinformatics experts, it is essential that chemoinformatics be included in chemistry courses to some extent. A select few universities must also provide training for experts in chemoinformatics. The initial steps have already been taken at several universities all across the world. More must be done in order to develop the chemoinformatics professionals that society so desperately needs.

2.4.1 CHEMINFORMATICS METHODS ARE REQUIRED IN CHEMISTRY

Chemistry has already produced a large amount of data, and this data avalanche is only growing. The number of known chemical compounds is around 45 million, and it grows by millions every year. Innovative methods like combinatorial chemistry and high-throughput screening generate enormous amounts of data. All of this data and information can only be regulated and made available by being properly stored in databases. Chemoinformatics is the sole method that makes that possible. On the other hand, a lot of issues lack crucial information. We are aware of the three-dimensional (3D) structures of around 300,000 organic molecules thanks to X-ray crystallography [6]. Alternatively, as another example, there are roughly 200,000 infrared spectra in the largest database of infrared spectra. Nevertheless, big as these figures may look, they are actually rather small.

Another justification for the use of informatics techniques in chemistry is this: Many chemistry issues are too intricate to be resolved using theoretical calculations or first-principles procedures. This holds true for both the relationships between the structure of a substance and its biological activity as well as the influence of reaction conditions on chemical reactivity [8].

In order to manage enormous amounts of chemical structures and data, to extract information from data, and to simulate complicated relationships, all these chemistry issues call for creative methods. Chemoinformatics techniques can be useful in this situation.

The review of chemoinformatics that follows has a focus on the issues and solutions that are shared by the different more specialized subfields.

2.4.1.1 Chemical Compound Representation

Matrix, connection tables, and linear codes are only a few of the techniques that have been created for the depiction of chemical structures and molecules on computers. To accurately describe a chemical structure and perceive properties, specialized methods must be developed to address stereochemistry, 3D structures, aromaticity, rings, and molecular surfaces.

2.4.1.2 Chemical Representation

Reaction starting ingredients, products, and reaction circumstances all serve as representations of chemical reactions. Moreover, the location of the reaction as well as the bonds that are formed and broken during a chemical reaction must be specified. Moreover, it is necessary to manage the stereochemistry of processes.

2.4.1.3 Chemistry Data

Our understanding of chemistry is mostly reliant on facts. A wide range of data on physical, chemical, and biological characteristics are provided by chemistry, including binary data for classification, real data for modelling, and spectrum data containing a lot of information. This information has to be transformed into a format that makes data analysis and information exchange straightforward.

2.5 DATA SOURCES AND DATABASES

Databases were developed very early on to store and exchange information electronically due to the enormous volume of data in chemistry. For compounds, 3D structures, reactions, and spectra in chemical literature and other topics, databases have been created. More and more, chemistry-related data and information are disseminated via the internet.

2.5.1 STRUCTURE SEARCH METHODS

Chemical structural data must be accessible in order to get data and information from databases. For complete structure, substructure, and similarity searches, methods have been devised.

2.5.2 CALCULATION TECHNIQUES FOR PHYSICAL AND CHEMICAL DATA

Many methods may be used to directly calculate the physical and chemical characteristics of compounds. First come computations of various degrees of complexity in quantum mechanics. For instance, simple methods like additive schemes may be applied to estimate a variety of variables with decent accuracy.

2.5.3 STRUCTURE DESCRIPTOR CALCULATION

The majority of the time, however, it is impossible to immediately determine a compound's physical, chemical, or biological properties from its structure. In this case, a deceptive strategy is required. Structure descriptors must first be used to describe the compound's structure, and then inductive learning techniques must be used to analyze a number of pairs of related attributes and structure descriptors in order to create a link between the two. There are several different structure descriptors that have been created to encode molecular surface characteristics or information about 1D, 2D, or 3D structures.

2.5.4 TECHNIQUES OF DATA ANALYSIS

Many inductive learning methods, including statistics, pattern recognition, artificial neural networks, and genetic algorithms, are used in the study of chemistry. These methods, which are suitable for quantitative modelling or categorization, can be divided into supervised and unsupervised learning techniques.

2.6 THE BASIC CONCEPTS OF CHEMOINFORMATICS

Graph theory and statistical learning are the two major mathematical approaches used by chemoinformatics to create its models for the objects in the chemical space. Although similar mathematical methods may be used to other fields, chemoinformatics has a special concept called the chemical space that specifies how to organize collections of chemical structures.

2.6.1 MOLECULAR MODELING

Very advanced systems for the detailed depiction of complicated molecular structures have been made possible by advancements, especially in hardware and software, with regard to graphics screens and graphics cards. Software for simulating molecular dynamics, modelling proteins, and constructing 3D structures has been developed. Molecular modeling has become a common practice. Argus Lab, Chimera, and Chemical are three of the most popular software programs for molecular modeling.

2.6.2 STRUCTURAL CLARIFICATION WITH COMPUTER ASSISTANCE

One of the main jobs of a chemist is to clarify the structure of a chemical substance, whether it be a reaction result or a compound isolated as a natural product. Structure elucidation must take into account a wide range of information, mostly from several spectroscopic approaches, as well as numerous structural possibilities.

2.6.3 Computer-Assisted Synthesis Design (CASD)

When planning a synthesis for an organic product, understanding chemical reactions and chemical reactivity is crucial. There are many alternatives to pick from when deciding how to assemble a molecule and which processes to use.

2.6.3.1 Chemical Space Paradigm

Chemical compounds, as opposed to stars, make up space; as noted by C. Lipinski and A. Hopkins, "chemical space might be considered as being akin to the cosmic universe in its immensity" [4]. Any effort to calculate the amount of chemical substances that may theoretically be created causes a combinatorial explosion, producing an estimate of more than 10^{60} [2], which is greater than the total amount of atomic particles in the cosmos. It is obvious that this number is so large that it is impossible to both synthesise and computationally create the structures of these molecules. Chemoinformatics seeks to navigate this figuratively limitless chemical universe by finding a logical representation for it. The creation of novel physiologically active chemicals and the design of novel pharmaceuticals depend critically on effective methods for navigating chemical space [4]. As opposed to being evenly dispersed over the entire chemical space, biologically active molecules of a particular type instead form relatively compact zones in it, much like galaxies do in the astronomical universe [4]. For any other chemical property, this is unquestionably true. Even a unique term, chemigraphy, which is similar to geography, has been proposed to describe the skill of navigating in chemical space [9]. While being frequently used in the literature on chemoinformatics, the term "chemical space" is still not clearly defined. In general, the concept of "space" refers to a collection of items having specific characteristics.

2.6.3.2 Chemical Compound Representation

Chemoinformatics identifies the structure and characteristics of molecules by treating them as informational objects. Graphs and descriptor vectors are the two basic types of objects that are typically employed. Undirected graphs with labelled vertices and edges stand in for atoms and chemical bonds, respectively. While the vertex labels provide symbols for chemical elements, the edge labels specify the sort of bond. The label either refers to the arrangement of bonds in molecules or, in more complicated systems, to specific bond types. For supramolecular systems, various "coordination" bond types can be defined, and chemical reactions can be encoded using "dynamic" bonds that correspond to chemical transformations [11]. Ensembles of graphs can be used to describe more chemical complexity, such as mixtures or polymers. The following are broader illustrations of chemical structures required for a number of practical applications. For instance, the vertexes of the graph identified for pharmacophore analysis as pharmacophoric centers (H-donors, H-acceptors, cation, etc.)

An edge that is marked with the 2D or 3D distance's value can be used to represent the distance between two centers (e.g., anion, aliphatic, aromatic). In Markush structures used for patent searches, a graph vertex can represent a range of various kinds of either individual atoms or full substructures. (e.g., substituents). Substructure queries used to search chemical databases exhibit the same behavior [10]. However, taking into account some complicated chemical substances reveals some restrictions on how graph theory may be used to represent chemical structures and the more suitable multicenter as a more efficient mathematical model to encode stereo chemical information bonding, hyper graphs [12] have been proposed. However, compared to graphs, hyper graphs are far more challenging objects to utilize, and as a result, their application is still quite restricted. Molecular descriptors-based, which Deschini and Consonni define as "the outcome of a logical and mathematical procedure which transforms chemical information encoded within a symbolic representation of a molecule into a useful number or the result of some standardized experiment," another well-liked representation of molecular structure is possible [12]. This molecular representation is widely used in chemoinformatics because (a) different descriptors can be generated from the same molecular graph, each of which describes a different aspect of the information concealed therein; (b) it is unaffected by any renumbering of the graph's vertices; and (c) the majority of the descriptors are molecular in nature and are simple to interpret; (d) descriptors can be used for inductive knowledge transmission; and (e) descriptors in addition to single molecules, more complex systems such as chemical processes [8] or multi-component mixtures [10] can also be described using descriptor vectors. More than 5,000 different description kinds have been reported as of late [12]. They are employed in clustering, similarity searching, SAR/QSAR/QSPR model construction, processing databases using (screens or fingerprints), and other processes. In the same manner, there are several molecular descriptors. Shortcomings should be mentioned: Two different molecules may be superposed on one point in the resulting chemical space if descriptors are poorly chosen for the following reasons: (a) There are many descriptors already in use, so there is always a chance of choosing ones that are redundant or irrelevant; (b) The loss of reciprocity between the descriptor and the molecular structure is a serious drawback of molecular descriptors. In fact, the "inverse" problem in QSAR refers to the extremely challenging and, in some situations, impossible process of reconstructing molecular graphs from descriptors. It covers the creation of molecular structures with desired property values from a practical standpoint. By assigning different chemical structures to different sets of molecular descriptor values, Gordeeva et al., Skvortsova et al., and Faulon et al. reported on attempts to tackle this challenge and found some degeneracy of solutions. This, as noted in [13], prohibits the construction of chemical structures from molecular descriptors, but it can also be helpful for securely exchanging several molecular descriptors.

2.7 MODELING BACKGROUND

Graph theory and computational learning theory are the two main mathematical techniques used in chemoinformatics. The use of graphs in chemistry is covered in a number of books and review articles; however, the latter is mostly covered in data mining literature. Here, we give a broad introduction of a number of computational learning theory's core concepts.

2.7.1 COMPUTERIZED LEARNING THEORY

The traditional statistical paradigm of "model parameterization" has recently given way to a new perspective on "predictive flexible modeling" in statistical modeling. According to the first paradigm, the statistical analysis' objective is to identify. The functional relationship between the input and output data is presumed to be established from some outside information, and a few independent free parameters are fitted to experimental data. For each free parameter, this often calls for a fixed number of experimental observations. Unfortunately, only a very small number of situations, such as the traditional Hansch-Fujita strategy based on just three descriptors, may satisfy this condition [14]. The second paradigm's goal is to create models with the highest degree of prediction power by adjusting to experimental results rather than versatile families. Evidently, this configuration is considerably more suitable for the majority of chemoinformatics studies. Due to the "curse of dimensionality," the early attempts to employ the second paradigm within the context of so-called nonparametric statistical analysis were unsuccessful because it required a large number of observations that grew exponentially with the number of free parameters [15]. Nonetheless, early attempts at predictive modelling were effective and used decision trees and artificial neural networks, both of which are totally heuristic methods [16, 17]. Vapnik's Statistical Learning Theory (SLT) provided the first theoretical foundation for creating statistical models using finite (even tiny) data sets. The foundation of contemporary computational learning theory is this strategy, along with those later developed as the MDL (Minimum Description Length) idea by Rissanen and the PAC (Probably Approximately Accurate) notion by Valiant. The objective of statistical analysis, according to SLT, is to select the "best" function (f(x, q*) from a group of functions with the lowest risk functional. R*f+ is the mean prediction performance on all conceivable test sets, and it is defined as an expected prediction error on fresh data obtained from the same distribution as the training set.

According to SLT, h is determined by the trade-off value used to simultaneously minimize both sides of Equation 2. Models with any (even a very large) number of variables can be created using kernel approaches, which imitate nonlinear functional dependencies of any kind by projecting descriptors onto a feature space of any (even infinite) dimensionality and building linear models in this feature space.

Computational learning theory is now a rapidly expanding field. A Bayesian learning technique to predictive flexible modeling has so just been described [18]. It takes into account the entire statistical distributions of models weighted by their capacity to fit data rather than just one particular model (as in STL), allowing one to average these distributions to produce probabilistic predictions. This method has gained a lot of chemoinformatics popularity; recent publications describe its implementations in Bayesian Neural Networks, Gaussian Processes, and Bayesian Networks.

2.7.2 VARIATIONS IN STATISTICAL MODELS

It should be noted that there is currently a fairly broad variety of applications for various statistical (machine learning) approaches in chemoinformatics. Currently used machine learning methods can be loosely divided into supervised and unsupervised machine learning, which make up the great majority of techniques. Other methods, such as partially supervised, active, and multi-instant learning, have only sporadically been used in chemistry up to this point.

Predicting the physicochemical characteristics and biological functions of chemical compounds is the objective of supervised learning in chemistry. Regression models carry out the quantitative forecasting of traits with actual value, while classification models evaluate the qualitative predictions (are they "active" or "inactive"?). Multiple linear regression (MLR), partial least squares (PLS), neural networks, support vector regression (SVR), and k-nearest neighbor (kNN) are the most widely used regression techniques at the moment in chemoinformatics applications, while Naïve Bayes, support vector machines (SVM), neural networks, and classification trees—particularly the Random Forest method [19]—are popular classification techniques. Additionally, there are models with structured output [16] and ranking models [20] which forecast values belong to classes of any complexity and which anticipate ranking order rather than property values. Specialized SVM modifications can be used to build the latter two types of models. By unsupervised learning, hidden patterns in the data are exposed. Unsupervised modelling approaches' main responsibilities are: a study of data clusters; dimensionality reduction; novelty (outlier) detection; and reduction in size. These are all distinct instances of data density estimation, in my opinion. The two types of clustering techniques—nonhierarchical (like k-means) and hierarchical—are frequently used. The most popular dimensionality reduction methods are ICA (Independent Component Analysis) and PCA (Principal Component Analysis). Kohonen Self-Organizing Maps (SOMs) [16] are frequently employed with the intention of visualising and analysing the chemical space. The application domains of QSAR/QSPR models are currently defined using a variety of machine learning techniques, including one-class SVM [54], as well as in virtual screening trials [16]. Two different sorts of models—primal and dual—can be distinguished in terms of data description. Primal models are based on the direct use of descriptors, whereas dual models are based on metrics describing similarity relationships between chemical structures. Kernels, which may be calculated from molecular descriptors and by directly comparing chemical structures, are the most beneficial kind of these measures. Both primal and dual strategies can be used for supervised and unsupervised modelling problems. Last but not least, statistical models for a network of mutually connected models may be built within the framework of the Multi-Task Learning and Feature Net approaches, and their predictive performance can be utilised due to the Inductive Learning Transfer phenomenon [8].

2.8   BIOINFORMATICS

The use of computational and analytical methods in bioinformatics allows for the collection and analysis of biological data. It is a multidisciplinary field that combines genetics, biology, statistics, mathematics, and computer science. In order to better understand genetic illnesses, bioinformatics is mostly utilized to discover genes and nucleotides. It has a strong connection to computational biology. The two names are frequently used interchangeably. Yet, they are two distinct locations literally.

2.8.1   CONTRAST COMPUTATIONAL BIOLOGY VS. BIOINFORMATICS

Bioinformatics and computational biology are interdisciplinary approaches to the study of life. They draw on empirical fields including computer science, physics, information science, and mathematics.

Both of these sectors, which are frequently employed in research facilities, labs, and universities, have developed as a result of the quick expansion of bio business around the world.

Despite the fact that the two professions may sound similar, the demands that they meet are different. For reference and clarification, the primary distinctions between them are listed below.

• Bioinformatics uses evaluation, analysis, and interpretation of bio data to address biological problems. The scripts, algorithms, and models used to record and preserve biological data are made by experts in bioinformatics. Computational biology's objective, on the other hand, is to resolve problems that result from bioinformatics research.

• Bioinformatics is employed in a variety of industries, including medication development, climate change research, molecular medicine, customized medicine, and microbial genome applications. Computational biology is used in a variety of fields, including stochastic models, molecular medicine, oncology, animal physiology, and genetic analysis.

2.8.2   USES OF MACHINE LEARNING IN BIOINFORMATICS

2.8.2.1   Genomics

The field of bioinformatics known as genomics is crucial since it focuses on the study of mapping, evolution, and editing of the genome. An organism's genome is its whole collection of genetic material. The primary divisions of genomics are described below.

- Regulatory genomics is the study of how to control the expression of the genome. Machine learning applications in this field of genomics include the creation of RNA-binding proteins and transcription factors, as well as the prediction and classification of gene expression.
- Structural genomics: By using computational and experimental methods, it aims to characterize genome structures. In this section, bioinformatics machine learning is used to categorize the three different types of protein structures (primary, secondary, and tertiary).
- Functional genome research: Scientists attempt to explain how genes interact in this field. In biology, machine learning can be used to categorize mutations and protein subcellular localization.

Natural language processing and machine learning techniques have made it possible for researchers to examine substantial biological data connected to genomics. They could quickly resolve issues like relation extraction and named entity recognition in this way.

2.8.3   GENOME SEQUENCING

Genome sequencing is essential to medical diagnostics. According to machine learning-enabled DNA sequencing techniques like next-generation sequencing, researchers can now read entire human genomes in a single day as opposed to the over 10 years it took with the previous Sanger Sequencing Technology [21–22].

2.8.4 GENE EDITING

The act of altering genes is called changing an organism's genetic make-up by deleting, adding, modifying a DNA sequence. It uses a technique known as CRISPR [23], which is a speedier and more affordable way to complete the surgery.

2.8.5 CLINICAL WORKFLOW

The clinical workflow process has been greatly changed by machine learning. For instance, accessing patient data that is kept in electronic records, paper charts, and other sources has never been simple for healthcare providers. But now that ML-enabled solutions like Intel's Analytics Toolkit are available, healthcare organizations can better utilize patient data.

2.8.5.1 Proteomics

Proteomics is the study of the components of proteins, how they interact, and how they work in an organism. Many human proteins have been examined thanks to mass spectrometry-enabled proteomics. Its development has been hampered by computational and experimental issues, necessitating informatics methods like machine learning to evaluate and understand vast biological data sets. Due to its high throughput capabilities, mass spectrometry is employed in the scientific fields associated with measuring such biological molecules in a high-throughput way investigation as an analytical method to characterize biological samples.

Proteins are not directly measured by mass spectrometry in their typical form. Instead, it divides them into smaller units made up of around 30 building block amino acid sequences. The amino acids are then assigned to particular proteins after being compared to the database. Because certain proteins are incorrectly identified, the results are not totally accurate.

A variety of proteins can be recognized by a given sample using machine learning techniques. They can be used with:

- Peaks in the mass spectrum, which allow samples to be examined without revealing the presence of specific proteins or peptides. Instead, potential biomarkers are compiled from peaks with high signal intensities.
- Proteins identified by examining sequence databases: Peptide masses are found in the material being studied, and they are then used to identify the proteins to which they belong.

In the identification of many illnesses, these technologies offer a significant superiority over more traditional ones as 2D gel electrophoresis, protein arrays, affinity separation, and enzyme-linked immunosorbent assays (ELISAs).

2.8.6 MICROARRAYS

Microarrays are laboratory techniques that monitor several gene expressions simultaneously. This approach is useful in investigating genome organization, gene expression, and chromatin architecture, which is becoming more prominent in animal, plant, and microbial genetic studies.

On a microarray, which is often on a silicon microchip or glass slide, different probes (DNA, RNA, tissues, proteins, and peptides) are organized in a certain pattern and match to different gene segments. This method is predicated on the idea that given the right circumstances, complementary sequences will attach to one another but non-complementary ones won't. The degree of hybridization between contemporary probes is indicated by fluorescence.

The complexity of microarray data sets is rising quickly. For big experiments, thousands of probes are needed.

- Using machine learning methods, Neural Designer, for example, has allowed researchers to find deep linkages as well as spot intricate patterns in microarray data. Furthermore, Array Express and other public databases record every detail of a microarray study, making it easily reusable by the research community.

Machine learning algorithms have been used on microarrays in the following ways:

- Gene evaluation: Examines alterations to gene patterns to see whether they are normal if they are caused by a disease.
- Recognize various gene stages: Identifies the circumstances under which genes change from a healthy to a diseased state.
- Foresee upcoming gene stages: Creates models for use previous biological data to anticipate future gene alterations.
- Disease prevention: Supports the discovery of associations between genes and illnesses and makes use of predictive modelling for early detection and preventative care.

2.8.7 TEXT MINING

Text analytics is another name for text mining. It is a machine learning-powered system that scans massive amounts of papers and unearths fresh data to assist in resolving research issues.

Because of the increased number of biological publications, it has become more challenging for researchers to comb through a variety of sources and compile helpful knowledge about a certain subject.

Using machine learning processes and evaluating data in databases using various forms of human-generated reports, saves labor expenses and accelerates the research process without sacrificing quality.

In bioinformatics, ML text analysis can be used for:

- Analysis of substantial protein and molecular interactions.
- Content conversion to other languages.
- Identification of new therapeutic targets (due to the extraction of material from biological journals and data sets).
- Automatic gene and protein activity annotation.
- DNA expression array analysis.

2.8.8 SYSTEMS BIOLOGY

Systems biology is the study of interactions and behaviors among biological building blocks including molecules, cells, organs, and organisms using numerical and statistical methods. In this area, computational modelling is a helpful tool. It simulates the overall behavior of the system and employs mathematical modelling to capture the interactions between the biological parts. It is challenging to create a solid mathematical model, nevertheless, due to the underlying systems' complexity and lack of adequate understanding.

Yet modeling complex relationships in areas like signal transduction networks, genetic networks, and metabolic pathways has been simpler with the introduction of data-driven machine learning techniques.

When there is sufficient biological data but insufficient biological insight to build theory-based models, machine learning can be helpful in biological systems. The finding of the relationship between *S. cerevisiae* genotype and phenotype is a nice illustration.

Despite the large number of strains with documented phenomes and genomes, there's no theory-based models that show how genotype differences influence strain phenotypes. Using a supervised model with genomes as input and phenomes as output allows us to identify the relationship between phenotypes and genotypes in this situation. The created model's interpretation reveals details about the organism's essential genetic makeup. It aids in the identification of the most important parameters that contribute to the prediction power of the model.

The probabilistic graphical model is a popular machine learning technique in systems biology. It establishes how many variables are organized and helps build genetic networks.

Using genetic algorithms is another well-liked method. It has been used to simulate genomic networks and regulatory systems and is based on natural evolution.

2.8.9 HEALTHCARE

Artificial intelligence and machine learning are frequently used in healthcare institutions to enhance patient care and quality of life. Hospitals may soon be able to improve treatment efficiency by collecting real-time data from various healthcare systems across numerous nations using machine learning-based technology. Among the most important uses of ML in healthcare are:

Drug development and production:

In the early stages of drug discovery, machine learning is commonly used. Precision medicine and next-generation sequencing are two of the research and development technologies used. They have demonstrated value in locating alternative treatments for complicated diseases.

Medical imaging and diagnosis:

Computer vision, a revolutionary technique, has been enhanced through the application of deep learning and machine learning. The technique has been widely adopted for a number of applications. For instance, the Microsoft Inner Eye project develops innovative technologies for the quantitative analysis of 3D medical images.

Individualized medical care:

Individualized treatment can benefit from predictive analytics on patient data. Now, doctors must make assumptions about a patient's risk of disease based on their health history and incomplete genetic information, or they are restricted to a certain spectrum of diagnoses. This could soon change since machine learning is making great strides in medicine by using patient data to help develop a variety of treatment options.

Diagnosis of stroke:

Machine learning uses pattern recognition to aid in the diagnosis, treatment, and prediction of complications in a variety of neurological illnesses. Acute Ischemic Stroke (AIS) treatment has made great progress in recent years. To anticipate motor abnormalities in stroke patients, machine learning techniques are currently being applied. Support Vector Machines (SVM) and 3D Conventional Neural Networks are the most often utilized approaches (CNN).

2.9 MACHINE LEARNING TOOLS USED IN BIOINFORMATICS

2.9.1 DEEPVARIANT

This is a tool for deep learning. It is used for mining genetic data. When compared to prior conventional approaches, it can predict common genetic changes more correctly. One of the first biological tools, DeepVariant, uses Google computers and machine learning to provide a scalable, cloud-based solution for the most challenging genomics data sets.

2.9.2 ALGORITHMS USING ATOMS

The first deep learning-based technique for rendering molecules as 3D pixels was developed by San Francisco-based biotech business Atomwise. This conversion helps with the atomic-level comprehension of the three-dimensional structure of molecules such as proteins.

Furthermore, it anticipates potential chemical reactions with a certain protein. The algorithms are mostly used in the creation of pharmaceuticals.

2.9.3 PROFILE OF A CELL

Formerly, software for biological imaging could only measure one variable from a collection of photos. Nevertheless, thanks to machine-learning algorithms, this has changed. With the Cell Profiler programme, scientists may now prepare and photograph an infinite number of samples every day.

Furthermore, the software can quantify particular characteristics such as the reflecting light of a cell in a microscope area. With the use of deep learning algorithms, it can also recognize thousands of cell traits.

Using deep learning techniques, a model is created based on the properties that were recovered from massive datasets like genomes or collections of photographs. The model is then used to analyze various biological datasets.

Processing large datasets created by newly emerging technology into usable information is one of the most important concerns in bioinformatics and biology in general. But as we go into the age of AI and big data, machine learning in bioinformatics is becoming a more significant force behind this transition.

Advantages:

- Very nicely justified theoretically.
- Because assumptions are explicit, they may be improved and reviewed.
- ML applications are consistently gentle.
- Sequence simulation trials have demonstrated that this strategy beats all others in the majority of circumstances.

Disadvantages:

- Bioinformatics is experimental; biology and bioinformatics are separate but complimentary fields of study.
- To generate raw data for analysis, bioinformatics relies on experimental research.
- The accuracy of bioinformatics predictions is dependent on the accuracy of the data and the sophistication of the algorithms being employed, which in turn provides helpful interpretation of experimental data and significant leads for additional experimental investigation.

2.10 ETHICS: BIOINFORMATICS AND CHEMOINFORMATICS

The field of bioinformatics, which studies the transmission, creation, and reception of information, developed in response to the overwhelming need for large-scale biological data management and storage. Both the public and corporate sectors have committed significant resources to the creation of bioinformatics tools and applications. It is therefore crucial to safeguard innovations by providing a legal framework.

The gathering, storage, analysis, and manipulation of chemical data are the main goals of the relatively new subject of information technology known as cheminformatics. Information about tiny molecule formulae, structures, characteristics, spectra, and activities is frequently included in the chemical data of interest (biological or industrial).

For a very long time, the main goal of bioinformatics has been the creation and maintenance of a database to store biological data. This type of database's development necessitated not only taking design challenges into account, but also developing an interface via which academics could access both already-existing data and submit new or updated data.

Some ethical issues on bioinformatics and chemo informatics are explored in the below subsections [24–25].

2.10.1 BIOETHICS

The term *Bioethics*: Greek *bios*, "life"; *ethos*, "moral nature, behavior" [26].

Bioethics is concerned with moral questions relating to health (mostly focusing on human ethics but also increasingly incorporating animal ethics), such as those arising from developments in biology, medicine, and technology. It suggests having a conversation about moral judgement in society (decisions that are "good" or "bad" and why), which is frequently tied to medical policy and practice but also to more general issues like the environment, wellbeing, and public health.

The ethical issues that occur in the interactions between life sciences, biotechnology, medicine, politics, law, theology, and philosophy are addressed by bioethics. The study of ethical principles in primary care, other medical specialties ("the ethics of the ordinary"), science education, animal, environmental, and public health ethics are all included.

2.10.2 BIOETHICS' DEVELOPMENT

Bioethics emerged as a distinct field of study in the early 1960s [27]. It was influenced not only by advances in life science, but also perfection of certain medicine and lifesaving technologies. Medical professionals must make difficult judgements concerning which patients receive treatment and which are permitted to pass away, such as organ transplantation and kidney dialysis [27]. The end consequence was broad opposition to traditional medical paternalism and the increasing acceptance of a patient's entitlement to full disclosure of his condition and some degree of control over what transpired to his body.

2.10.3 BIOETHICS AND TECHNOLOGY

Bio with technology means biotechnology. Bio means natural amenities and technology means technical ventures, with these ventures solving medical purposes. Two major subfields of what is sometimes referred to as "applied ethics" are bioethics and computer ethics.

In its broadest sense, applied ethics is an effort to use philosophical terminology, theories, and argumentative techniques to address "real world" issues like nuclear proliferation, pollution, the economy, and crimes related to these, as well as issues that are typically related to bioethics and computer ethics, like abortion.

Computer ethics is a dynamic and complicated subject of study that takes into account how concepts, laws, and values relate to the always evolving realm of computer technology. Computer ethics is a discipline in and of itself that offers both conceptualizations for understanding and rules for using computer technology, even though it is a field between science and ethics and depends on them.

The premise is that there seems to be a need for a unique kind of tools that can effectively handle cases when computers are actually engaged. Discussions on the nature of computers and information technology may be included in such a set. Computer ethics do not have an issue when a thief breaks into a home and steals a computer, but they do have a problem when a hacker utilizes their knowledge to steal a lot of personal information. One needs a set of conceptual tools that are unavailable in the former situation in order to fully consider the ethical implications of the latter conduct.

The main ethical concerns raised by bioinformatics relate to the use of computers to modify genetic data. Although privacy is arguably the most pressing issue, other issues include discrimination, genetic profiling, the creation of medications that specifically target particular people and could result in stratification, and more. There will be an ethics that addresses these problems if they are genuine and have an impact on our lives. What should it be, though? Should it be bioethics or computer ethics? Computer ethics differs from traditional or theoretical ethics in that it calls for a unique lexicon and set of instruments. This particular set is necessary due to the nature of the computer itself. If this is the case, then because bioinformatics involves extensive computer use, its ethics should be seen as a subset of computer ethics and hence call for a particular set of rules.

2.10.4 LEGAL ISSUES IN BIOETHICS

A study of key case laws concerning biotechnological inventions and discoveries, appraisal, and case studies are all part of the field of bioethics. Case studies and instances of successful patent grants are also studied. Industrial designs, plant breeder's rights/protection of plant varieties, IC layout designs, trademarks, geographical indications, and trade secrets are some examples of further IPRs. [28]

Chem-bioinformatic models relate the chemical makeup of a medication and/or its target (a protein, gene, RNA, microbe, tissue, or disease, for example) to the biological activity of the drug against that target. A comprehensive judicial framework, on the other hand, is required to give suitable and relevant direction for handling diverse computer approaches as applied to scientific research in bioscience frontiers.

Chem-bioinformatics model predictions for regulatory reasons, as well as to legally safeguard molecular system models and the software used to find them.

Large-scale computational biology's use of software tools raises a number of legal difficulties, including those relating to the purchase and licensing of commercial software systems, in-house and outside software development, and open-source software.

Bioinformatics Law: Legal Issues for Computational Biology in the Post-Genome Era begins with an overview of the rapidly developing field of bioinformatics before examining the legal problems related to the software tools that enable extensive computational biology, such as the purchase and licensing of commercial software systems, in-house and outside software development, and open source software problems. The focus then turns to the legal protection of models and software, along with a quick summary of the themes and strategies for the legal defense of computer software and any discovery made in science utilizing that application. It also affects the potential patentability of any substance or product, such as when computational methods wholly or largely take the place of an experimental finding.

All of the assessments we've reviewed on the legal protection of chem-bioinformatics models and software address essentially the same legal problems in relation to chem-bioinformatics studies.

Problem 1 – Copyright protection: It is widely understood that intellectual property is the field of law that protects software through the copyright system. Methodologies based on computer software or software-related innovations, on the other hand, should be patentable. But there is a separate conceptual foundation in favor of multinational copyright laws [29]. On a practical level, software companies employ a variety of safeguards, including trade secret rights, the publication of "objects code," contractually obligated users, and, increasingly, patent protection [30].

Problem 2 – Patent protection: A patent is an exclusive right awarded for an invention, which is a product or technique that offers a new way of doing something or a new technological solution to a problem [31]. Although many nations have welcomed the patentability of software-related inventions, international legislation on the patentability of software is still not standardized. The "technical effects" theory, initially promulgated by the European Patent Office, is the most generally accepted doctrine defining the extent of patent protection for software-related discoveries

(EPO). According to this concept, software is typically patentable if its application has a "technical impact." [29]

Problem 3 – Trade secret protection: The necessity for trade secret legislation is immediately clear given that access to or lack of access to source code is such a crucial computing issue and that the majority of proprietary software owners go to great pains to keep such data private (non-disclosed) [31].

Problem 4 – Proteomics-related computer programmes are covered by trademark protection, which also covers the programmes' external components such distinctive trade dress (trademark). A trademark is described as a brand or medium that may be graphically expressed and can distinguish the goods or services of one enterprise from those of another enterprise. A trademark may consist of both words and designs, characters, numerals, or a product's or package's shape [32].

TRIPS provide legal consistency among member countries by demanding the harmonization of applicable legislation [33, 34]

Problem 5 – Contractual matters: The terms of a contract may be stricter or less strict than those of a copyright law [35]. Software creators and publishers have long attempted to strengthen the implicit safeguards provided to software programmes by adding additional obligations through contract, frequently in the form of a licence agreement [36]. The so-called "click- wrap" licence may be used for online contracts. The licencing conditions displayed on the screen are accepted by the buyer when they click the appropriate button or type "Agree," "Yes," or "I Accept" in the appropriate fields of an online registration form [36]. This seems to be a common practice to ensure that the user is aware of any contractual terms and conditions at the time the contract is actually entered into [37].

Problem 6 – Difficulties with software taxes: The taxation structure reflects the expanding significance of software on the global economy. Taxation of cross-border transactions involving software has long been a contentious issue because tax systems are reliant on domestic nation rules. A "royalty" is described as compensation for the use of software. Royalty payments are acquired through the use or transfer of intellectual or industrial property rights, according to the OECD Model Tax Treaty. Most tax conventions tend to expand or reduce the range of payments recognized as royalties based on a set of agreed-upon interests rather than entirely adhering to this classification.

2.10.5 LEGAL SAFEGUARDS FOR COMPUTER SOFTWARE

Researchers in the field of bioinformatics who are looking for knowledge on any legal problems that may affect their work have several challenges in acquiring appropriate information. This is mostly due to the fact that there is no one means of legally securing software. In truth, there is no single worldwide framework addressing software protection. In actuality, there isn't a solitary global framework for protecting software. Despite a global trend towards explicit legislative protection for software, the breadth and viability of enforcement from this type of protection differ greatly by jurisdiction, i.e., through copyright laws [29].

2.10.5.1 Copyright Protection

There is no question that the area of law known as intellectual property deals with the protection of software, particularly through the copyright system, and that methods involving computer software or discoveries linked to software should be patentable. It is worth emphasizing, however, that a separate conceptual foundation resides in favor of international copyright laws. Although it is legally correct to distinguish between the copyright model and the author's right model, we will refer to both regimes as "copyright" throughout the work because it is a term that is often used in the English language. Common law copyright and continental author's rights are the two traditions that make up copyright. The rights model (from the French: droit d'auteur), the Anglo-American tradition, the first copyright custom, the common law copyright system are recognized in both US and UK legal systems. Based on economic trends, this regime's emphasis is thought to be more utilitarian in nature. Common law copyright is supported by the practical necessity to maintain incentives for individuals and businesses to make fresh investments that will lead to social advancement [39]. Another model, author's rights, is used in some form by the majority of European Community continental members and the majority of the rest of the globe. While these countries' approaches differ in specifics, such as perpetuity of rights, they all share a common focus that is centered on the idea that the individual retains the right to control concerning their creations; both the economic exploitation of the work (such as reproducing and distributing), but also the "moral rights.", such as deciding whether to publish the work or not, attributing authorship, and preserving the integrity of the work [39–40]. Yet, it is important not to exaggerate the differences between the two systems. There are certain similarities between these systems that are crucial for comprehending how copyright might be handled in the context of software [30].

In fact, one of the main difficulties in creating a coordinated multi-party intellectual property strategy for every new technology is bridging the gaps between these two systems' disparate methods. This problem affects creating international agreements on copyright as well as creating copyright legislation for the European Commission (EC). [29]. The territoriality principle governs the application of copyright rules. They frequently restrict protection to either works that were first published in a country or to its residents. The availability of protection is addressed by conventions and bilateral agreements, which extend it to foreign authors in accordance with formal reciprocity or national treatment standards [29]. Generally speaking, a nation will accord works by foreign writers the same protection it accords to works by its own citizens under the concept of national treatment. Similar to formal reciprocity, a nation will protect works by writers from other nations just as it does works by its own citizens, but only if it judges that the works of its own citizens are given at least some level of protection in the other nation [29]. The Berne Convention and the Universal Copyright Convention are the two most important international agreements pertaining to copyright protection [42], as well as some of the provisions of the TRIPS (Trade Related Aspects of Intellectual Property Rights) agreement [43]. In nations that have ratified the Berne Convention, copyright protection is formality-free, meaning that it is not contingent on formalities like registration or copy deposit. Software is protected by the Berne Convention in the same way as literary creations are, to the fullest degree possible. Recent developments in the European Union have demonstrated potential for the creation of a more coordinated and effective global copyright regime. The European Commission was compelled to provide a standardized approach to intellectual property, spanning computer software, semi-conductors, and biotechnology products [44]. This was accomplished amongst nations that did not even adhere to the same common law copyright principles or author's rights framework. Finally, a functional software copyright regime for all member nations was developed [41]. According to the Community Directive on Software Copyright (91/250/EEC Directive), software is covered by copyright across the EU. According to European law, the programmer who created the programme is the rightful owner of the intellectual property, and running, copying, altering, or disseminating the software without the owner's consent is prohibited. When there are many programmers, the Directive also allows for co-ownership. Unless there exists a specific agreement to the contrary between the programmer and the employer, the rights belong to the employer when a programmer writes a programme while employed. The "moral rights," which are freely assignable, nevertheless belong to the coder. Several Member States demand written assignments [45]. A practical and viable answer to the copyright issue is the Software Directive. It establishes a framework to offer effective anti-piracy defence for software. It achieves a compromise between enabling actions that promote innovation and market competitiveness while also preserving the right holder's investment in developing software. However, several concerns that may have been resolved were left unresolved, which might lead to future uncertainty [41]. Freedom of action is also guaranteed for the software sector to the extent that software copyright only grants exclusivity to their expression and does not grant exclusivity to ideas or functions. Practically speaking, software providers employ many layers of protection, including trade secret rights, disclosing "objects code," enforcing contracts with users, and increasingly looking for patent protection [30].

2.10.5.2 Patent Protection

An innovation, such as a product or a technique that gives a novel approach to a problem or a new technological solution, is given an exclusive right known as a patent. Patents must have an innovative step, be evaluated for commercial viability, and go through an examination process [31]. There is currently no international harmonization of the laws governing software patentability, despite the fact that many countries have to some extent recognized the patentability of advances relating to software. Since a patent can be used to sue anyone who uses or sells a protected innovation within that country, even if the infringer independently created it, there is a global trend towards accepting patent protection for software-related concepts [29]. If the underlying concept falls within one idea is protected by patent law if it falls under one of the statutory categories of patentable subject matter and is not so fundamental as to constitute a law of nature [30]. Copyright law, however, merely defends the expression of ideas. The "technical effects" concept, initially introduced by the European Patent Office, is the most commonly regarded theory defining the range of patent protection for innovations relating to software (EPO). According to this approach, software is typically patentable if it is used in a way that has a "technical impact." Hence, for instance, software that manages an electrical engine's timing is Software that recognizes and corrects contextual homophone mistakes (such as changing "there" to "their") may not be patentable under this concept [4]

2.10.5.3 Trade Secret Protection

Given that having access to the source code or not is such a key computing issue and that the majority of proprietary software owners take great care to maintain such code as private (non-disclosed) information, the significance of trade secret legislation is immediately apparent [7]. This type of defence encompasses any attempt to take advantage of the firm's accomplishments in software development, whether it results from the dishonesty of workers or authorities, the buyer of the programme who is obligated by a contract with the company, or any other comparable behaviour. Computer repair and maintenance manuals may be protected under the trade secrets regime, allowing for easier software reverse engineering. Trade secrets of any complexity are covered under the TRIPS agreement. In accordance with Article 39 of TRIPS, all signatory nations are required to implement trade secret law, which would encompass software if conditions like those relating to secrecy were satisfied [32].

2.10.5.4 Trademark Protection

The law tackles issues with computer programmes' exterior elements, such as a distinctive commerce, and safeguards their internal content (trademark). A trademark is defined as a brand or media that can be visually displayed and may differentiate the goods or services of one enterprise from those of another enterprise. Words, images, characters, numbers, or even the form of a product or packaging can be used as a trademark [33]. Internationally, trademark registration as a method of protection is well-established. Most nations first acquire rights by registration, then retain them through continued usage of the country. In general, trademark protection is not offered by use of the mark without registration. Some nations, such as the United States, Canada, and the Philippines, require use of the mark prior to registration, while others grant some rights priority based purely on usage (e.g., common law countries). Many intellectual property treaties, most notably the WTO Agreement on Trade-Related Aspects of Intellectual Property Rights, have reduced the fundamental restrictions of the geographical applicability of trademark regulations. TRIPS establishes legal compatibility between member jurisdictions by requiring harmonization of applicable laws. In many nations around the world, the definition of a "sign" used as a "trademark" or contained therein is provided under TRIPS. International trademark issues are also governed by the Madrid Agreement, Madrid Protocol, and the European Community Trademark (ECT) [34, 35]. Only per-country protection for trademarks is possible. Depending on the jurisdiction, it might take many years or just a few months to get from application to registration. Other nations check the current register for potentially contradicting earlier decisions and uncommon occurrences. In the end, the enforceability is determined by the specific facts and the appropriate legislation of the nation [38].

Software distribution under shareware licences is another popular strategy. It can be downloaded from an internet site or offered through a CD, which is typically given away with a computer magazine. The way that such a licence operates is by offering a free trial of the software, usually for a set amount of time, after which the user can choose whether to pay a registration fee to obtain a licence for the software's continued use, subject to the license's terms and conditions, or to stop using it. The shareware software programme typically has a tendency to stop functioning after the trial time. Software may also be sold in limited-feature versions that may be upgraded after registering. A variation of the shareware system, this is referred to as the "crippleware" system. In this case, software is made available under a shareware licence, but in an unfinished or "crippled" state; for example, the software is unable to save or print files. The copyright holder will send a buyer a "uncrippled" version of the programme in exchange for a price. "Nagware" is a frequent variation. This consists of software used to remind users on a regular basis to pay the copyright owner. After paying the licence price for a "crippleware" system, the user will receive a "nag-free" version [38]. All of these licence types are often referred to as "proprietary" licences. The term "non-proprietary licences" refers to a number of less restricted licence types as well as unconventional business structures. Both "public domain" software and "all-permissive licences," which may be regarded as software not protected by any copyright, fall under this category. Certain licences provide unrestricted use, distribution, and copying but may forbid modification; this class of licences comes very near to being "all-permissive." A "semi-free" licence is also quite lenient in terms of non-commercial users (academics, governmental organisations, etc.), but it forbids usage for profit. Typically, the business plan is to offer the same software under two separate licences: a proprietary software licence for corporations and a semi-free licence intended to promote widespread non-commercial usage of the product (which, of course, businesses have to pay for). In addition, many pieces of "open source" software fall under "copyleft" licences. Anybody may use, copy, and distribute the programme as they see fit under the terms of the licence. The programme may also be altered for an individual's personal usage. Under the condition that both the source code and the modified object code are distributed, modified versions of the programme may be distributed. Also, this version has to be shared under a "copyleft" licence [38]. Using licenced software and being aware of the permissible usage as outlined by the licencing conditions are crucial in the field of medicinal chemistry. This establishes the legitimacy in law and the potential for publishing of any scientific discovery produced using that programme. It has an effect on any product or substance's potential patentability as well, such as when computational algorithms completely or substantially replace an experimental finding.

2.11 INTELLECTUAL PROPERTY

2.11.1 INTELLECTUAL PROPERTY PROTECTION FOR BIOINFORMATICS DATABASES: TRANSNATIONAL PERSPECTIVES

Nowadays, rules that safeguard bioinformatics databases also safeguard other databases. While most nations employ copyright laws to safeguard bioinformatics datasets, the European Union adopts a *sui generis* approach. Databases were protected by copyright, but there was also a *sui generis* right that could be used to safeguard the creator's investment in some unique but unoriginal databases.

Copyright protection for databases was made accessible to nations and parties under the Berne Convention or Trade-Related Aspects of Intellectual Property Rights (TRIPS) Agreement as a consequence of the finding of a balance [46] *Sui generis* rights were also exclusively applicable to those who created the European Union. In the US, databases are shielded by copyright legislation. It was examined in 1991 to see if databases might be protected by copyright or by using the principles of "industrious collecting" or "sweat of the brow" [47]. The disparity between the rules of the European Union and the United States has a big influence, making it unlikely that the two would work together on anything. As a result, the IP rights system may have unintended effects. Complex databases, such the bioinformatics databases, are not properly covered by database law as a result of these conflicts [47].

2.11.2 PATENTABLE TECHNOLOGIES IN BIOINFORMATICS

Although databases themselves are not patentable, inventions linked to databases may be covered by patents. If a software innovation meets all the requirements for software patents [45], it is patentable in the US. Software was not previously covered by IP protection on its own since they did not fall under the definition of patentable subject matter.

Nevertheless, the U.S. Federal Circuit Court of Appeal significantly altered the law in this field of software patents in State Street Bank & Trust Co. v. Signature Financial Group Inc. [46], which may have a significant effect on the bioinformatics sector [47]. According to the court's decision in the State Street case, mathematical algorithms are patentable if they result in a "useful, concrete, and tangible result," and the mere fact that a claimed invention involves entering, calculating, and storing numbers [48] does not make it non-statutory subject matter unless it fails to produce a tangible outcome. According to this case law, algorithms and software that satisfy the requirements are patentable in the United States.

In Schlumberger Canada Ltd. v. Canada [48] (Commissioner of Patents), the main court case in Canada, it was determined that algorithms employed in software programmes to compute protein sequences, shapes, locations, and functions may also be relevant. Hence, it can be demonstrated that bioinformatics patents are compatible with present patent law [48].

2.11.3 PATENT PROTECTION FOR CHEM-BIOLOGICAL DATABASES

Collections of biological sequences make up biological databases. Biological databases are also not patentable if biological sequences are not. They must be connected to a statutory subject matter in order to be patentable. The method used to create the database may qualify as a patented technique, even if the database itself may not qualify as patentable subject matter.

It was stated in the State Street case [49] that even while information itself is not patentable as a physical good, a method of creating the information may be. Second, the database itself would not be covered by the patent; only the database creation method would. That would reduce the patent's value since a rival might easily create the patented product in a method that does not violate the law [49].

2.11.3.1 Patent Protection for Bioinformatics Software and Hardware

In contrast to biological sequences and databases, computer software qualifies as patentable subject matter if it produces an actual, usable good. Because the output of bioinformatics software has biological applications, it is definitely perceptible subject matter. Since that bioinformatics software can be used to manufacture drugs and conduct medical diagnosis, it would be difficult to deem it unpatentable.

Bioinformatics hardware is similarly patentable. A patent cannot cover a biological sequence or database that is only a portion of a covered machine or apparatus since the patentee will only be protected from infringement if they employ the machine or apparatus that fully encompasses the claimed innovation.

2.11.3.2 Challenges to Bioinformatics

Bioinformatics patents raise a number of problems, some of which are obvious while others are less apparent. The obvious issue is that the involved technologies are transdisciplinary. The less obvious challenges relate to the variety of business models employed in bioinformatics and the corresponding variety of patent claim types that may be necessary to ensure the maximum amount of patent protection.

A challenge for many bioinformatics inventions is locating a patent attorney who is knowledge-able about both the IT and biotechnology components of the breakthrough. Many have suggested using two patent attorneys—an IT patent attorney and a biotech patent attorney—to ensure that the technical aspects of a bioinformatics patent are addressed.

This "tag team" approach is comparable to how the US Patent and Trademark Office assesses bioinformatics patents. The Patent and Trademark Office has put together a special team of exam-iners with interdisciplinary training to assess bioinformatics patents. These examiners have back-grounds in biotechnology, computer science, physics, math, and other subjects.

The "tag team" approach might be helpful in overcoming technical challenges, but it is insuf-ficient to maximize the benefits of a bioinformatics patent on its own. A significant challenge is comprehending the various bioinformatics business models and the various patent claim types that may be pertinent to those enterprises.

Data-selling businesses have significant challenges as a result of patent protection. Data itself is not patentable, but data structures are. One of the most important aspects of bioinformatics is managing the large amount of data generated by genomic and proteomic investigations, which can amount to terabytes and beyond. Among the improvements related to the technology for processing this data are novel ways of storing data. It is possible to make statements about "data structure" to describe how data is stored.

ASP models and other client/server technologies present a number of specific international con-cerns. If an ASP system claim contains communication between a client terminal and a server, territoriality considerations may hinder a finding of infringement. If the server is operated by a competitor outside of the US and accessed from a client terminal inside the US, a system claim might not be applicable (or vice versa).

This is due to the fact that not all claim-covered activity occurred in the US. One way around this is to write separate client-side and server-side assertions. In this way, if either the client or the server is located in the United States, the action might be protected by the patent [50].

A number of bioinformatics business models involve licencing research data or tools in addi-tion to receiving royalties on any products made as a result of the use of the research tool as part of the licence fee. A product-by-process claim can be used to support this form of patent licence. A product-by-process claim protects a good made employing a particular technique. If the patent sufficiently explains and states how to make a product using the data or research tool, protection for this source of income might be acquired [51].

2.11.4 Database

These are databases that include structured biological data, such as protein sequencing, molecular structure, DNA sequences, etc.

The biological data may be altered using a number of computer tools, such as update, remove, insert, etc. To make their experiment data and findings more widely accessible, scientists and researchers from all over the world register them in biological databases.

The vast majority of different biological data may be found in free-to-use biological databases.

2.11.4.1 Types of Biological Databases

There are basically three types of biological databases.

As the experimental findings that the scientists submit are archived, it is also possible to refer to this database as an archive. The main database is filled with information that was obtained through experimentation, such as the genomic sequence and macromolecular structure. The information placed here is un-curated (no modifications are performed over the data).

It gathers distinctive data from the lab and makes these unchanged data available to regular consumers.

As the data are added to the database, accession numbers are assigned to them. The accession number can be used to subsequently obtain the same data. Accession number identifies each data individually and it never changes.

Examples:

Examples of primary database/Nucleic Acid Databases are GenBank and DDBJ.

Examples of protein databases are PDB, SwissProt, PIR,TrEMBL,Metacyc, etc.

2.11.4.1.1 Secondary Database

The primary database's analysis output is the data that is saved in these sorts of databases. The primary database is subjected to computational techniques, and the secondary database contains useful and instructive data.

The data are carefully vetted here (processing the data before it is presented in the database). A secondary database is superior to a primary database and has more useful information.

Examples of secondary databases are as follows.

- InterPro (protein families, motifs, and domains).
- UniProt Knowledgebase (sequence and functional information on proteins).

2.4.11.1.2 Composite Databases

These databases compare the data that is input before filtering it according to the specified criteria.

The primary database is used to pull the original data, which is then combined depending on certain criteria.

It facilitates quick sequence searches. Non-redundant data is present in composite databases.

Examples of composite databases are as follows.

- OWL, NRD and Swiss port +TREMBL.

2.11.4.2   Limitations of Bioinformatics Databases [52]

Overdependence on sequence information and related annotations without comprehending the accuracy of the information is one of the issues with biological databases.

The fact that there are numerous inaccuracies in sequence databases is frequently overlooked. The core sequence databases have substantial amounts of redundancy as well.

Gene annotations can occasionally be incorrect or lacking. Any of these mistakes can spread to other databases and affect other databases.

2.12   DATA ACCESS AND SOCIAL ISSUES

In recent years, the development of numerous bioinformatics applications has played a significant role in the health industry. Although there are many benefits to using bioinformatics, there are also certain moral, legal, and social concerns that arise with each use. Using a thorough evaluation of the literature that is currently available, in this chapter, bioinformatics applications in forensic databases with DNA fingerprinting, genetically engineered biological weapons, pharmacogenomics, personalized medicine, and mutation detection with NGS are examined. Due to the computational analysis of genetic data, which reveals certain private information about persons, ethical, legal, and social challenges are emerging. Because the outcome can disclose some unpleasant and terrible truths that society sometimes cannot accept. Privacy is the fundamental concern in bioinformatics because most applications work with genes, which are private and particular to each person. The security and safety of any bioinformatics application should be attested to, along with that of the input and output data, which are more private and sensitive. This report makes various recommendations for positive actions and regulations to mitigate the aforementioned non-technical impacts.

In this decade, computational technology advancements have greatly improved. These significant advances in computer and biology have enabled the realization of many scientific aspirations. Using computational methods also reveals a number of hidden and undiscovered facts in biology. The biological domain has greatly benefited from the development of data science, particularly in the areas of big data analysis, data mining, and pattern recognition. This development results in the creation of a brand-new field called bioinformatics, which modifies both biology and computing. The health sector benefits greatly from bioinformatics in many ways, including treatment methods, monitoring, predicting, and other things. Nonetheless, a lot of problems could develop because of human social, cultural, and ethical ties. If we have a computational pipeline and a person's genetic information, it is simple to learn about their physical and psychological behavior. This could potentially hurt the specific person by disclosing private medical information or other information. The purpose of this study is to identify the ethical, legal, and social difficulties that arise from the use of bioinformatics in the health sector.

In YouTube videos and several online materials given by scientists and biologists, the effect of information technology and issues possessed on the functioning of these bioinformatics apps were explored. These documents and videos were examined in terms of ethical and cultural issues as well as potential solutions. According to a survey conducted by Deniz and Canduri, the top ten bioinformatics applications are DNA and protein sequencing, protein modelling, evolutionary studies, pharmacogenomics, genetic engineering, and biological weapons, personalized medicine with IBM Watson, mutation detection using next-generation sequencing (NGS), use of genomics with NGS, forensic databases with DNA fingerprinting, and proteomic technology [53]. Below are five applications that were chosen based on the survey's findings in order to examine the ethical, legal, and societal problems that using modern technology raises.

The five applications selected among the ten by Diniz and Canduri [53] include forensic databases with DNA fingerprinting, genetic engineering and biological weapons, pharmacogenomics, personalised medicine, and mutation detection using NGS.

2.12.1 FORENSIC DATABASES WITH DNA FINGERPRINTING

DNA forensics is a branch of forensic science that focuses on using genetic evidence in criminal investigations. The procedure first entered the court system in 1986, when police in England asked molecular biologist Alec Jeffreys to use DNA to verify a 17-year-old boy's confession in two rape-murder cases in the English Midlands [54]. The software programmes used in DNA fingerprinting include GelJ.

Images from a 1D gel electrophoresis are used to perform an automated DNA diagnosis using a bio-image processing technique (GELect).

2.12.2 BIOLOGICAL WEAPONS AND GENETIC MANIPULATION

The cornerstone of genetic engineering is gene transfer. Today, a large portion of the meals we consume are either entirely genetically modified foods or contain genetically modified ingredients. The United States exports food worth billions of dollars as a result of the sale of GM seeds and crops [55]. The threat of a public health emergency seems to be the greatest concern because there is a chance that some genetically altered germs could accidently enter the general population [56]. The goal of biological weapons is to spread disease among people, plants, and animals by introducing poisons and pathogens like viruses and bacteria [57]. The bioweapons business can use genetic engineering to modify genes to create new pathogenic features that are meant to increase the efficiency of the weapon by increasing its survivability, infectivity, virulence, and medicine resistance. Despite the fact that it is obvious that improved biotechnology will be advantageous to society, one of the major threats we will likely face in the near future is the "black biology" of the creation of bioweapons.

2.12.3   PHARMACOGENOMICS

Pharmacogenomics is a field that seeks to identify the genetic causes of individual variability in medication response. It is described as the use of whole-genome technology to forecast a patient's disease's sensitivity to or resistance to chemotherapy. Another significant application for pharmacogenomics profiling is the assessment of the safety profile of prescription medications that have already been approved for usage. It has been established that DNA microarrays, sometimes referred to as microchips or microarrays, are a state-of-the-art technique for high throughput, detailed analysis of thousands of genes simultaneously.

2.12.4   MEDICAL PERSONALIZATION USING IBM WATSON

The ability to provide the appropriate drug, suitable patient, suitable disease, suitable time, and suitable dose. Personalized medicine comprises locating the genetic data that paves the way for such forecasts in order to foresee a person's sensitivity. IBM Watson is utilizing cognitive computing to complement clinical judgement in personalized medicine. The discovery of biomarkers that function as auxiliary diagnostics for the targeted medicine is essential to the development of this cutting-edge treatment strategy. For patients, medical practitioners, regulatory organizations, pharmaceutical and diagnostic firms, and healthcare professionals in general, personalized medicine holds great potential benefits.

2.12.5   DETECTION OF MUTATION WITH NGS

DNA sequencing, frequently in a large number of patients, can be used to find undiscovered mutations. This prompted the creation of techniques for both detecting and screening DNA for mutations. A quick sequencing method is called next-generation sequencing (NGS). It creates a patient's whole molecular profile. The finding of altered genes responsible for oncogenic characteristics in cancers was made by targeted high throughput sequencing.

A promising technology for diagnostic applications is next-generation sequencing, which allows the simultaneous detection of several mutations in various genes in a single test. Today, a number of nations maintain DNA databases of criminals. Sadly, there have been instances where the DNA of individuals who were detained but not found guilty was mistakenly added to the database. It is possible to view DNA fingerprinting in this context as a method that invades people's privacy and makes their personal information easily accessible to others.

Genetically engineered new species might cause ecological issues. The need to produce more food products has led to the genetic modification of plants, which has sparked intense debates over political, ethical, and societal issues. The effects of genetically altered organisms on the ecosystem are unpredictable. Also, a lot of genetically modified foods use donors that are microorganisms whose potential to cause allergies is either unstudied or unknown. New gene combinations and genes derived from non-food sources may also trigger allergy reactions in some individuals or exacerbate pre-existing ones [55]. The main disadvantage of using genetic engineering in human existence is the misuse of this technology in the development of biological weapons or warfare. Yet they also raise important questions about how far we should push the limits of genetic engineering. Transgenic animals and plants offer immense promise. This raises political issues, which have been debated in legislative hearings, regulatory actions, and court cases [58]. The first gene-edited infants, named Lulu and Nana, who are naturally immune to the human immunodeficiency virus (HIV) were recently generated by Chinese scientist Jian-kui.

He altered the kids' germline genes using the CRISPR-Cas9 technology. Because of ethical and scientific issues, China's guidelines and regulations forbid germline genome editing on human embryos for clinical application. In addition to breaking other ethical and regulatory standards, Jian-kui HE's human experiments broke various Chinese laws. One hundred and twenty-two

Chinese experts concluded that CRISPR-Cas should not be employed to create genetically altered offspring because of substantial off-target hazards and related ethical concerns. So, this gene alteration may not provide the newborns with many significant advantages while bringing unknown and uncontrollable risks to them and their future generations [59]. Individual pharmacogenomics profiles are being used increasingly frequently, putting privacy at risk. The U.S. Senate and House of Representatives passed the *Genetic Information Nondiscrimination Act of 2008* to protect people from genetic discrimination in the employment and health insurance industries [60]. If patients consult Watson directly, they run the risk of receiving the incorrect diagnosis before ever seeing a doctor. What party is at fault in this situation? Watson's output is incorrect because of malicious human activity, and people who receive inaccurate diagnoses about them experience depression. Dataset selection after ethical practices can lead informatic application to more reliable results. The ethics enabled in informational applications include virtual chemical library screening, system chemical biology networks, and the prediction of molecular features pertinent in better approach.

2.13   CONCLUSION

There are many issues to address for implementing any bioinformatics or chemoinformatics systems. The basic development in the maintenance of such systems needs to be adequate, as there are direct associations with human life. Chemists need to plan their experiments more effectively and get more information out of their data where cheminformatics play a vital role for transforming raw data to model fitted datasets. Patentable bioinformatics system and application are popular after bioethics is carried out in appropriate way.

REFERENCES

1. Barnard JM. *J. Chem. Inf. Comput. Sci.* 1991;31: 64–68.
2. Dobson CM. *Nature* 2004;432: 824–828.
3. Gordeeva EV, Molchanova MS, Zefirov NS. *Tetrahedron Comput. Methodol.* 1990;3: 389–415.
4. Zhao Y, Truhlar DG. *Theor. Chem. Acc.* 2008;120: 215–241.
5. Breiman L, Friedman J, Stone CJ, Olshen RA. *Classification, Regression Trees*, Chapman & Hall/CRC, Wadsworth, CA, 1984.
6. Oloff S, Zhang S, Sukumar N, Breneman C, Tropsha A. Chemometric analysis of ligand receptor complementarity: identifying Complementary Ligands Based on Receptor Information (CoLiBRI). *J. Chem. Inf. Model.* 2006;46: 844–851.
7. Laggner C, Wolber G, Kirchmair J, Schuster D, Langer T. *Chemoinformatics Approaches to Virtual Screening*, Eds. A Varnek, A Tropsha. RSC Publisher, Cambridge, 2008, pp. 76–101.
8. Varnek A, Gaudin C, Marcou G, Baskin I, Pey AK, Tetko I V. Inductive transfer of knowledge: application of multi-task learning and feature net approaches to model tissue-air partition coefficients. *J. Chem. Inf. Model.* 2009;49: 133–144.
9. Faulon JL, Churchwell CJ, Visco DP Jr. The signature molecular descriptor. 2. Enumerating molecules from their extended valence sequences. *J. Chem. Inf. Comput. Sci.* 2003;43: 721–734.
10. Johnson AM, Maggiora GM. *Concepts, Applications of Molecular Similarity*, Wiley, New York, 1990.
11. Cherkassky V, Mulier F. *Learning from Data: Concept, Theory, Methods*, 2nd ed., Wiley, Hoboken, NJ, 2007.
12. Zupan J, Gasteiger J. *Neural Networks in Chemistry*, Wiley-VCH, Weinheim, 1999.
13. Baskin II, Gordeeva EV, Devdariani RO, Zefirov NS, Palyulin VA, Stankevich MI. *Dokl. Akad. Nauk. SSSR.* 1989;307: 613–617.
14. Cover TM, Thomas JA, Abramson NM. *Information Theory and Coding*, McGraw-Hill, New York, 1963.
15. Burden FR, Winkler DA. Robust QSAR models using Bayesian regularized neural networks. *J. Med. Chem.* 1999;42: 3183–3187.
16. Agarwal S, Dugar D, Sengupta S. Ranking chemical structures for drug discovery: a new machine learning approach. *J. Chem. Inf. Model.* 2010;50: 716–731.

17. Joachims T, Hofmann T, Yue Y, Yu CN. Predicting structured objects with support vector machines. *Commun. ACM* 2009;52: 97–104.
18. Rupp M, Schneider G. Graph kernels for molecular similarity. *Mol. Inf.* 2010;29: 266–273.
19. Kohonen T. *Self-Organizing Maps*, Springer, 2001.
20. Schuffenhauer A, Ertl P, Roggo S, Wetzel S, Koch MA, Waldmann H. *J. Chem. Inf. Model.* 2007;47: 47–58.
21. Marketsandmarkets.com. Genomics Market by Product & Service. https://bit.ly/3Am0N6I. Accessed Sep 20, 2021.
22. NCBI.gov. What Is Next Generation Sequencing? https://www.ncbi.nlm.nih.gov/pmc/articles/PMC3841808/. Accessed Sep 20, 2021.
23. Nature.com. CRISPR, the Disruptor, Issue 522, 2015. https://www.nature.com/articles/522020a. Accessed Sep 20, 2021.
24. ScienceDaily.com. Artificial Intelligence Boosts Proteome Research. https://www.sciencedaily.com/releases/2019/05/190529113044.htm. Accessed Sep 20, 2021.
25. https://www.pharmskool.com/2022/02/bioinformatics-objectives-advantages-disadvantages-applications.html#:~:text=Bioinformatics%20tools%20aid%20in%20comparing,important%20part%20of%20systems%20biology
26. www.merriam-webster.com
27. https://www.britannica.com/topic/bioethics/Approaches
28. https://physics.snu.edu.in/node/8176#:~:text=Bioethics%3A%20ethical%20concerns%20of%20biotechnology,on%20successful%20grant%20of%20patents
29. International legal protection for software. Fenwick &West LLP, 2007.
30. Steering committee for intellectual property issues in software computer science and telecommunications board commission on physical sciences, M., and applications national research council: Intellectual property issues in software. National Academy Press, Washington, DC, 1991.
31. Westkamp GN. Protección del material biológico mediante derechos de autor. ¿Vueltadela Bioinformática a los Derechos de autor en la biotecnología? *IPR- Helpdesk Bulletin* 2005.
32. Story A. Intellectual property and computer software. In *Intellectual Property Rights and Sustainable Development*, Ed. ICTSD-UNCTAD. Imprimerie Typhon, Chavanod, 2004.
33. Morcon C, Roughton A, Gaham J. *The Modern Law of Trade Marks*. Butterworth, London, 2005.
34. WIPO. Madrid agreement concerning the international registration of marks of April 14, 1891, as revised at Brussels on December 14, 1900, at Washington on June 2, 1911, at The Hague on November 6, 1925, at London on June 2, 1934, at Nice on June 15, 1957, and at Stockholm on July 14, 1967, 1 and as amended on September 28, 1979. WIPO Database of Intellectual Property and Legislative Texts.
35. WIPO. Protocol relating to the Madrid agreement concerning the international registration of marks adopted at Madrid on June 27, 1989 and amended on October 3, 2006. World Intellectual Property Organization.
36. Rowland D, Campbell A. Supply of software: Copyright and contract issues. *International Journal of Law and Information Technolog.* 2002;10:23–40.
37. Davidson SJ, Bergs SJ Kapsner M. Open, click, download, send ... What have you agreed to? The possibilities seem endless. Computer Associations Law Conference (CLA20012001).
38. PR-Helpdesk. Software copyright licensing. B. Papers, Ed., 2–3. 2003.
39. OECD. Organization for Economic Co-operation and Development. Articles of the model convention with respect to taxes on income and on capital. OECD, 2005.
40. Stewart SM. *International Copyright and Neighbouring Rights*. Butterworths, London, 1989.
41. Horton KL. The Software Copyright Directive and the Internet: Collision on the Information Superhighway? Jean Monnet Chair, New York School of Law, 1997.
42. WIPO. Berne Convention for the protection of literary and artistic works of September 9, 1886, completed at PARIS on May 4, 1896, revised at BERLIN on November13, 1908, completed at Berne on March 20, 1914, revised at Rome on June2, 1928, at Brussels on June 26, 1948, at Stockholm on July 14, 1967, and at Paris on July 24, 1971, and amended on September 28, 1979. WIPO Database of Intellectual Property and Legislative Texts.
43. WTO. Trade-related aspects of intellectual property rights. Annex 1C of the Marrakech agreement establishing the world trade organization, signed in Marrakech, Morocco on 15April 1994, World Trade Organization.
44. EC: Commission of the European Council White Paper: Completing the internal market. 1985.
45. IPR-Helpdesk. Software copyright. T.B. Papers, Ed. 2004.

46. University of Buffalo Center of Excellence in Bioinformatics, Introduction to bioinformatics. University of Buffalo Centre of Excellence in Bioinformatics. http://www.bioinformatics.buffalo.edu/current_buffalo/primer.html

47. Murashige KH. Genome research and traditional intellectual property protection–A Bad Fit? (1996) 7 Risk: Health, Safety & Environment 231, online: Franklin Pierce Law Centre http://www.piercelaw.edu/risk/vol7/summer/murashig.htm

48. National Research Council, A question of balance: private rights and the public interest in scientific and technical databases. Washington, DC: National Academies Press, 1999. National Academies Press http://books.nap.edu/html/question_balance/index.html

49. Howard K. The bioinformatics gold rush (2000) 282:1 Scientific American 58 at 63. See also Teresa K. Attwood, "The Babel of Bioinformatics" (2000) 290:5491 Science 471.

50. David PA. Will building 'good fences' really make 'good neighbours' in science? 2001. Stanford University Economics Department.

51. National Research Council, A question of balance: Private rights and the public interest in scientific and technical databases. Washington, DC: National Academies Press, 1999.

52. https://omicstutorials.com/data-errors-in-bioinformatics-databases/

53. Canduri WJSDF. Bioinformatics: An overview and its applications. *Genet. Mol. Res.* 2017;16 (1): gmr16019645.

54. Cho MK, Sankar P. Forensics genetics and ethical, legal and social implications beyond the clinic. *Nature Genetics.* 2004 Nov;36 (Suppl 11): S8–12.

55. Uzogara SG. The impact of genetic modification of human foods in the 21st century: A review. *Biotechnology Advances.* 2000;18: 179–206.

56. Greely HT. Legal, ethical, and social issues in human genome research. *Annu. Rev. Anthropol.* 1998 Oct;27(1): 473–502.

57. van Aken J, Hammond E. Genetic engineering and biological weapons: New technologies, desires and threats from biological research. *EMBO Reports.* 2003 Jun;4(S1):S57–60.

58. Blank RH. Politics and Genetic Engineering. *Politics Life Sci.* 1992 Feb;11(1): 81–85.

59. Li J-R, Walker S, Nie J-B, Zhang X-Q. Experiments that led to the first gene-edited babies: The ethical failings and the urgent need for better governance. *J Zhejiang Univ Sci B.* 2019: 32–38.

60. Fletcher JC. Moral problems and ethical issues in prospective human gene therapy. *Symposium on Biomedical Ethics* 1983: 515–546.

3 The Ethical Responsibility and Reasoning for Using AI in Big Data

Priyanka Banerjee, Amartya Sen, and Saptarshi Sanyal

3.1 INTRODUCTION

Information and communication technologies (ICTs), which require regulatory supervision and ethical and social assessment, have a long-term significant impact on social and economic life. Nowadays, two types of interlinking are observed which can develop the benefits of ICTs in a great sense, but there also have been opposite impacts on ethics and human rights. ICTs are the way of production and collection of big data and using the data for analysing. The key technological drivers of Smart Information System (SIS) are AI and big data. Some examples of intelligent sociotechnical systems that range from social network data analysis for helping the advertising of data prediction are Google search engine, Google Translate, Amazon Alexa home assistant, predictive policing systems, healthcare surgery robots, personal fitness application, and many others.

SIS promise to solve many social problems and challenges, which is the actual reason for the current prominence. To overcome the financial crisis of 2008, the European Commission [1] formulated a strategy to promote smart, sustainable, inclusive growth, which is still visible and driving European policy decisions. Various changes like demographic change to migration, social inclusion, healthcare, skills, and education are the effects of the strategy in Europe. Many of these challenges require intelligent and specific solutions as per the accepted public narrative. To achieve these policy goals, the only way is SIS. To improve the workflow and user satisfaction, SIS generate new sources of income, improve processes, and provide bespoke solutions. SIS raised some significant concerns at that time. Privacy and data protection are the most obvious issues, but they are far from the only ones. Concerns range from questions of fairness and hidden biases in big data all the way to the possibility of truly autonomous machines that may harm or kill people, but that may also be subjects of ethical rights. A key topic of debate is the social consequence that SIS may have in the future of work and employment.

The advantages and downsides of SIS get profited from integrating various viewpoints and concerns which forward the fragmented debate of the thesis. For such integration, the perspective offered by Responsible Research and Innovation (RRI). For addressing a broad range of current and emerging ethical and social issues, the needed scope and flexibility is possessed by RRI. From the above discussed issues, we come to a point that a single approach is not sufficient for a single focused issue even though it is discussed well about privacy and data protection. RRI implemented to support both AI and big data analytics.

3.2 SMART INFORMATION SYSTEMS: THE COMBINATION OF AI AND BIG DATA ANALYTICS

Using the ethical and social consequences of artificial intelligence like hardware, software, application for performing big analytical data and for mimicking human cognition is the topmost debate

DOI: 10.1201/9781003353751-3

issue. Due to the developments of these artificial intelligence in current years, the progress of ICT is pushed up. Using the term Smart Information Systems as a shorthand for technologies which involve artificial intelligence, machine learning, big data are much easier to understand the social consequences of such technologies. Enabling technologies is an environment that helps to generate and collect data and act on the world and interact with humans. SIS tends to be enveloped in the broader technical infrastructure, but they are not in isolation. To understand the SIS requires broader reflection and oversight; the technology of social and ethical relevance is raised in a broader socio-economic environment. Science and technology research and development are framed to resolve social challenges like economic growth, environmental sustainability, security, social inclusion as per the political context. The research organizations and companies accept and replicate the political rhetoric. The political rhetoric produces the scene of SIS for positive expectations.

In Figure 3.1, the existence of SIS in the institutional and societal ecosystem, addressing challenges and raising ethical concerns are shown. The key technical drivers of SIS – AI and big data, which provide core capabilities – are located in the top rectangle. These technical drivers of SIS provide the required data and enabling functionality from their social relevance. Such enabling functions are social networks and the Internet of Things (IoT), which provide datasets and affective devices that interact with humans. This branch of technologies in the figure is symptomatic. There are also other technologies present with enabling functions. SIS mainly combined the grand societal challenges and the ethical and human rights concerns in a broader sense.

3.3 THE ETHICS OF SMART INFORMATION SYSTEMS

In this part, the ethical concerns as well as current proposals is discussed with the established mechanisms for directing them.

3.3.1 ETHICAL ISSUES

Discussion of ethics and information technology has a long history, started from the very early times of modern computing technology [2]. There are numerous discussed issues found in the recent comprehensive review of ethics in ICT [3]. Ethical issues that are applicable for ICT are also applicable for ICT because SIS is a subcategory of ICT. After collecting some documents from a paper on ethics in ICT, it is found that 177 conveyed the privacy issue and data protection, making it the most distinguished one. And also, numerous other issues are discussed, including users, their agency, trust, identity, inclusion, security, misuse, etc.

Besides the ethical issues, there is some argument on specific issues of SIS like artificial intelligence, machine learning, big data etc. The forum FAT ML (Fairness, Accountability, and transparency in Machine Learning) is mainly developed on Principles for Accountable Algorithms and a Social Impact Statement for Algorithms [4]. Another forum, the World Economic Forum, has raised nine ethical issues to ask about AI systems in Davos, Switzerland [5].

In an article in *MIT Technology Review*, we found that the AI app-making engineers are unable to explain correctly the "mind-boggling" questions of Will Knight [6]. Apart from the questions of Knight, there are also other questions about the ethics of AI.

To develop the future of AI, the Association for Advancement of Artificial Intelligence (AAAI) [7] co-sponsored a Presidential Panel to look over the long-term study of future AI. This study panel started in California as the Asilomar Study, which mainly give concerns about control of AI and the actions that enhance the societal outcomes. In 2017, the Asilomar council evolved 23 new principles to AI Study [8]. Even in 2018, the AAAI council collaborated with the ACM on the AI, Ethics, and Society conference in New Orleans [9].

There are some measures that have addressed the ethical issues arising from AI: A methodology for AI research "Ethics by Design" proposed by an Irish researcher [10]. Some American researchers developed a way to determine the bias in black-box model [9, 11]. Every AI practitioner

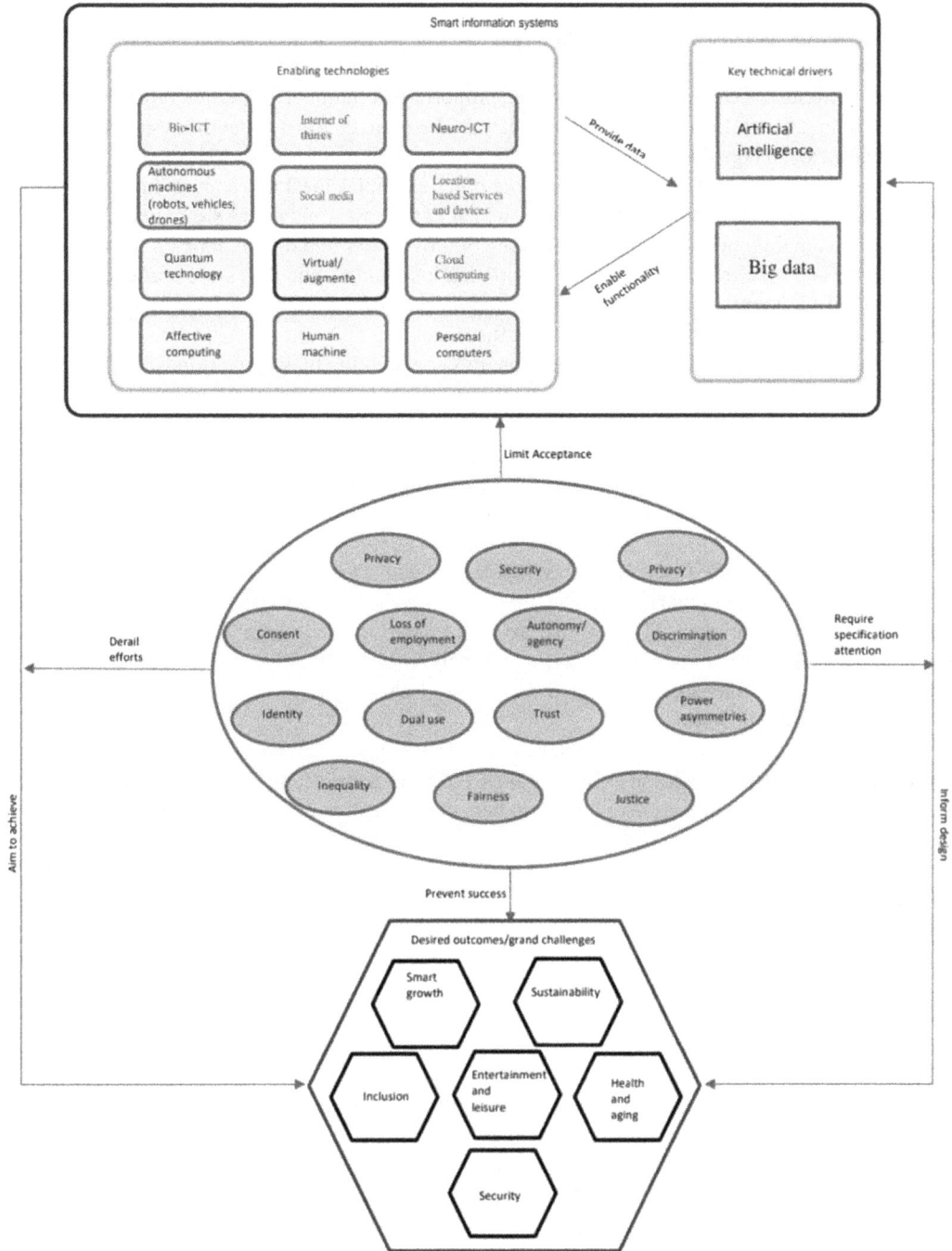

FIGURE 3.1 The ecosystem of Smart Information Systems.

and student should take part in ethical training. Students researching AI learning, data science, or computer science should discuss related ethics and security topics [9].

An algorithm's creator should be accountable for its outcome [12]. When the creators are held accountable, a good outcome is more likely. Enforcement is a regulatory function. The AI system is regulated by academics [13–15] and also by the parliamentarians [16] due to the insufficiency of current ethical and legal principles in this zone.

AI systems become complex in consequences, for example, some algorithms predicting the scene of a crime for the police, which can help to prevent the crime within time, and also being able to deploy appropriate numbers of police force for the right situation. But this type of algorithm raises some ethical issues. One example is that a disproportionate number of people might be arrested from a certain area while other areas are neglected. The main concern about the system is that such technologies can change the whole law and enforcement, which may be undesirable.

From the *US News and World Report*, some other examples are known about how the AI system is used in the Universities league table, which can boost a magazine's circulation; also, the algorithm was mainly powered for the weighting of universities and did not look on the factors that are arguable and took into account some factors that it should avoid. This system affects the study system in US by doing some unwanted increasing in the cost of education [17]. These algorithms should have been designed better to overcome the issues.

From the study of so many experts, we come to know that the biases and mindset of the creator are reflected by the algorithms. An experiment was organized by Facebook to study the effect of content on their users by posting some news feeds with positive and negative sights. They were reprimanded for not informing their users earlier [18].

The Cambridge Analytica scandal is a case which is related to the prediction of algorithms, like how an algorithm can predict someone's profile by using data analysis and tell different things about different electors.

In the book *How Big Data Increases Inequality and Threatens Democracy*, Cathy O'Neil describes negative side of algorithms and artificial intelligence – that it not only raises ethical issues but also threatens democracy. [17] Another article of *Scientific American* also raises a concern on the same topic: Will Democracy Survive Big Data and Artificial Intelligence? [19]. Nowadays these AI systems make a "filter-bubble" around the society like a digital prison, which can destroy the democracy in different ways.

In the current development of technology, SIS can increase some existing issues, but they also present essential matters. In these existing issues, there are some secret biases in algorithmic decision making [20]. The implication of large numbers of people in new technology can raise questions of liability for policymakers [21].

In Figure 3.1, some ethical issues caused by SIS are discussed. We make an overview on these various issues and denote how SIS influences society and also address data protection and privacy along with ethical and social issues. This type of broad approach is made on reliable research and innovation. Before proceeding in further detail, now it's time to understand how to address ethics in current SIS.

3.3.2 CURRENT RESPONSES TO ETHICAL ISSUES

To indicate a novel development, we used the term SIS, a technology involved with a long history, and raised some questions like privacy and data protection. SIS mostly use personal data. Therefore, it is an important step to protect that data to avoid negative outcomes. The impact of SIS is addressed in so many novel features of the European General Data Protection Regulation (GDPR). In GDPR, novel features relevant to SIS are breach notifications, hefty financial penalties, data protection impact assessments [22], privacy by design [16], and the so-called right to be forgotten.

Some specific action targeting privacy can form to predict the issues raised by SIS. Value-sensitive design are aimed at the technical level to address the ethical questions [23, 24]. To support privacy by design in GDPR, this value-sensitive design idea is started.

Standardization is the other method to solve some problems raised by SIS, and this is mostly used to address ethical and social issues, as in the ISO 27000. BSI8611 is the most advanced standards on Robots and Robotic Devices: Guide to the Ethical Design and Application of Robots and Robotic Systems [25]. The recent IEEE Global Initiative is to complement this standard for Ethical Considerations in Artificial Intelligence and Autonomous Systems [26].

Professionalism is the traditional method to solve the issues in ICT with ethics and human rights. To make some guidelines and codes of ethics for addressing ethical issues, some bodies are involved such as the British Computer Society, the Association for Computer Machinery (ACM), and the IEEE. Sometimes these bodies should have rethought their stances due to the impulsion of the emergence of SIS. As an example, the ACM is currently consulting on an updated version of its general code [27] which is more specifically supported the ACM Statement on Algorithmic Transparency and Accountability [28]. Researchers are also adopting bottom-up approaches to make a change in guidelines which are dealing with the capabilities of emerging Smart Information Systems [8].

To deal with ethics and human rights in SIS, numerous activities and attempts denote that one can perceive a significant challenge in calling for a concerted effort to address them. This perception has spread among policymakers like various governments and political bodies to produce reports on Smart Information Systems, notably AI [21, 29, 30] and robotics [31]. The European Parliament is proposing the introduction of a registration system, a new regulator, and other measures.

3.4 COMBINING VIABLE APPROACHES: RESPONSIBLE RESEARCH AND INNOVATION

Till now we have argued on the complexation of ethical and social issues raised by SIS and the numerous initiatives to solve the issues. These seem to be missing the way of alignment of various remedies and ensuring them to join up and create synergies. This is a determining effort to develop SIS in a globalized way. To achieve the goal, there are requirements of so many stakeholders and decision-makers.

Responsible Research and Innovation (RRI) is the concept to deal with the issues in a realistic way. RRI is a venture to rethink research and innovation which ensures that the outcomes of research and innovation are acceptable, desirable, and sustainable [32]. Though the RRI is a new term, it is built on the previous system of activities like technology ethics, technology assessment, technology studies, etc. The European Commission is one of the key proponents of RRI which emphasizes public engagement, ethics, science education, gender equality, open access, governance, the six keys of RRI. The UK Engineering and Physical Science Research Council has chosen another view of RRI which is based on anticipation of future outcomes, reflection of the research, engagement of stakeholders, and action and responsiveness. Stakeholders need to be engaged at the early stage of research, which is agreed by all proponents of RRI. RRI needs to be joined into projects in the broader funding and supportive environment and requires transparency and willingness to be flexible.

3.4.1 IMPLEMENTING RESPONSIBLE RESEARCH AND INNOVATION IN SMART INFORMATION SYSTEM RESEARCH

RRI is currently integrated into the research of the Human Brain Project (HBP) and two current flagship research projects under the European Union. The HBP targets making a development in ICT infrastructure for neuroscience [33]. This project needs huge amounts of neuroscientific data for the generation of new perceptions about the brain. To overcome the limitation of current AI, RRI develop some novel computational architectures based on the neuroscience.

From the starting time of the project, people expected that HBP can raise ethical and social queries. One of the 12 HBP subprojects is going under the heading Ethics and Society to program of RRI [34]. The four main components of HBP subproject are: the foresight lab of the HBP, which investigates possible outcomes of the work undertaken by the HBP; a work package that undertakes philosophical research on complex issues; public engagement, which reaches out to stakeholders and the public to discuss issues of relevance to the HBP and the public; and ethics management, which deals with the administration of ethics-related issues. A researcher awareness program is included by the HBP's RRI, which helps to reach scientists and engage with them on specific issues.

The partners of HBP are dealing with various ethical issues. These include both human and animals as traditional biomedical research ethics issues. The European rules on informed consent of research subjects can act as a compiling body for the project. The HBP needs to develop suitable data to ensure that data is treated appropriately because this is a data integration project.

It is now possible to integrate principles of RRI in complex research structures due to the help of HBP. HBP is a project which can influence future development of SIS. The HBP has commenced on a journey to reflect on ethical and social issues by incorporating RRI into its structure. This proves that integrating RRI into SIS research is possible.

Here we select the HBP as an example for this article to prove that RRI can be put into practice. HBP is the significant influence for next stage development of SIS because of the large number of participants and its status as an EU flagship project.

3.4.2 RESPONSIBLE RESEARCH AND INNOVATION AS A RESPONSE TO ETHICS AND PRIVACY IN SMART INFORMATION SYSTEMS

After the complexity of SIS technologies and their impacts on the social and ethical sides, an approach is needed which is able to learn, include external voices, incorporate reflection, and bring together different stakeholder groups. RRI offers this type of approach, and it is already in use in the HBP. All the authors are EU-funded SHERPA project partners. SHERPA is the acronym for Shaping the Ethical Dimensions of Information Technologies – a European Perspective, which mainly works on the social and ethical issues arising from SIS. The partners of SHERPA will apply the RRI approach to a wide range of AI-embedded technologies.

Nowadays RRI is more focused in universities and research work, which is funded by society, whereas SIS systems are more focused in big technology companies. Also, it is noticed that well-established ethical reflections are finding that they are driven to technical companies. The governance boards and ethic panels are now supervised the AI investments in industry [35].

Our aim is to come to a broader consensus on what it would mean to RRI in the field of SIS. We do not believe that there will be an algorithmic solution to the ethics of SIS. The range of technologies, applications, and uses is simply too broad. The goal of both the HBP and SHERPA is to develop a culture of responsibility where stakeholders are ready and willing to accept responsibility for the processes and outcomes of their research and innovation work.

3.5 CONCLUSION

From this chapter, we observe that the activities of RRI is totally failsafe. The prediction of ethical concerns leads by the development of SIS. RRI is an unexpected approach of SIS which identifies and deal with consequences. We know that the future is not predictable, so we need to realize the current consequences of research and innovation. Novel technologies like AI and big data analytics may cause problems which we have to deal with retrospectively. Knowing that there are so many imperfections, we can accept the challenge and proactively engage with the ethics of SIS. To overcome all these things, some initiative is taken by the RRI.

REFERENCES

1. "COM(2010), 2020: Europe, 2020—A Strategy for Smart, Sustainable and Inclusive Growth," European Commission, Mar. 2010.
2. N. Wiener, *The Human Use of Human Beings*, Doubleday, 1954.
3. B.C. Stahl, J. Timmermans, and B.D. Mittel Stadt, "The Ethics of Computing: A Survey of the Computing Oriented Literature," *ACM Computing Surveys*, 48(4), 2016, pp. 1–55.
4. "Principles for Accountable Algorithms and a Social Impact Statement for Algorithms," FAT/ML; https://www.fatml.org/resources/principles-for-accountable-algorithms.

5. J. Bossmann, "Top 9 Ethical Issues in Artificial Intelligence," World Economic Forum, 21 Oct. 2016; https://www.weforum.org/agenda/2016/10/top-10-ethical -issues-in-artificial-intelligence.

6. W. Knight, "The Dark Secret at the Heart of AI," *MIT Technology Review*, 11 Apr. 2017; https://www.technologyreview.com/s/604087/the-dark-secret-at-the-heart-of-ai.

7. "AAAI Presidential Panel on Long-Term AI Futures," AAAI, 2009; https://www.aaai.org/Organization /presidential-panel.php.

8. "Asilomar AI Principles," Future of Life Institute, Asilomar Conference, 2017; https://futureoflife.org/ai -principles.

9. C. Sandvig et al., "Auditing Algorithms: Research Methods for Detecting Discrimination on Internet Platforms," *Data and Discrimination: Converting Critical Concerns into Productive Inquiry*, 2014, pp. 1–23.

10. M. d'Aquin et al., "Towards an 'Ethics in Design' Methodology for AI Research Projects," *AAAI/ACM Conference Artificial Intelligence, Ethics, and Society*, 2018.

11. S. Tan et al., "Detecting Bias in Black-Box Models Using Transparent Model Distillation," Oct. 2017; https://arxiv.org/abs/1710.06169.

12. J.A. Kroll et al., "Accountable Algorithms," *University of Pennsylvania Law Review*, 165, 2016, pp. 633–705.

13. O.J. Erdélyi and J. Goldsmith, "Regulating Artificial Intelligence Proposal for a Global Solution," *AAAI/ACM Conference on Artificial Intelligence, Ethics and Society*, 2018.

14. A. Mantelero, "Regulating Big Data. The Guidelines of the Council of Europe in the Context of the European Data Protection Framework," *Computer Law and Security Review*, 33(5), 2017, pp. 584–602.

15. K. Yeung, "Algorithmic Regulation: A Critical Interrogation," *Regulation and Governance*, 12(4), 2017, pp. 503–523.

16. *Privacy by Design*, Information Commissioner's Office, 2008.

17. C. O'Neil, *Weapons of Math Destruction: How Big Data Increases Inequality and Threatens Democracy*, Penguin, 2016.

18. C. Flick, "Informed Consent and the Facebook Emotional Manipulation Study," *Research Ethics*, 12(1), 2016, pp. 14–28.

19. D. Helbing et al., "Will Democracy Survive Big Data and Artificial Intelligence," *Scientific American*, 25(Feb.), 2017.

20. V. Mayer-Schonberger and K. Cukier, *Big Data: A Revolution That Will Transform How We Live, Work and Think*, John Murray, 2013.

21. "Preparing for the Future of Artificial Intelligence," *Executive Office of the President National Science and Technology Council Committee on Technology*, Oct. 2016.

22. "Privacy Impact Assessment (PIA) Good Practice," CNIL, 2015.

23. B. Friedman, P. Kahn, and A. Borning, "Value Sensitive Design and Information Systems," in *The Handbook of Information and Computer Ethics*, K. Himma and H. Tavani, eds., Wiley Blackwell, pp. 69–102, 2008.

24. N. Manders-Huits and J. van den Hoven, "The Need for a Value-Sensitive Design of Communication Infrastructures," in *Evaluating New Technologies: Methodological Problems for the Ethical Assessment of Technology Developments*, P. Sollie and M. Düwell, eds., Springer, pp. 51–62, 2009.

25. "BS8611—Robots and Robotic Devices: Guide to the Ethical Design and Application of Robots and Robotic Systems," *BSI Standards*. Publication BS8611:2016, Apr. 2016.

26. "The IEEE Global Initiative on Ethics of Autonomous and Intelligent Systems," IEEE, 2017; https://standards.ieee.org/develop/indconn/ec/autonomous_systems.html.

27. B. Brinkman et al., "Listening to Professional Voices: Draft 2 of the ACM Code of Ethics and Professional Conduct," *Communications of the ACM*, 60(5), 2017, pp. 105–111.

28. "Statement on Algorithmic Transparency and Accountability," ACM US Public Policy Council, Dec. 2017.

29. "Artificial Intelligence, Automation, and the Economy," *Executive Office of the President National Science and Technology Council Committee on Technology*, Dec. 2016.

30. "Robotics and Artificial Intelligence," House of Commons Science and Technology Committee, Sept. 2016.

31. "Report with Recommendations to the Commission on Civil Law Rules on Robotics," 2015/2103(INL), Committee on Legal Affairs, European Parliament, A8–0005/2017, Jan. 2017.

32. J. Stilgoe, R. Owen, and P. Macnaghten, "Developing a Framework for Responsible Innovation," *Resources Policy*, 42(9), 2013, pp. 1568–1580.

33. K. Amunts et al., "The Human Brain Project: Creating a European Research Infrastructure to Decode the Human Brain," *Neuron*, 92(3), 2016, pp. 574–581.

34. C. Aicardi, M. Reins Borough, and N. Rose, "The Integrated Ethics and Society Programme of the Human Brain Project: Reflecting on an Ongoing Experience," *Journal of Responsible Innovation*, Jun. 2017, pp. 1–25.

35. D. Etherington, "Microsoft Creates an AI Research Lab to Challenge Google and DeepMind," *Tech. Crunch*, 2017.

4 Ethical Theories in AI
A Machine Learning Context

Savan Patel, Roshani J Patel, Rina K Patel, and Jigna Prajapati

4.1 INTRODUCTION

In our increasingly interconnected society, artificial intelligence is gaining ground. AI is gradually but steadily altering how we live and work, from self-driving cars to automated customer service representatives. The ethical ramifications of using AI are becoming more complicated as it becomes more advanced (1, 2).

4.1.1 ARTIFICIAL INTELLIGENCE

In its most basic form, artificial intelligence is a subject that integrates computer science with big datasets to aid in problem-solving. In addition, it covers the aspects of artificial intelligence known as deep learning and machine learning, which are frequently discussed in conjunction. Building expert systems that anticipate outcomes or classify data based on input is a common practice in these domains (3, 4). There are various perspectives where systems mimic human thought and machines behave like people. Another strategy are systems with logic and reasoned behavior.

4.1.2 MACHINE LEARNING

In computer science and artificial intelligence (AI), a field known as machine learning, the goal is to simulate human learning by using data and algorithms to gradually increase a system's accuracy (5). Data science, a field that is expanding quickly, is fundamentally dependent on machine learning. Algorithms are trained using statistical techniques in order to provide classifications or predictions and unearth critical insights in data mining projects. Ideally, decisions made as a result of these insights have an impact on significant growth metrics in applications and businesses. Data scientists will be sought after more and more as big data develops and gets better. They will be expected to contribute to the identification of the most crucial business queries and the information needed to respond to them (6). Machine learning algorithms are typically created using frameworks that accelerate solution development, such as TensorFlow and PyTorch (7).

4.2 AI ETHICS CONCERNS

Moral guidelines that help us distinguish between good and wrong are a part of ethics. A collection of rules called AI ethics offers guidance on the creation and results of artificial intelligence (8, 9). Recent results and confirmation bias are just two examples of the many cognitive biases that affect humans and are reflected in our actions and, subsequently, in our data. Since data is the foundation of all machine learning algorithms, we must keep this in mind as we design experiments and algorithms (10). These human biases can be amplified and scaled by artificial intelligence at a previously unheard-of rate. Businesses are paying more attention to adopting automation and data-driven decision-making throughout their operations as a result of the proliferation of big data. The purpose of most businesses' AI applications is to improve business outcomes, but there are some cases where

DOI: 10.1201/9781003353751-4

this isn't the case. This is especially relevant because the initial study design was flawed and the datasets were skewed (11).

4.3   AI PRINCIPLES

I. Instructions: The Agent Does According to Instruction

The myth of King Midas serves as a warning, but as Russell has shown, the propensity for extreme literalism offers enormous issues for AI and the principle that oversees it (12). In this fabled scenario, the hero gets exactly what he asks for—everything he touches turns to gold—instead of what he really wanted. However, it may be difficult to prevent such effects in practice. In the context of the video game *CoastRunners*, an artificial agent that had been programmed to maximize its score looped endlessly. obtaining a high score without actually completing the race, which is what it was intended to do (13–15).

II. Expressed Intentions: The Agent Does What Intended It to Do

In light of the aforementioned concern, many researchers have stated that the challenge is to ensure that AI actually does what we actually intend it to do. A machine that was capable of understanding the principle's intention in this way would be able to understand the subtleties of language and meaning that perhaps less advanced forms of AI couldn't. As a result, when performing routine tasks, the agent would be aware not to endanger human life or property. In high-stakes situations, it would also be able to weigh trade-offs logically. This is a serious challenge. AI may need a thorough model of human language and interaction, including knowledge of the culture, institutions, and practices that allow people to understand the underlying meaning of phrases, in order to truly understand the intention behind instructions. Therefore, it makes sense that the scientific community is paying close attention to the issue of closing the instruction–intention gap (16–18).

III. Revealed Preferences: The Agent Does What My Behaviour Reveals Is Preferred

This strategy has certain benefits. It could, for starters, assist an artificial agent in responding to circumstances in real time correctly. Additionally, it might enable the agent to successfully negotiate social situations by becoming sensitive to other people's preferences and avoiding actions that would b eunpopular. Furthermore, an emphasis on revealed preferences is applicable to facts that the agent has access to, and the strategy has received much research in the field of welfare economics (19).

IV. Informed Preferences or Desires: The Agent Does What One Would Want It to Do If It Were Rational and Informed

AI could prevent a significant proportion of mistakes that result from incomplete information and faulty reasoning by concentrating on the subset of preferences that an informed and instrumentally rational person would hold. We could also get closer to people's true preferences by adopting a set of criteria that more accurately reflects their true aspirations (20).

V. Interest or Well-Being: The Agent Does What Is in My Interest or What Is Best for Me Objectively

This viewpoint holds that AI would be created to support whatever is advantageous for a person's wellbeing. It would be focused on pursuing the kinds of things that fulfil our desires, enhance our lives, and support overall well-being. We are still able to gather information about what people believe constitutes or contributes to their well-being as well as specifics about the kinds of things that make a human life go smoothly even though interest, viewed in this way, is not susceptible to direct scientific observation. The fields of philosophy, psychology, and economics have helped us understand well-being in this sense (21, 22).

VI. Values: The Agent Does What It Morally Ought to Do, as Defined by the Individual or Society

What does the term "values" mean in this situation? Values can be thought of as natural or artificial realities about what is right or wrong and what kinds of things should be supported. When applied to things or commodities in market situations, the notion of value is very different from this normative sense of worth. Many items have value that is either not recognized by markets or that is clearly distorted in how it is priced (20). For things like love, friendship, the environment, justice, freedom, and equality, this is true. While this is going on, there is a lot of metaethical discussion surrounding the existence of values that are both normative and objective (23). The premise of metaethical realists is that they are. The opposing viewpoint asserts that our evaluations eventually lack this factual support.

4.3.1 GOAL ACHIEVEMENT

In order for AI to be in line with human values, this section aims to define guiding principles for AI. Before delving deeper into the solutions, we must first be clear about the issue at hand. Because, contrary to what we might initially think, it is not our responsibility to identify the best or most accurate moral theory, then apply it to robots. Instead, it entails determining how to make appropriate decisions in light of the fact that we live in a diverse world where people hold a wide range of valid and diametrically opposed value beliefs (24, 25).

4.3.2 AI PRINCIPLES AND HUMAN VALUES

These approaches question what values people in that circumstance might rationally agree upon after starting with the premise that all people are free and equal. A key tenet of this argument is that people are not required to set aside their different values and viewpoints. Instead, the parties only need to agree on broad principles to manage a specific subject matter or set of relationships. Even though they must agree on some fundamental principles, they may choose to support these principles for a variety of ethical, intellectual, or even religious reasons. Their agreement thus appears to be a "overlapping consensus" between various points of view (25).

Political liberals assert that, starting from the premise that people are free and equal, it is possible to find justice principles that are supported by an overlapping consensus of opinion. This tradition also recognizes that non-liberal societies might uphold just values in accordance with their own overlapping internal consensus. In both cases, people uphold the moral standards they have set for themselves, largely avoiding the problem of dominance or value imposition (26).

4.4 ARTIFICIAL MORALITY

Making autonomous creatures learn to behave morally, that is, in accordance with moral principles, is the challenge facing AI research. Here, we offer a novel two-step method for resolving the value alignment problem. The first step entails using philosophical foundations to formalise moral principles and behaviour that complies with those principles. The (Multi-Objective) Reinforcement Learning paradigm, which makes it simpler to manage an agent's moral and personal goals, is compatible with our formalisation. In the second stage, an environment where an agent can practice moral behaviour while pursuing its own objectives must be created (27, 28).

We present our interpretation of the value alignment procedure, which is depicted in Figure 4.1, in order to accomplish our goal. According to this perspective, a reward specification phase integrates the personal and moral objectives to create a multi-objective environment. Through an ethical embedding phase, the multi-objective environment is then transformed into a single-objective ethical environment, which is the environment in which an agent learns.

This work aims to develop a value alignment method that leads to an agent learning environment where the agent can learn to behave in a value-aligned manner while pursuing its own personal goal. We believe that a value-aligned agent behaves morally, upholding moral principles at all times,

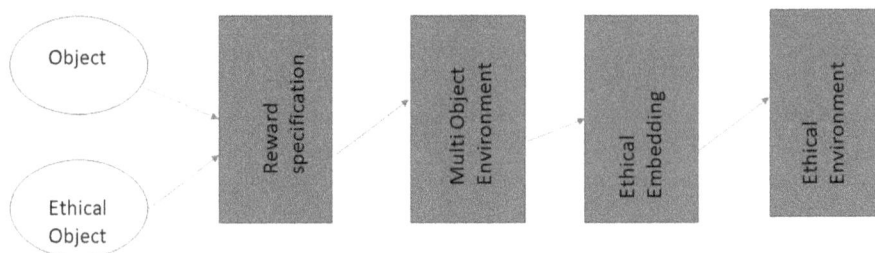

FIGURE 4.1   Ethical environment.

and acting in the most admirable way imaginable. In this study, we also assume that the agent can behave ethically in accordance with our definition. These are the fundamental presumptions for all of our subsequent contributions (29).

4.5   ETHICAL THEORIES

The ethics of AI are discussed in the social sciences, humanities, media, and policy fields—all of which are interested in these technologies. Concerns range from sentient machines enslaving humans to bias in datasets leading to discrimination. The social impact of AI-based technology provides the context and justification for the flurry of activity in public discourse and policy changes concerning whether and how AI should be regulated or whether other solutions should be sought to solve the drawbacks of AI (30).

4.5.1   Primary Ethical Issues

1) Discrimination, harassment, and safety

Discrimination and harassment are two of the most serious ethical issues that managers and human resources professionals face (31).

Some of the anti-discrimination topics are:

- Age: Companies and domestic laws are not permitted to discriminate against senior employees.
- Disability: In order to combat disability discrimination, it is necessary to make adjustments for and treat employees with mental or physical disabilities equally.
- Pay equity: Pay equity seeks to make sure that every employee, regardless of religion, sexuality, or culture, are paid equally for comparable labour.
- Pregnant workers are safeguarded from being unfairly targeted due to their situation.
- Race: Workers ought to be handled similarly regardless of race or ethnicity.
- Religion: Nobody at the workplace must treat a worker differently due to their religious beliefs.
- Gender and sex: A worker's gender and sexual identity should not influence how they're handled while they work for a company.

2) Safety and health in the workplace

Each worker has the privilege of a safe workplace and adequate working conditions. The following are some of the most common employee safety concerns (32):

- **Fall protection:** Fall protection entails precautions, such as guard rails, for protecting workers from crashes.

- **Hazards interaction:** This involves evaluating hazardous compounds that workers work with and discussing how and when to manage these hazardous items properly.
- **Scaffolding:** Human resources in development or service companies are required to inform workers about the highest weight capacities of scaffolds.
- **Breathing prevention:** If applicable, offer recommendations for urgent situation procedures as well as guidelines for the use of breathing equipment.
- **Locking, tagged:** This entails defining specific control mechanisms for hazardous equipment and hazardous power sources like petroleum and natural gas.
- **Commercial vehicles:** This is critical to make sure that the appropriate truck safety measures are all in place to safeguard workers.
- **Staircases:** When operating stairs, staff must be aware of the weight capacity of such staircases.
- **Techniques of power connections:** This involves developing standards of electronics and cables activities. Such rules, for example, could describe when staff should design a circuit that prevent electrostatic discharge.
- **Machinery protection:** When appropriate, provides operational protecting guidelines of equipment like rotary blades, powerful press, clippers, and other equipment.
- **Energized maintenance guidelines:** Developing general electrical regulations for employees is critical for safety in work environments that require the frequent use of electrical equipment. For example, employees should never place conductors or equipment in damp or wet locations.

3) Blowing the whistle or ranting on media platforms

Employers have been paying more attention to an employee's online behaviour as social media has grown in prominence. It is still illegal to reprimand employees for offensive social media posts, and the ramifications of a negative social media post may affect the employee's treatment. You may choose to fire an employee if their social media posts harm the company's reputation or cause a reduction in business (33).

4) Accounting practices

Businesses must adhere to the law's requirements for proper bookkeeping procedures. Accountancy that is unethical is a big concern problem, especially for companies that are traded on the open market. To protect consumers and investors, the law sets standards for financial reports. In order to attract investors and business partners, no matter how big the company is, every institution is obligated to maintain accurate accounting records and tax filings (34).

5) Corporate espionage and nondisclosure

There is a risk for many businesses that information, like client information, will be stolen by current or previous workers and exploited by rivals. Business espionage encompasses both theft of intellectual property belonging to a business and the secret collection of details about the client. Hence, requiring confidentiality agreements may be helpful. To deter this type of ethical wrongdoing, you as a supervisor or HR professional may wish to apply harsh financial consequences (35).

6) Cyber security and privacy procedures

Customers and employees may grow increasingly concerned about their privacy as a result of improvements in a company's technological security mechanisms. You may observe what employees are doing on computer systems or other task devices given by the business. You, as either a manager or HR expert, can utilise this form of electronic monitoring to maintain productivity and efficiency as long as it does not violate the worker's privacy (36).

4.5.2 Human Rights Issues of AI

1) Artificial intelligence as a discriminating tool

As AI has grown in our organic communities, the issue of discrimination and institutional racism has acquired relevance in political discussions concerning technological growth. According to Article 2 of the UDHR and Article 2 of the ICCPR, everyone has the right to exercise all freedoms and rights without restriction. Given the wide spectrum of prejudiced attitudes and oppressive behaviours that characterise human contact, putting this into practice is difficult. Although some people wrongly believe that AI is the solution to this dilemma, a technological innovation that would free us from the bias of human decision-making, such beliefs fail to consider the traces of human intellect in AI technology (37–39).

2) The impact of technology on joblessness

According to Article 23 of the UDHR, Article 6 of the ICESCR, and Article 1(2) of the ILO, everyone has the right to work and to be protected against unemployment. The efficacy of tools and services has altered how businesses and individuals live, but the rapid growth of AI has also resulted in a period of unemployment as it has displaced human labour. In his book *Work in the Future*, Robert Skidelsky claims that "sooner or later, we shall run out of work," referring to the rapid growth of technology. This point was also addressed in seminal research by Oxford professors Carl Frey and Michael Osborne, who predicted that 47% of US occupations will be automated as a result of artificial intelligence (40–42).

3) Controlling populations and movement

The ability to travel freely, which is widely regarded as a fundamental human right, is supported by a number of international agreements. The way AI is used as a tool for surveillance will have a significant impact on how much this privilege can be restricted. According to research by Stanford Foundation for International Security, border management is just one of the security-related uses of AI that at least 75 of the 176 countries in the world are currently actively pursuing. As predictive policing systems end up considering "dirty data" reflecting conscious and subconscious bias, there have been concerns related to the disproportionate impact of surveillance on already vulnerable communities, for example, police discrimination against Black people, refugees, and migrants. According to the *Guardian*, dozens of towers with laser-enhanced cameras have been erected in Arizona, close to the US–Mexico border, in an effort to halt illegal immigration. The US government also put in place a facial recognition system to take pictures of people driving into and out of the country.

4.5.3 Problems with Machine Learning

1) Understanding which processes need automation

Separating fact from fiction is getting harder and harder in today's machine learning world. Consider the issues you want to solve before selecting an AI platform. The tasks that are routinely performed manually and have a predetermined output are the simplest to automate. Before automation, complicated processes need to be examined more closely. Machine learning can undoubtedly help with some operations, but not all automation issues require it.

2) Inadequate data quality

The lack of high-quality data is machine learning's biggest issue. High-quality data is crucial for the algorithms to work as intended, despite the fact that in AI the majority of developers' time is typically spent on optimising algorithms. Ideal machine learning is stymied by clean data, noisy data, and incomplete data. This issue can be solved by carefully assessing and scoping data through data governance, data integration, and data exploration until you obtain unambiguous data. This should be completed before beginning.

3) Inadequate infrastructure

For machine learning, enormous amounts of data processing power are required. Legacy systems frequently collapse as a result of the burden. You should determine if your infrastructure is prepared to handle machine learning. If it can't, you should upgrade, especially to a version with accelerated graphics and customizable memory.

4) Put into action or implementation

By the time an organisation decides to upgrade to machine learning, it frequently already has installed analytical systems. It can be challenging to combine more advanced machine learning techniques with more traditional ones. Implementation is greatly facilitated by maintaining accurate interpretation and documentation. Implementing services like anomaly detection, predictive analysis, and ensemble modelling can be made much simpler by collaboration with a partner for implementation.

5) A scarcity of skilled personnel

Machine learning and deep analytics are still comparatively young academic fields. Managing and producing analytical data materials for automated learning is difficult due to a lack of qualified staff. Data scientists typically need to have both deep knowledge of science, technology, and mathematics in addition to specialised knowledge in their field. Paying high compensation at the time of hiring will be necessary because these professionals are in high demand and aware of their value. Additionally, since many managed service providers always have a list of qualified data scientists on hand, you can ask them for assistance with staffing (43).

4.5.4 ETHICS AT THE ALGORITHM LEVEL

Though bias in machine learning algorithms continues to outpace machine ethics, it is growing in acceptance and momentum. In the context of algorithmic prediction making, the word *bias* can have both positive and negative connotations. Particularly in situations where they might have legal or moral repercussions, we need to check the results of these machines to make sure they are fair. This study makes an effort to address algorithmic ethics in autonomous robots. The circumstances of the particular system determine the best strategy to prevent biased decisions while maintaining the highest algorithmic functionality, which cannot be resolved by a single method. To assist us in making the best decision, we turn to the ethics of machines (44–46).

4.6 CONCLUSION

Ethics are essential for implementing or managing any artificial intelligence system. Ethics are fitted with wide range of purposes, such as to offer safety guidelines capable of preventing existential threats to humanity, to resolve any bias issues, and to construct amiable artificial intelligence systems that will comply to ethical norms and improve the wellbeing of humanity. Institutions and scientists working on AI ultimately hope to solve the majority of issues and tasks that humans are unable to directly complete. The ethical practice in a subset of artificial intelligent such as machine learning covers some concrete concentration on various ethical theories to produce more reliable predictions.

REFERENCES

1. Rodrigues R. Legal and human rights issues of AI: Gaps, challenges and vulnerabilities. *Journal of Responsible Technology* 2020;4:100005.
2. Chu CH, Nyrup R, Leslie K, Shi J, Bianchi A, Lyn A, et al. Digital ageism: Challenges and opportunities in artificial intelligence for older adults. *Gerontologist* 2022;62(7):947–55.

3. Winston PH. *Artificial Intelligence*: Wesley Longman Publishing Co., Inc.; 1984.
4. Charniak E. *Introduction to Artificial Intelligence*: Pearson Education; 1985.
5. Carbonell JG, Michalski RS, Mitchell TM. An overview of machine learning. *Machine Learning* 1983:3–23.
6. Mohammed M, Khan MB, Bashier EBM. *Machine Learning: Algorithms and Applications*: CRC Press; 2016.
7. Witten I, Ian H, Frank, Eibe. *Data Mining: Practical Machine Learning Tools and Techniques with JAVA Implementations*. 2002;31(1):76–7.
8. Floridi L, Cowls J, Beltrametti M, Chatila R, Chazerand P, Dignum V, et al. AI4People—An ethical framework for a good AI society: Opportunities, risks, principles, and recommendations. *Minds and Machines* 2018;28(4):689–707.
9. Herschel R, Miori VM. Ethics & big data. *Technology in Society*. 2017;49:31–6.
10. Jobin A, Ienca M, Vayena E. The global landscape of AI ethics guidelines. *Nature Machine Intelligence* 2019;1(9):389–99.
11. Olson DL, Delen D. *Advanced Data Mining Techniques*: Springer Science & Business Media; 2008.
12. Da Silva FL, Warnell G, Costa AHR, Stone P. Agents teaching agents: A survey on inter-agent transfer learning. *Autonomous Agents and Multi-Agent Systems* 2020;34:1–17.
13. Peddyreddy S. Applications of artificial intelligence for public cloud. *The American Journal of Science and Medical Research* 2021;7(4):1–4.
14. Roff HM. Artificial intelligence: Power to the people. *Ethics and International Affairs* 2019;33(2):127–40.
15. Brundage M, Avin S, Clark J, Toner H, Eckersley P, Garfinkel B, Dafoe A, Scharre P, Zeitzoff T, Filar B, Anderson H. The malicious use of artificial intelligence: Forecasting, prevention, and mitigation. arXiv preprint arXiv:1802.07228. 2018.
16. Tuomela, R. We-intentions revisited. *Philosophical Studies* 2005;125:327–69. https://doi.org/10.1007/s11098-005-7781-1.
17. Falvey K. Knowledge in intention. 2000;99(1):21–44.
18. Hadfield-Menell D, Hadfield GK, editors. Incomplete contracting and AI alignment. In: *Proceedings of the 2019 AAAI/ACM Conference on AI, Ethics, and Society*; 2019.
19. Vredenburgh K. Philosophy. A unificationist defence of revealed preferences. *Economics and Philosophy* 2020;36(1):149–69.
20. Gabriel I. Artificial intelligence, values, and alignment. *Minds and Machines* 2020;30(3):411–37.
21. Axelrod R. Agent-based modeling as a bridge between disciplines. *Handbook of Computational Economics*, 2006;2:1565–84.
22. Sumner LW. *Welfare, Happiness, and Ethics*: Clarendon Press; 1996.
23. Dyck AJ. Ethics and medicine. *The Linacre Quarterly* 1973;40(3):17.
24. Han S, Kelly E, Nikou S, Svee E-O. Aligning artificial intelligence with human values: Reflections from a phenomenological perspective. *AI and Society* 2021:1–13.
25. Enholm IM, Papagiannidis E, Mikalef P, Krogstie J. Artificial intelligence and business value: A literature review. *Information Systems Frontiers* 2022;24(5):1709–34.
26. Gabriel I. Artificial intelligence, values, and alignment. *Minds and Machines*. 2020;30(3):411–37.
27. Misselhorn CJS. Artificial morality. *Concepts, Issues and Challenges*. 2018;55:161–9.
28. Allen C, Smit I, Wallach W. Artificial morality: Top-down, bottom-up, and hybrid approaches. *Ethics and Information Technology* 2005;7:149–55.
29. Rodriguez-Soto M, Serramia M, Lopez-Sanchez M, Rodriguez-Aguilar JA. Instilling moral value alignment by means of multi-objective reinforcement learning. *Ethics and Information Technology*. 2022;24(1):9.
30. Stahl BC, Andreou A, Brey P, Hatzakis T, Kirichenko A, Macnish K, et al. Artificial intelligence for human flourishing – Beyond principles for machine learning. *Journal of Business Research*. 2021;124:374–88.
31. Brown J, Drury L, Raub K, Levy B, Brantner P, Krivak TC, et al. Workplace harassment and discrimination in gynecology: Results of the AAGL member survey. *Journal of Minimally Invasive Gynecology* 2019;26(5):838–46.
32. Newman CJ, De Vries DH, d'Arc Kanakuze J. Ngendahimana G. Workplace violence and gender discrimination in Rwanda's health workforce: Increasing safety and gender equality. *Human Resources for Health* 2011;9(1):1–13.
33. Dungan JA, Young L, Waytz A. The power of moral concerns in predicting whistleblowing decisions. *Journal of Experimental Social Psychology* 2019;85:103848.

34. Nambukara-Gamage B, Rahman S. Ethics in accounting practices and its influence on business performance. *People: International Journal of Social Sciences* 2020;6(1):331–48.

35. Fitzpatrick WM, DiLullo SA, Burke DR. Trade secret piracy and protection: Corporate espionage, corporate security and the law. *Advances in Competitiveness Research* 2004;12(1):57.

36. Liu J, Xiao Y, Li S, Liang W, Chen CLP. Cyber security and privacy issues in smart grids. *IEEE Communications Surveys & Tutorials* 2012;14(4):981–97.

37. Zuiderveen Borgesius FJ, *Discrimination, Artificial Intelligence, and Algorithmic Decision-Making*: Council of Europe; 2018.

38. Tischbirek A. Artificial intelligence and discrimination: Discriminating against discriminatory systems. *Regulating Artificial Intelligence* 2020:103–21.

39. Jebara T. *Machine Learning: Discriminative and Generative*: Springer Science & Business Media; 2012.

40. Virgillito ME. Rise of the robots: Technology and the threat of a jobless future. *Labor History* 2017;58(2):240–2.

41. Chessell D. The jobless economy in a post-work society: How automation will transform the labor market. *Psychosociological Issues in Human Resource Management* 2018;6(2):74–9.

42. Frey CB, Osborne MA. Technology at work v2. 0: The future is not what it used to be. *Citi GPS: Global Perspectives & Solutions*. 2016.

43. Acemoglu D, Restrepo P. *Artificial Intelligence, Automation, and Work. The Economics of Artificial Intelligence: An Agenda*: University of Chicago Press; 2018, pp. 197–236.

44. Shadowen N. Ethics and bias in machine learning: A technical study of what makes us "good". In: Lee N, editor. *The Transhumanism Handbook*: Springer International Publishing; 2019, pp. 247–61.

45. Kazim E, Koshiyama AS. A high-level overview of AI ethics. *Patterns (N Y)* 2021;2(9):100314.

46. Dörr KN, Hollnbuchner K. Ethical challenges of algorithmic journalism. *Digital Journalism* 2017;5(4):404–19.

5 Leave-One-Out Cross-Validation in Machine Learning

Radhika Sreedharan, Jigna Prajapati, Pinalkumar Engineer, and Deep Prajapati

5.1 INTRODUCTION

Machine learning has become so popular for many reasons such as large volumes of available data to manage; the large-scale companies are using large-scale datasets for the various types to generate huge phenomenal rate (1). The cost of storage is reducing day by day with capturing, processing, distributing, and transmitting digital information. The multidimensional algorithm in supports to machine learning and deep learning available for large scale data analysis (2). Machine learning is one of the most popular sub-branches of artificial intelligence (AI) (3). ML mechanisms are working on the paradigm about thinking like humans in the reference of data scrutiny and analysis (4). The big data set is examined using various techniques which enable computers to learn without being explicitly programmed (5).

Machine learning is popular in each and every field, especially in the healthcare sector where trial and error directly deal with human lifestyle. Bioinformatics and cheminformatics are very specific domains of the healthcare sector (6, 7). Bioinformatics is an interdisciplinary domain where molecular biology and genetics with computational statistics are functioning for better advancements (8). The basic learning pattern for biological data with the concern features various scientific research identified for further studies (9). Artificial neural networks and deep learning are particularly useful in exploring, analyzing, and extracting useful features in biological data to predict and classify gene and protein-oriented data (10). Machine learning with computational statistics helps in the identification of genes and nucleotides to identify a disease based on genes. They are also used to determine biological sequence and examine patterns like RNA, protein-sequence, DNA, etc (11).

Chemo-informatics is also a multidimensional discipline involving extracting information, processing data, and inferring meaningful outcomes from various chemical structures (12). Along with rapid detonation of chemical big data from HTS and combinatory synthesis, ML is becoming an indispensable tool for drug designers. The chemical information from large-scale platforms with combinational data-driven analysis is executing for designer drugs by crucial biological parameters (13).

Leave-one-out cross-validation is being used for estimating the working effectiveness of machine learning algorithms. Leave-one-out cross validation is used to predict the desired outcomes when there is no train model (14).

5.2 MACHINE LEARNING TECHNIQUES

Supervised learning: Machine learning broadly categories supervised and unsupervised learning (15). It's also known as labeled or non-labeled datasets. In supervised learning, an instructor or supervisor is involved. In order to build the model and provide test data, a supervisor gives labelled data. Learning happens twice in a supervised learning algorithm. The instructor presents the material to the student at the first stage so that the student can learn it. The pupil gets and comprehends the material. At this point, the teacher is unaware of the student's understanding of the material. The

 DOI: 10.1201/9781003353751-5

teacher will next pose a series of questions to the student to gauge their understanding of the material. The student is put to the exam based on these inquiries, and the teacher notifies the student of his evaluation. Supervised learning is the name given to this type of learning (16, 17).

a. **Classification:** This is a method of supervised learning. The classification algorithms' input characteristics are referred to as independent variables. Label or dependent variable are other names for the target attribute (18). A classification model is a framework that depicts the relationship between the input and target variables (19). Forecasting the "label," which is in a discrete form, is the goal of classification. There are two levels of learning when it comes to categorization. The learning algorithm uses a labelled dataset in the first stage, known as the training stage, to begin learning. Samples are processed after the training set, and the model is then created. The created model is evaluated using a test or unidentified sample and given a label in the second stage. This is how things are classified. The classification learning algorithm builds the model while learning from the set of labelled data. The model then chooses a test case and labels it (15, 19).

b. **Regression:** Regression learns from and forecasts responses with continuous values. With training data, it makes a single output value prediction. The real estate process may be predicted using regression (20). Locality, housing size, and other factors might be input variables (21). The regression model uses the input x to produce a model that looks like a fitted line with the formula y=f (x). Here, y is the dependent variable, and x is an independent variable that might be one or more qualities.

Unsupervised learning: There are no components for a supervisor or teacher. Self-instruction is the most typical sort of learning process when a teacher or supervisor is not present. Trial and error is the foundation of this self-education process (22). Objects are provided to the application, but no labels are established. Based on grouping principles, the algorithm itself examines the instances and detects patterns. Similar things are grouped together when forming groups. Unsupervised learning is analogous to techniques used by people to identify items or events as belonging to the same class, such as assessing how similar two things are to one another (22). Unsupervised learning algorithms are classified into: cluster analysis, association, and dimensional reduction (23, 24).

a. **Cluster analysis:** To organise things into distinct clusters or groups, cluster analysis is used. Based on an object's properties, a cluster analysis groups the items. All of the partition's data items have some characteristics and differ dramatically from those of the other partitions in other areas (25). Each grouping results in a forecasting error for the predictions. The grouping that lessens inaccuracy is the best. For instance, an intelligent tutoring system would seek to group students according to how they learn so that other students in the class might benefit from a student's successful learning methods (26). The segmentation of an image's region of interest, the detection of aberrant growth in a medical imaging, and the identification of signature clusters in a gene database are all examples of clustering processes (27). Among the most important clustering algorithms are: hierarchical algorithms and the k-means algorithm (28).

b. **Association:** In huge databases, relationships between data items are possible thanks to association rules. The goal of this unsupervised method is to discover intriguing connections between variables in sizable databases (29, 30).

c. **Dimensionality reduction:** By utilising the variance of the data, it takes data of a higher dimension as input and produces the data in a lower dimension. The challenge is to condense the dataset to a small number of characteristics without sacrificing generality (31, 32).

Semi-supervised learning: In certain cases, the dataset contains both a sizable amount of unlabeled data and some labelled data. Labeling is a pricey and challenging task for people to complete.

When developing a completely labelled dataset is expensive and labelling a limited portion is more feasible, semi-supervised learning is used. Semi-supervised algorithms apply a pseudo-label to unlabeled data. The data may be merged as the labelled and pseudo-labelled dataset (17, 33).

Reinforcement learning: Reinforcement learning is linked to applications where the algorithm must make choices, so the end result is prescriptive, not merely descriptive, as in unsupervised learning, and the choices have an impact (34). That is similar to learning from mistakes in the human world. Accuracy helps you learn since mistakes come with a cost, which teaches you that a certain line of action has a lower success rate than others. Playing video games is an intriguing example of reinforcement learning that occurs with computers. When an algorithm is given instances without labels, as in supervised learning, reinforcement learning takes place. Yet, depending on the algorithm's suggested remedy, a particular case may come with either positive or negative feedback (34, 35).

This type of learning is linked to applications for the algorithm seeking decisions. The product is treated as prescriptive and not as descriptive, likely happening in unsupervised learning. The choices for the predictive parameters have repercussions. Sometimes it works like trial and error, especially when the data is executing for certain predictions and the outcomes followed by such trials. Certain inaccuracies happen in datasets, which leads to a penalty for certain courses of action. Such action can be successful after applying the desired consequence of inputs/instances (36, 37).

Dataset: Overfitting and underfitting are problems that arise in machine learning when training a model. Our dataset has to be split into three distinct portions in order to address the problems: the data are broadly categorized as training dataset, validation dataset and test dataset (38). The dataset is divided into the three groups mentioned above in the following ratio: 60:20:20 and as per need.

a. Training Dataset
   These datasets are used to update the model's weight once this data set has been used to train the model.
b. Validation Dataset
   These kinds of datasets are used to cut down on overfitting. The model with the data that is not included in the training is evaluated in order to confirm that the rise in accuracy of the training dataset is truly increased.
c. Overfitting, or excessive variance, occurs when the accuracy over the training dataset rises while the accuracy over the validation dataset falls.
d. Test Dataset
   The majority of the time, when we attempt to modify the model based on the results of the validation set, we accidentally cause the model to look into our validation set, which may cause our model to become overfit on the validation set as well.

   To solve this problem, we have a test dataset that is exclusively used to validate the correctness of the model's final output.

The different qualities, such as the attributes or features, establish the dataset's structure and properties. The process of creating a dataset often involves manual observation; however, occasionally an algorithm may be used for testing an application. The dataset includes text, time series, categorical, numerical, and categorical data. For example, numerical information will be used to anticipate automobile prices. Each row in the dataset represents an observation or a sample.

Bias: When data is sampled in a way that causes the distributions of the samples to diverge from those of the population you are pulling them from, this is what is known as sampling bias. Consider a country where, according to a poll, the northern half prefers the colour yellow and the southern half prefers the colour green. You would have a totally yellow preferred colour distribution if you conducted a poll that only included respondents from the southern half of the country, and vice versa. Your sample would be strongly skewed in one direction or the other. The degree to which a sample statistic deviates from the population is known as sample variance. By choosing the best method for sampling our data, we can control both of them (39).

Generalization: The capacity to learn via data the universal rules that may be applied to all other data is known as generalisation. In order to determine if and to what degree learning from data is possible, out-of-sample data becomes crucial (40).

No matter how large the in-sample dataset, bias brought about by certain selection criteria still makes it exceedingly improbable that identical events would be seen frequently and consistently in reality. For instance, there is a story about drawing conclusions from skewed samples in statistics. In order to predict the outcome of the 1936 U.S. presidential election between Alfred Landon and Franklin D. Roosevelt, the *Literary Digest* used skewed poll data.

Training, Validating, and Testing: In an ideal scenario, you would test your machine learning system on data that it has never encountered before. Yet, it is not always feasible in terms of time and money to wait for new data. You can divide your data into training and test sets at random as a first easy fix. The typical ratio is between 25% and 30% for testing and the rest 75–70% for training. You simultaneously split the data that is made up of your responses and characteristics while preserving correspondence between each response and its corresponding characteristics.

The second remedy happens when have to tune your learning algorithm. In this case, the test split data is not a good practice as it causes another kind of overfitting known as snooping. A validation set, or third split, is required to combat spying. Your instances should be divided into thirds and used as follows: 70% for training, 20% for validation, and 10% for testing.

You could split the data arbitrarily, regardless of how the data were initially ordered. If there is no meaningful ordering, ordering might result in either an overestimation or an underestimation, which would make your test unreliable. You must ensure that sequential ordering exists in the split data and that the test set distribution does not deviate significantly from the training distribution as a solution. Check, for instance, that your sets' identification numbers are continuous when they are accessible.

When there are many instances (n), say, n>10000, you may reliably generate a dataset that is randomly divided. When the dataset is smaller, you may determine whether the test set is inappropriate by analyzing basic statistics like mean, mode, median, and variance between the response and features in the training and test sets. Just recalculate a revised split if you are unsure of the original.

LOOCV model evaluation: k-fold cross-validation: Cross-validation, also known as k-fold cross-validation, is a technique used to gauge how well a machine learning algorithm performs while generating forecasts on data that was not used during model training (41).

The single hyperparameter "k" in the cross-validation determines how many subsets of a dataset are created. Each subset is given the opportunity to be used as a test set after being divided, and all other subsets are combined to be used as a training dataset (42).

This means that k models must be fitted and assessed during k-fold cross-validation. With the use of summary statistics like mean and standard deviation, this in turn offers k estimations of how well a model performed on the dataset. Then, using this score, a model and configuration may be compared and finally chosen as the "final model" for a dataset.

The most frequent value of k is 10, with typical values being 3, 5, and 10. This is due to the fact that, when compared to various k values and a single train-test split, 10-fold cross-validation offers a reasonable mix of cheap computational cost and minimal bias in the estimation of model performance.

In contrast to holdout cross-validation, k-fold cross-validation is a considerably more popular method. You must first divide the dataset into k equal portions. The data is then divided into smaller train and test sets for each of these chunks, after which the error of each chunk is assessed. You just take the average after you have all the mistakes for all the chunks. Instead of only testing on one particular subset of your data, this technique allows you to see the issue in all elements of it.

The basic "holdout" cross-validation strategy is referred to as the simple 70/30 train/test split. But there are other additional statistical cross-validation methods, and because R is based on statistical design, you may model a wide variety of cross-validation methods.

Resorting to cross-validation: With the train/test set split, you really introduce bias into your testing since the quantity of your in-sample training data is being reduced (43). There could be some important instances kept out of training when you divide your data. Also, if data is very complicated, a test set that appears to be identical to the training set may not actually be, since different combinations of values are used in the test set. Since you don't have many instances, these problems increase the volatility of the sampling findings. The train/test split is not the preferred option among machine learning practitioners when you have to review and optimise a machine learning solution because of the possibility of having your data divided in an undesirable way (44).

The solution is cross-validation using k-folds. It is based on random splitting, except this time it divides your data into k equal-sized folds (portions). Then, a test set for each fold is held out one at a time while training folds are employed on the others. Each iteration uses a different fold as a test to estimate the degree of correctness. In actuality, a succeeding fold that differs from the prior fold is held out after the test on one-fold against the others used as training, and the method is repeated to provide another assessment of inaccuracy. Once each of the k-folds has been used as a test set once, the procedure is complete and you have k numbers of error estimates that you can calculate into mean and standard error estimates.

LOOCV procedure in scikit-learn: LOOCV is executable using various classes as LeaveOneOut class and some others also using scikit-learn Python library. Such a library provides a basic implementation facility for the LOOCV via ML class Lib. As there is no configuration, methods do not need arguments (45). The instance of class will be created without argument as per the predefined case.

```
# creating_the_ procedure _for_ loocv
cv1 = LeaveOneOut()
```

after creating the procedure, the function will be called as split() and the datset is provided as enumerate. The train and test sets is given the row as per the retrieval occurred on the specified iteration.

```
...
for train_ix1, test_ix1 in cv1.split(X):
...
```

The input/ output X1 /Y1 are the indices columns from the given dataset arrys for splitting the dataset...

```
# split _data
Xx_train1, Xx_test1 = Xx[train_ix1, :], Xx[test_ix1, :]
yy_train1, yy_test1 = yy[train_ix1], yy[test_ix1]
```

The model will be identified using the training set, and the same test dataset will be used to assess or evaluate. Such evaluation will be for prediction about the specified problem by calculating a performance metric. Such calculation can process using the predicted values and its relevant expected values in effective ways.

```
...
# fit model1
model1 = RandomForestClassifier(random_state1=1)
model1.fit(X_train1, y_train1)
# evaluate model1
Yhat1 = model.predict(X_test1)
```

The derived scores will be saved from each computation or evaluation and a final mean estimate of model performance can be presented. This can tie and demonstrate the use of LOOCV to evaluate a RandomForestClassifier model. This is from the synthetic binary classification dataset which is created with the make_blobs() function.

The following example will present the same condition:

LOOCV : for_ evaluating the performance _ random forest classifier _ mannually

Importing make_blobs from sklearn.datasets

Importing LeaveOneOut from sklearn.model_selection

Importing RandomForestClassifier from sklearn.ensemble

Importing accuracy_score from sklearn.metrics

```
X1, y1 = make_blobs(n_samples1=100, random_state1=1)
# create loocv procedure
cv1 = LeaveOneOut()
# enumerate splits
y_true1, y_pred1 = list(), list()
for train_ix1, test_ix1 in cv.split(X1):
# split data
X_train1, X_test1 = X[train_ix1, :], X[test_ix1, :]
y_train1, y_test1 = y1[train_ix1], y[test_ix1]

# fit model
model11 = RandomForestClassifier(random_state=1)
model11.fit(X_train1, y_train1)
# evaluate model
yhat = model.predict(X_test1)
# store
y_true.append(y_test1[0])
y_pred.append(yhat1[0])
# calculate accuracy
Acc1 = accuracy_score(y_true1, y_pred1)
print('Accuracy: %.3f' % acc1)
```

This is the example of estimates the performance of the random forest classifier on our specified dataset which is synthetic in nature. If a dataset has 100 examples, the test and train dataset split will be created as 100 for evaluation. The mentioned example may lead classification accuracy from mentioned predictions and may derive around 99 percent.

Accuracy: 0.990

Enumerating the folds manually is slow and contains more emphasis on code, and such bulky code may lead bugs. The cross_val_score() function is an alternative to evaluating a model using LOOCV. This instantiated LOOCV object set via the "cv1" argument. The mean and standard deviation will be returned as summarized by calculating as accuracy scores.

The next example will illustrate evaluating the RandomForestClassifier using LOOCV on the same dataset using the cross_val_score() function.

loocv : random forest classifier for automatically evaluate the performance

```
# loocv : random forest classifier for  automatically evaluate the performance
Impoting mean from numpy
Importing std from numpy
Importing  make_blobs from sklearn.datasets
Importing  LeaveOneOut from sklearn.model_selection
Importing  cross_val_score from sklearn.model_selection
Importing  RandomForestClassifier from sklearn.ensemble
# create dataset
```

```
X1, y1 = make_blobs(n_samples1=100, random_state1=1)
# create loocv procedure
cv1 = LeaveOneOut()
# create model
model11 = RandomForestClassifier(random_state1=1)
# evaluate model
scores = cross_val_score(model11, X1, y1, scoring='accuracy', cv1=cv1, n_jobs=-1)
# report performance
print('Accuracy: %.3f (%.3f)' % (mean(scores), std(scores)))
```

The mean classification Accuracy: 0.990 (0.099) across all folds is found adequate with our manual estimate previously.

5.3 LOOCV TO EVALUATE MACHINE LEARNING MODELS

Evaluating machine learning models on common classification and regression predictive modelling datasets using the LOOCV approach.

a. LOOCV for classification

   With the sonar dataset, we will show how to utilise LOOCV to assess a random forest method. The sonar dataset is a typical machine learning dataset with 208 rows of data and a goal variable with two class values, such as binary classification. Predicting whether sonar1 returns point to a rock or a simulated mine is part of the dataset.

Sonar1 Dataset (sonar.csv)

0.0200,0.0371,0.0428,0.0207,0.0954,0.0986,0.1539,0.1601,0.3109,0.2111,0.1609,0.1582,0.2238,0.06
45,0.0660,0.2273,0.3100,0.2999,0.5078,0.4797,0.5783,0.5071,0.4328,0.5550,0.6711,0.6415,0.7104,0.
8080,0.6791,0.3857,0.1307,0.2604,0.5121,0.7547,0.8537,0.8507,0.6692,0.6097,0.4943,0.2744,0.0510
,0.2834,0.2825,0.4256,0.2641,0.1386,0.1051,0.1343,0.0383,0.0324,0.0232,0.0027,0.0065,0.0159,0.00
72,0.0167,0.0180,0.0084,0.0090,0.0032,R

0.0453,0.0523,0.0843,0.0689,0.1183,0.2583,0.2156,0.3481,0.3337,0.2872,0.4918,0.6552,0.6919,0.77
97,0.7464,0.9444,1.0000,0.8874,0.8024,0.7818,0.5212,0.4052,0.3957,0.3914,0.3250,0.3200,0.3271,0.
2767,0.4423,0.2028,0.3788,0.2947,0.1984,0.2341,0.1306,0.4182,0.3835,0.1057,0.1840,0.1970,0.1674
,0.0583,0.1401,0.1628,0.0621,0.0203,0.0530,0.0742,0.0409,0.0061,0.0125,0.0084,0.0089,0.0048,0.00
94,0.0191,0.0140,0.0049,0.0052,0.0044,R

0.0262,0.0582,0.1099,0.1083,0.0974,0.2280,0.2431,0.3771,0.5598,0.6194,0.6333,0.7060,0.5544,0.53
20,0.6479,0.6931,0.6759,0.7551,0.8929,0.8619,0.7974,0.6737,0.4293,0.3648,0.5331,0.2413,0.5070,0.
8533,0.6036,0.8514,0.8512,0.5045,0.1862,0.2709,0.4232,0.3043,0.6116,0.6756,0.5375,0.4719,0.4647
,0.2587,0.2129,0.2222,0.2111,0.0176,0.1348,0.0744,0.0130,0.0106,0.0033,0.0232,0.0166,0.0095,0.01
80,0.0244,0.0316,0.0164,0.0095,0.0078,R

0.0100,0.0171,0.0623,0.0205,0.0205,0.0368,0.1098,0.1276,0.0598,0.1264,0.0881,0.1992,0.0184,0.22
61,0.1729,0.2131,0.0693,0.2281,0.4060,0.3973,0.2741,0.3690,0.5556,0.4846,0.3140,0.5334,0.5256,0.
2520,0.2090,0.3559,0.6260,0.7340,0.6120,0.3497,0.3953,0.3012,0.5408,0.8814,0.9857,0.9167,0.6121
,0.5006,0.3210,0.3202,0.4295,0.3654,0.2655,0.1576,0.0681,0.0294,0.0241,0.0121,0.0036,0.0150,0.00
85,0.0073,0.0050,0.0044,0.0040,0.0117,R

0.0762,0.0666,0.0481,0.0394,0.0590,0.0649,0.1209,0.2467,0.3564,0.4459,0.4152,0.3952,0.4256,0.41
35,0.4528,0.5326,0.7306,0.6193,0.2032,0.4636,0.4148,0.4292,0.5730,0.5399,0.3161,0.2285,0.6995,1.
0000,0.7262,0.4724,0.5103,0.5459,0.2881,0.0981,0.1951,0.4181,0.4604,0.3217,0.2828,0.2430,0.1979
,0.2444,0.1847,0.0841,0.0692,0.0528,0.0357,0.0085,0.0230,0.0046,0.0156,0.0031,0.0054,0.0105,0.01

10,0.0015,0.0072,0.0048,0.0107,0.0094,R

0.0286,0.0453,0.0277,0.0174,0.0384,0.0990,0.1201,0.1833,0.2105,0.3039,0.2988,0.4250,0.6343,0.81
98,1.0000,0.9988,0.9508,0.9025,0.7234,0.5122,0.2074,0.3985,0.5890,0.2872,0.2043,0.5782,0.5389,0.
3750,0.3411,0.5067,0.5580,0.4778,0.3299,0.2198,0.1407,0.2856,0.3807,0.4158,0.4054,0.3296,0.2707
,0.2650,0.0723,0.1238,0.1192,0.1089,0.0623,0.0494,0.0264,0.0081,0.0104,0.0045,0.0014,0.0038,0.00
13,0.0089,0.0057,0.0027,0.0051,0.0062,R

0.0317,0.0956,0.1321,0.1408,0.1674,0.1710,0.0731,0.1401,0.2083,0.3513,0.1786,0.0658,0.0513,0.37
52,0.5419,0.5440,0.5150,0.4262,0.2024,0.4233,0.7723,0.9735,0.9390,0.5559,0.5268,0.6826,0.5713,0.
5429,0.2177,0.2149,0.5811,0.6323,0.2965,0.1873,0.2969,0.5163,0.6153,0.4283,0.5479,0.6133,0.5017
,0.2377,0.1957,0.1749,0.1304,0.0597,0.1124,0.1047,0.0507,0.0159,0.0195,0.0201,0.0248,0.0131,0.00
70,0.0138,0.0092,0.0143,0.0036,0.0103,R

0.0519,0.0548,0.0842,0.0319,0.1158,0.0922,0.1027,0.0613,0.1465,0.2838,0.2802,0.3086,0.2657,0.38
01,0.5626,0.4376,0.2617,0.1199,0.6676,0.9402,0.7832,0.5352,0.6809,0.9174,0.7613,0.8220,0.8872,0.
6091,0.2967,0.1103,0.1318,0.0624,0.0990,0.4006,0.3666,0.1050,0.1915,0.3930,0.4288,0.2546,0.1151
,0.2196,0.1879,0.1437,0.2146,0.2360,0.1125,0.0254,0.0285,0.0178,0.0052,0.0081,0.0120,0.0045,0.01
21,0.0097,0.0085,0.0047,0.0048,0.0053,R

0.0223,0.0375,0.0484,0.0475,0.0647,0.0591,0.0753,0.0098,0.0684,0.1487,0.1156,0.1654,0.3833,0.35
98,0.1713,0.1136,0.0349,0.3796,0.7401,0.9925,0.9802,0.8890,0.6712,0.4286,0.3374,0.7366,0.9611,0.
7353,0.4856,0.1594,0.3007,0.4096,0.3170,0.3305,0.3408,0.2186,0.2463,0.2726,0.1680,0.2792,0.2558
,0.1740,0.2121,0.1099,0.0985,0.1271,0.1459,0.1164,0.0777,0.0439,0.0061,0.0145,0.0128,0.0145,0.00
58,0.0049,0.0065,0.0093,0.0059,0.0022,R

The example below downloads the dataset and summarizes its shape.

```
# summarize the sonar1 dataset
Importing read.csv from pandas
# load dataset
url1 = 'https://raw.githubusercontent.com/jbrownlee/Datasets/master/sonar1.csv'
dataframe = read_csv(url1, header=None)
# split into input and output elements
data = dataframe.values
X1, y1 = data[:, :-1], data[:, -1]
print(X1.shape, y1.shape)
```

The mentioned example sets the dataset and splits it into input and output elements as 208 rows of data with 60 input variables; after this we can now evaluate a model using LOOCV. Initially the dataset is being loaded and slipt input and output components.

```
# split into inputs and outputs
X1, y1 = data[:, :-1], data[:, -1]
print(X1.shape, y1.shape)
…. Procedure creation …
# create loocv procedure
cv1 = LeaveOneOut()
….model to evaluating….
# create model
model1 = RandomForestClassifier(random_state=1)
```

Then use the cross_val_score() function to enumerate the folds, fit models, then make and evaluate predictions. We can then report the mean and standard deviation of model performance.

```
# evaluate model
scores = cross_val_score(model, X, y, scoring='accuracy', cv=cv, n_jobs=-1)
# report performance
print('Accuracy: %.3f (%.3f)' % (mean(scores), std(scores)))
# loocv evaluate random forest on the sonar dataset
Importing mean from numpy
Importing std from numpy
importing read_csv from pandas
importing  LeaveOneOut from sklearn.model_selection
importing  cross_val_score from sklearn.model_selection
importing  RandomForestClassifier  from sklearn.ensemble


# load dataset
url1 = 'https://raw.githubusercontent.com/jbrownlee/Datasets/master/sonar.csv'
dataframe1 = read_csv(url1, header=None)
data = dataframe1.values
# split into inputs and outputs
   X1, y1 = data[:, :-1], data[:, -1]
print(X1.shape, y1.shape)
# create loocv procedure
Cv1 = LeaveOneOut()
# create model
model = RandomForestClassifier(random_state=1)
# evaluate model
scores = cross_val_score(model, X, y, scoring='accuracy', cv=cv, n_jobs=-1)
# report performance
ı
```

print('Accuracy: %.3f (%.3f)' % (mean(scores), std(scores)))

The outcomes first load the dataset and confirm the number of rows in the input and output elements. The model is then evaluated using LOOCV, and the estimated performance when making predictions on new data has an accuracy of around 82.2 percent. (208, 60) (208,)

Accuracy: 0.822 (0.382)

5.4   LOOCV FOR REGRESSION

For the housing dataset, we will illustrate how to utilise LOOCV to assess a random forest method.

The housing dataset, which has 506 rows of data and 13 numerical input variables and a numerical target variable, is a typical machine learning dataset. Housing Dataset (housing.csv) (Sample data)

```
0.15086,0.00,27.740,0,0.6090,5.4540,92.70,1.8209,4,711.0,20.10,395.09,18.06,15.20
0.18337,0.00,27.740,0,0.6090,5.4140,98.30,1.7554,4,711.0,20.10,344.05,23.97,7.00
0.20746,0.00,27.740,0,0.6090,5.0930,98.00,1.8226,4,711.0,20.10,318.43,29.68,8.10
0.10574,0.00,27.740,0,0.6090,5.9830,98.80,1.8681,4,711.0,20.10,390.11,18.07,13.60
0.11132,0.00,27.740,0,0.6090,5.9830,83.50,2.1099,4,711.0,20.10,396.90,13.35,20.10
0.17331,0.00,9.690,0,0.5850,5.7070,54.00,2.3817,6,391.0,19.20,396.90,12.01,21.80
0.27957,0.00,9.690,0,0.5850,5.9260,42.60,2.3817,6,391.0,19.20,396.90,13.59,24.50
0.17899,0.00,9.690,0,0.5850,5.6700,28.80,2.7986,6,391.0,19.20,393.29,17.60,23.10
0.28960,0.00,9.690,0,0.5850,5.3900,72.90,2.7986,6,391.0,19.20,396.90,21.14,19.70
0.26838,0.00,9.690,0,0.5850,5.7940,70.60,2.8927,6,391.0,19.20,396.90,14.10,18.30
0.23912,0.00,9.690,0,0.5850,6.0190,65.30,2.4091,6,391.0,19.20,396.90,12.92,21.20
0.17783,0.00,9.690,0,0.5850,5.5690,73.50,2.3999,6,391.0,19.20,395.77,15.10,17.50
0.22438,0.00,9.690,0,0.5850,6.0270,79.70,2.4982,6,391.0,19.20,396.90,14.33,16.80
```

0.06263,0.00,11.930,0,0.5730,6.5930,69.10,2.4786,1,273.0,21.00,391.99,9.67,22.40
0.04527,0.00,11.930,0,0.5730,6.1200,76.70,2.2875,1,273.0,21.00,396.90,9.08,20.60
0.06076,0.00,11.930,0,0.5730,6.9760,91.00,2.1675,1,273.0,21.00,396.90,5.64,23.90
0.10959,0.00,11.930,0,0.5730,6.7940,89.30,2.3889,1,273.0,21.00,393.45,6.48,22.00
0.04741,0.00,11.930,0,0.5730,6.0300,80.80,2.5050,1,273.0,21.00,396.90,7.88,11.90

The example below downloads and loads the dataset as a Pandas DataFrame and summarizes the shape of the dataset.

```
# load and summarize the housing dataset
from pandas import read_csv
# load dataset
url = 'https://raw.githubusercontent.com/jbrownlee/Datasets/master/housing.csv'
dataframe = read_csv(url, header=None)
# summarize shape
print(dataframe.shape)
```

It confirms the 506 rows of data and 13 input variables and single numeric target variables (14 in total). (506, 14), now model evaluation using LOOCV.

First, the loaded dataset must be split into input and output components.

```
# split into inputs and outputs
X1, y1 = data[:, :-1], data[:, -1]
print(X1.shape, y1.shape)
```

Next, we define the LOOCV procedure.

```
...# create loocv procedure
Cv1 = LeaveOneOut()
We can then define the model to evaluate....
# create model
model11 = RandomForestRegressor(random_state=1)
```

Then use the cross_val_score() function to enumerate the folds, fit models, then make and evaluate predictions. We can then report the mean and standard deviation of model performance.

In this case, we use the mean absolute error (MAE) performance metric appropriate for regression....

```
# evaluate model
scores = cross_val_score(model, X, y, scoring='neg_mean_absolute_error', cv=cv, n_jobs=-1)
# force positive
scores = absolute(scores)
# report performance
print('MAE: %.3f (%.3f)' % (mean(scores), std(scores)))
```

```
# loocv evaluate random forest on the housing dataset
from numpy import mean
from numpy import std
from numpy import absolute
from pandas import read_csv
from sklearn.model_selection import LeaveOneOut
from sklearn.model_selection import cross_val_score
from sklearn.ensemble import RandomForestRegressor
# load dataset
url = 'https://raw.githubusercontent.com/jbrownlee/Datasets/master/housing.csv'
dataframe = read_csv(url, header=None)
data = dataframe.values
```

```
# split into inputs and outputs
X, y = data[:, :-1], data[:, -1]
print(X.shape, y.shape)
# create loocv procedure
Cv1 = LeaveOneOut()
# create model
Model1 = RandomForestRegressor(random_state=1)
# evaluate model
scores = cross_val_score(model, X, y, scoring='neg_mean_absolute_error', cv=cv, n_jobs=-1)
# force positive
scores = absolute(scores)
# report performance
print('MAE: %.3f (%.3f)' % (mean(scores), std(scores)))
```

The model is evaluated using LOOCV, and the performance of the model when making predictions on new data is a mean absolute error of about 2.180 (thousands of dollars).

(506, 13) (506,)

MAE: 2.180 (2.346)

LOOCV in R: Each observation is used as the validation set in the LOOCV (leave-one-out cross-validation) method, while the remaining (N-1) observations are treated as the training set. One observation validation set is used in LOOCV to fit the model and forecast. Moreover, doing this for every observation as the validation set N times. A model is fitted, and it is then utilised to forecast an observational value. The number of folds in this particular instance of K-fold cross-validation is equal to the number of observations (K = N). This technique helps to lessen bias and randomness. The approach seeks to prevent overfitting and lower the Mean-Squared error rate. LOOCV is really simple to execute in R programming.

How closely the model's predictions match the actual data must be measured in order to assess a model's performance on a dataset. In leave-one-out cross-validation (LOOCV), a popular technique for performing this, the following strategy is used:

1. Divide a dataset into training and testing sets, with the training set consisting of all but one observation.
2. Create a model simply utilising the training set of data.
3. Calculate the mean squared error by using the model to forecast the response value of the one observation that was left out of the model (MSE).
4. Continue doing this n times. Compute the test MSE such that it is equal to the sum of all the test MSEs.

Using the trainControl() method from the caret package in R is the simplest approach to execute LOOCV in R. Suppose we have the following dataset in R:

```
#create data frame
df <- data.frame(y=c(6, 8, 12, 14, 14, 15, 17, 22, 24, 23),
                 x1=c(2, 5, 4, 3, 4, 6, 7, 5, 8, 9),
                 x2=c(14, 12, 12, 13, 7, 8, 7, 4, 6, 5))

#view data frame
df
```

y	x1	x2
6	2	14
8	5	12
12	4	12
14	3	13
14	4	7
15	6	8
17	7	7
22	5	4
24	8	6
23	9	5

The following code shows how to fit a multiple linear regression model to this dataset in R and perform LOOCV to evaluate the model performance:

```
library(caret)

#specify the cross-validation method
ctrl <- trainControl(method = "LOOCV")

#fit a regression model and use LOOCV to evaluate performance
model <- train(y ~ x1 + x2, data = df, method = "lm", trControl = ctrl)

#view summary of LOOCV
print(model)

Linear Regression

10 samples
 2 predictor

No pre-processing
Resampling: Leave-One-Out Cross-Validation
Summary of sample sizes: 9, 9, 9, 9, 9, 9, ...
```

```
Resampling results:


 RMSE        Rsquared    MAE
 3.619456   0.6186766   3.146155
```

Tuning parameter "intercept" was held constant at a value of TRUE.
Here is interpretation of output:

- Ten distinct samples were used to create ten models. Each model made use of two predictor variables.
- There was no pre-processing. In other words, before fitting the models, we did not scale the data in any manner.
- We generated the 10 samples using the leave-one-out cross-validation resampling technique.
- The sample size for each training set was 9.
- The root mean squared error (RMSE). This calculates the typical discrepancy between the model's predictions and the actual data. A model may predict real observations more accurately the lower the RMSE.
- R-squared: This is a measurement of the relationship between the model's predictions and the actual data. The more precisely a model can predict the actual observations, the greater the R-squared.
- MAE: This is the typical absolute difference between the model's predictions and the actual data. A model may predict real observations more accurately the lower the MAE.
- The three metrics (RMSE, R-squared, and MAE) in the output each offer us a sense of how well the model performed on untested data.
- To determine which model yields the lowest test error rates and is thus the optimal model to utilise, we often fit a number of different models and compare the three metrics offered by the output shown above.

LOOCV in Python: We must gauge how well a model's predictions match the actual data in order to assess a model's performance on a dataset. Leave-one-out cross-validation (LOOCV), a popular technique for doing this, takes the following approach:

- Divide a dataset into a training set and a testing set, using all but one observation as part of the training set.
- Build a model using only data from the training set.
- Utilize the model for predicting the response value of the one observation left out of the model and calculate the mean squared error (MSE).
- Repeat this process n times. Calculate the test MSE to be the average of all of the test MSE's.

Example of how to perform LOOCV for a given model in Python.
    Step _1: Load Necessary Libraries
    Importing all necessary libraries.
    Step_2: Creating the Data
    Creating a pandas DataFrame that contains two predictor variables, x1 and x2, and a single response variable y.
    df1= pd.DataFrame({'y': [6, 8, 12, 14, 14, 15, 17, 22, 24, 23],
    'x1': [2, 5, 4, 3, 4, 6, 7, 5, 8, 9],
    'x2': [14, 12, 12, 13, 7, 8, 7, 4, 6, 5]})
    Step_3: Perform Leave-One-Out Cross-Validation

Setting a multiple linear regression model to the dataset and performing LOOCV to evaluate the model performance.

#define predictor and response variables

```
X1 = df1[['x1', 'x2']]
Y1 = df1['y']
#define cross-validation method to use
Cv1 = LeaveOneOut()
#build multiple linear regression model
model1 = LinearRegression()
#use LOOCV to evaluate model
scores = cross_val_score(model, X, y, scoring='neg_mean_absolute_error',cv=cv, n_jobs=-1)
```

#view mean absolute error
mean(absolute(scores))
3.1461548083469726

From the output we can see that the mean absolute error (MAE) was 3.146. That is, the average absolute error between the model prediction and the actual observed data is 3.146.

In general, the lower the MAE, the more closely a model is able to predict the actual observations.

Another commonly used metric to evaluate model performance is the root mean squared error (RMSE). The following code shows how to calculate this metric using LOOCV:

```
#define predictor and response variables
X1 = df[['x1', 'x2']]
Y1 = df['y']
#define cross-validation method to use
cv1 = LeaveOneOut()
#build multiple linear regression model
model1 = LinearRegression()
#use LOOCV to evaluate model
scores = cross_val_score(model, X, y, scoring='neg_mean_squared_error',
                         cv=cv, n_jobs=-1)
#view RMSE
sqrt(mean(absolute(scores)))
```

3.619456476385567

We can see from the result that the RMSE, or root mean square error, was 3.619. A model can predict real observations more accurately with a smaller RMSE.

In actual practice, we usually fit a number of different models and compare the RMSE or MAE of each model to determine which model yields the lowest test error rates and is, thus, the ideal model to employ.

5.5 CONCLUSION

Machine learning is used widely in each and every field for data analysis and data-related prediction activities. There are many types of machine learning algorithms, especially supervised and unsupervised. Leave-one-out cross-validation is being used to estimate the working effectiveness of machine learning algorithms. Leave-one-out cross-validation is used to predict the desired outcomes when there is no training model. The major focus of such procedures is on bias and randomness. Such methods also focus on minimising the Mean-Squared error rate. As experiments on data using LOOCV, the LOOCV is preferable when there is a small dataset. It is expensive when there is large dataset.

REFERENCES

1. Jordan MI, Mitchell TM. Machine learning: Trends, perspectives, and prospects. *Science.* 2015;349(6245):255–60.
2. Mahesh B. Machine learning algorithms-a review. *International Journal of Science and Research.* 2020;9:381–6.
3. Mitchell TM. *Machine Learning*: McGraw-Hill; 2007.
4. Bi Q, Goodman KE, Kaminsky J, Lessler J. What is machine learning? A primer for the epidemiologist. *American Journal of Epidemiology.* 2019;188(12):2222–39.
5. Sagiroglu S, Sinanc D, editors. Big data: A review. *International Conference on Collaboration Technologies and Systems (CTS)*; 2013. IEEE.
6. Baxevanis AD, Bader GD, Wishart DS, *Bioinformatics*: John Wiley & Sons; 2020.
7. Engel T. Basic overview of chemoinformatics. *Journal of Chemical Information and Modeling.* 2006;46(6):2267–77.
8. Luscombe NM, Greenbaum D, Gerstein M. What is bioinformatics? A proposed definition and overview of the field. *Methods of Information in Medicine.* 2001;40(4):346–58.
9. Koch A, Meinhardt H. Biological pattern formation: From basic mechanisms to complex structures. *Reviews of Modern Physics.* 1994;66(4):1481.
10. Kumar M, Bhasin M, Natt NK, Raghava GPS. BhairPred: Prediction of β-hairpins in a protein from multiple alignment information using ANN and SVM techniques. *Protein Science.* 2005;33(Suppl_2):W154–W9.
11. Wassenegger M. RNA-directed DNA methylation. *Plant Molecular Biology.* 2000;43:203–20.
12. Bharati D, Jagtap R, Kanase K, Sonawame S, Undale V, Bhosale AV. Chemo informatics: Newer approach for drug development. *Asian Journal of Research in Chemistry.* 2009;2(1):1–7.
13. Vandewiele N, Van Geem K, Reyniers M-F, Marin G, editors. Automatic reaction network using chemoinformatics. *AIChE Annual Meeting: American Institute of Chemical Engineers (AIChE).* 2011
14. Wong T-T. Performance evaluation of classification algorithms by k-fold and leave-one-out cross validation. *Pattern Recognition.* 2015;48(9):2839–46.
15. Cunningham P, Cord M, Delany SJ. *Machine Learning Techniques for Multimedia.* Supervised learning. 2008:21–49.
16. Hastie T, Tibshirani R, Friedman J. Overview of supervised learning. *The Elements of Statistical Learning.* 2009:9–41.
17. Van Engelen JE, Hoos HH. A survey on semi-supervised learning. *Machine Learning.* 2020;109(2):373–440.
18. Castelli M, Vanneschi L. Largo ÁR. Supervised learning: Classification. *Encyclopedia of Bioinformatics and Computational Biology.* 2018;1:342–9.
19. Kotsiantis SB, Zaharakis I, Pintelas PJ. Supervised machine learning: A review of classification techniques. *Informatica.* 2007;160(1):3–24.
20. Stöter F-R, Chakrabarty S, Edler B, Habets EA, editors. Classification vs. regression in supervised learning for single channel speaker count estimation. *IEEE International Conference on Acoustics, Speech and Signal Processing (ICASSP)*; 2018. IEEE.
21. Nasteski VJ. An overview of the supervised machine learning methods. Horizons. 2017;4:51–62.
22. Ghahramani Z. Canberra, Australia, February 2–14, Tübingen, Germany, August 4–16, Revised Lectures. Unsupervised learning. 2004:72–112.
23. Sathya R, Abraham A. Comparison of supervised and unsupervised learning algorithms for pattern classification. *International Journal of Advanced Research in Artificial Intelligence.* 2013;2(2):34–8.
24. Dike HU, Zhou Y, Deveerasetty KK, Wu Q, editors. Unsupervised learning based on artificial neural network: A review. *IEEE International Conference on Cyborg and Bionic Systems (CBS)*; 2018: IEEE.
25. Greene D, Cunningham P, Mayer RJ. Unsupervised learning and clustering. *Machine Learning Techniques for Multimedia: Case Studies on Organization and Retrieval.* 2008:51–90.
26. Grira N, Crucianu M, Boujemaa NJ. Unsupervised and semi-supervised clustering: A brief survey. *A Review of Machine Learning Techniques for Processing Multimedia Content.* 2004;2004:9–16.
27. Caron M, Bojanowski P, Joulin A, Douze M, editors. Deep clustering for unsupervised learning of visual features. In *Proceedings of the European Conference on Computer Vision (ECCV).* 2018.
28. Sinaga KP, Yang M-S. Unsupervised K-means clustering algorithm. *IEEE Access.* 2020;8:80716–27.
29. Pham M, Cheng F, Ramachandran KJDS. A comparison study of algorithms to detect drug–adverse event associations: Frequentist, Bayesian, and machine-learning approaches. *Drug Safety.* 2019;42(6):743–50.

30. Mannila H, editor. Data mining: Machine learning, statistics, and databases. In *Proceedings of the 8th International Conference on Scientific and Statistical Data Base Management*. 1996: IEEE.

31. Stone CJJT. The dimensionality reduction principle for generalized additive models. *Annals of Statistics*. 1986:590–606.

32. Ayesha S, Hanif MK, Talib RJIF. Overview and comparative study of dimensionality reduction techniques for high dimensional data. *Information Fusion*. 2020;59:44–58.

33. Ouali Y, Hudelot C, Tami M. An overview of deep semi-supervised learning. arXiv preprint arXiv:2006.05278. 2020.

34. Moerland TM, Broekens J, Plaat A, Jonker CMJF. Model-based reinforcement learning: A survey. *Foundations and Trends® in Machine Learning*. 2023;16(1):1–118.

35. Botvinick M, Ritter S, Wang JX, Kurth-Nelson Z, Blundell C, Hassabis D. Reinforcement learning, fast and slow. *Trends in Cognitive Sciences*. 2019;23(5):408–22.

36. Laskin M, Lee K, Stooke A, Pinto L, Abbeel P, Srinivas AJ. Reinforcement learning with augmented data. *Advances in Neural Information Processing Systems*. 2020;33:19884–95.

37. Chen L, Lu K, Rajeswaran A, Lee K, Grover A, Laskin M, et al. Decision transformer: Reinforcement learning via sequence modeling. 2021;34:15084–97.

38. Triantafillou E, Zhu T, Dumoulin V, Lamblin P, Evci U, Xu K, Goroshin R, Gelada C, Swersky K, Manzagol PA, Larochelle H. Meta-dataset: A dataset of datasets for learning to learn from few examples. arXiv preprint arXiv:1903.03096. 2019.

39. Vabalas A, Gowen E, Poliakoff E, Casson AJ. Machine learning algorithm validation with a limited sample size. *PLOS One*. 2019;14(11):e0224365.

40. Subasi A. *Practical Machine Learning for Data Analysis Using Python*: Academic Press. 2020.

41. Pozo Montero F. *Validation Procedures. Leave-One-Out Cross-Validation (LOOCV) and K-Fold Cross-Validation*. 2019.

42. Nti IK, Nyarko-Boateng O, Aning JJ. Performance of machine learning algorithms with different K values in K-fold cross-validation. *International Journal of Information Technology and Computer Science*. 2021;13(6):61–71.

43. Aparicio J, Esteve M. How to peel a data envelopment analysis frontier: A cross-validation-based approach. *Journal of the Operational Research Society*. 2022:1–15.

44. Mila C, Mateu J, Pebesma E, Meyer HJ. Nearest neighbour distance matching leave-one-out cross-validation for map validation. *Methods in Ecology and Evolution*. 2022;13(6):1304–16.

45. Hao J, Ho TK. Machine learning made easy: a review of scikit-learn package in python programming language. *Journal of Educational and Behavioral Statistics*. 2019;44(3):348–61.

6 Ethical Issues and Artificial Intelligence Technologies in Bioinformatics Concerning Behavioural and Mental Health Care

Divya Sheth, Param Patel, and Yashwant Pathak

6.1 INTRODUCTION

Artificial intelligence, or AI for short, is a development of various computer systems with the goal of being able to perform tasks that can be normally performed by human intelligence. Within this, AI would be able to complete tasks dependent on human intelligence such as in fields that require visual perception, speech recognition, decision-making, and language acquisition. This technology uses mathematical algorithms to imitate human intelligence and their cognitive abilities while addressing a variety of healthcare challenges. These algorithms hold great promise in expanding the knowledge of potential use in the healthcare system. Not only can this AI technology be used in patient care but also in collaborating in continuous research efforts in the efforts to heal millions and add to preventative medicine [1]. More specifically, AI technology holds great potential to be a milestone in the transformation of mental healthcare and improve the pitfalls within them. AI is currently used in healthcare and facilitates detection of disease in the early state before it is too late to cure.

Bioinformatics in conjunction with AI can have endless positive implications on the healthcare system. Bioinformatics consists of a development of systems and software tools dedicated to understanding and storing biological information in data sets that can be later manipulated for in-depth multidisciplinary analysis. Thus, by combining computer programing and algorithms, basically AI, data can be analysed and help scientists understand specific elements in the data which can be translated into future research. In essence, bioinformatics takes portions of data and interdisciplinary topics from biology, medicine, maths/physics, and computer science to be able to construct meaningful outcomes as a reflection of the data.

Currently, the costs of healthcare are on the rise and will continue to rise. Challenges such as increases in chronic illness, life expectancy, and the high cost of new therapies are major reasons why artificial intelligence will start to emerge. The main premises of AI in medicine is to alleviate the high cost and to make the tasks more efficient. AI is used in clinical medicine for disease diagnosis and treatment outcomes; however, there are drawbacks such as accuracy and natural error due to the fact that in medicine not everything can be definitive, along with the fact that there is a good amount of overlap in disease diagnosis [2]. Some additional concerns with AI technology include healthcare liability, privacy concerns, and data compromise. Healthcare administrations across the world have raised questions regarding patient data security, which continues to be the major player in the decisions for AI advancement. Depending on the illness of a patient, the complexity

DOI: 10.1201/9781003353751-6

of care may vary, and without proper empathy between the doctor and patient, there cannot be a true human-level approach to patient interaction. The need for a proper patient–physician interaction continues to be a major factor in determining the quality of care a patient has. With AI at the forefront, it is difficult to tell if AI can truly supplement all aspects of a patient's care [3].

6.2 IMPLICATIONS OF BIOTECHNOLOGY IN THE CLINICAL SETTING

While bioinformatics is about the computer biotechnology used to aid with clinical applications, there are always questions regarding computer bioethics and how we can integrate nanotechnology into our own human bodies. Recent discoveries have been made on the potential for nanodevices to be able to go inside the body and destroy cancer cells specifically. This would be a major breakthrough, as we have never been able to have proper targeted drug delivery in cancer treatment. If the efficacy of these devices is seen only from the patient benefit standpoint, they offer benefits such as longevity, quality of life, and cell-specific targeting. AI has now come to a point where it is able to predict diseases using lab work such as radiology, bloodwork, and other serum specific tests much sooner than we can detect [4].

The majority of the methods used in a clinical setting have a one-size-fits-all approach with adjustments made as needed by a doctor. With AI, we are able to analyse a larger volume of data very quickly and create personalised treatments for patients. With treatments catered towards each patient's specific needs, there is a very high chance that patient prognosis will increase too. Mental health care is a sensitive area for AI because many of the diseases that encompass the field are sensitive in nature. In order for AI to be beneficial, there needs to be a proper integration of both the physician and the AI technology in use. Studies have shown that AI can be used in mental illness as a method of monitoring and supervising patients. This enables AI to relay any changes out of the norm in a patient's brain activity, vitals, or any other relevant information to the physician quickly so that the physician can make the best decision for the patient. With this type of complex integration, we could potentially see a rise in the use of AI in more fields than were thought possible [5].

6.3 ARTIFICIAL INTELLIGENCE IN MEDICINE

Artificial intelligence uses computer systems and robotic instruments to mimic human behaviour and the thought process. In clinical situations, AI helps in drug formulations and assists in clinical diagnosis and robotic surgeries [1]. Thus, AI has both a virtual and physical impact on medicine and clinical expectation and interventions. In the virtual impacts, AI requires information management tools that can interpret information in large data sets for electronic healthcare. Some of the virtual impacts AI can have is in designing drug compositions, running simulations, and analysing lab results. While in the physical impacts, AI is used in robotic-assisted surgeries, like the da Vinci Surgical System, medical devices using robotic technology to monitor patients in critical condition, and helping in clinical presentation analysis.

More specifically, AI use in virtual healthcare can be broadened and taken in many avenues. Areas within medicine that have seen the most success are found in leveraging pattern recognition in specialties such as cancer detection, radiology, and ophthalmology, where AI algorithms can perform at a higher accuracy and work better compared to experienced clinicians in evaluating medical images for abnormalities undetectable to the human eye [6]. These slight differences unnoticeable to the human eye can be monitored by prior algorithms made and can check for slight subtleties indicating various diseases. In the future, intelligent machines will be heavily emphasised and equipped through healthcare clinics, allowing the technology to flourish while human learning is limited by its capacity to learn and access knowledge. However, it is unlikely that medical practices will completely forget human clinicians and physicians. Instead, artificial intelligence machines will be used in support of physicians in clinical decision making.

While AI has been vastly implemented in certain specialities such as surgery and internal medicine, some soft practices like psychiatry and mental health care have had more challenges in being adopted. Figure 6.1 shows a trend of the number of publications found when searching "Artificial intelligence and Mental Health" [6]. This is not a systematic means of analysis and is merely searched through the PubMed and Google Scholar databases while avoiding studies including neurocognitive disorders like dementia and mild cognitive impairment, in spite of the relation to mental health care [6]. This data is not an overall or extensive view of the number of publications; however, the trend can be extrapolated to show that there is a rising trend in the number of publication related to artificial intelligence and mental health, where in recent years, 2015–2019, there was a significant increase for those same publications. Thus, the figure shows a recent spike in publications and studies in the relation to the use of AI in mental health care.

However, while AI is slowly integrating into mental health care, ethical issues must be taken into considerations to show improvement in the effectiveness and safety of this slow integration. A critical ethical issue is the effectiveness of computer algorithms used to predict and diagnose mental health illnesses. This requires accuracy and precision such that is does not lead to an increased risk to the patient. It is important to prioritise certain guiding principle like autonomy, beneficence, and justice such that there is a mitigation in certain data sets that could be biased. The ethical issue of biased data sets can link unexpected disease to certain factors and make unrealistic predictions and associations, such as a certain mental illness being linked to a specific ethnic group if the algorithm is not carefully monitored or accurate.

In terms of accuracy in diagnosis, AI has been utilised in ophthalmology to check for fundus signs of patients with diabetes. The AI was able to take these results and then relay which screenings are out of the norm so that an ophthalmologist can review those cases personally. In neurology, a learning AI was used to check 12-month-old infants to determine the possibility of autism symptoms developing at 24 months. The results were promising in that the AI had a positive predictive value of 81%. In some cancers, AI is able to find single isolated markers through PET scans and identify very specific serum markers that can accurately predict cancers at a much earlier stage with higher efficacy [7]. All of these capabilities of AI is due to the ability to accurately identify minor structural abnormalities in protein structures. We have the ability to find these changes; however, the methods are far too expensive and consume a significant amount of time. With AI assistance,

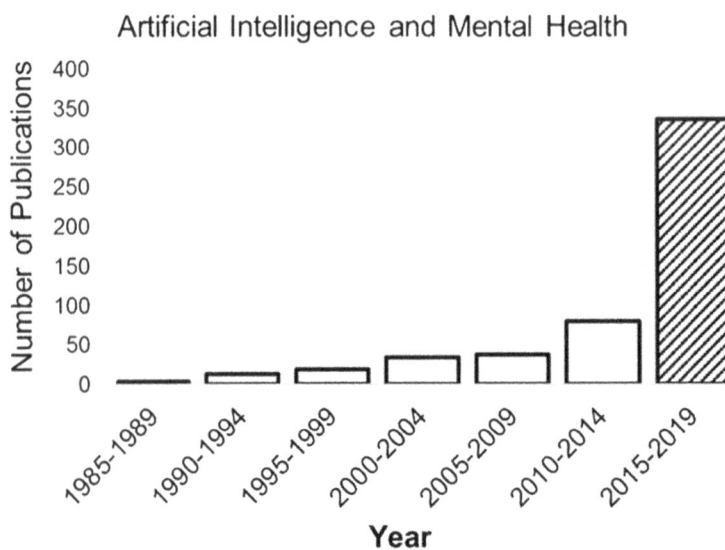

Artificial Intelligence and Mental Health

FIGURE 6.1   Frequency of publications by year in PubMed using search terms "Artificial Intelligence and Mental Health" [6].

we would be able to bypass this major hurdle and help patients be seen by the appropriate doctors sooner than the first onset of symptoms [8].

6.4 ETHICAL IMPLICATIONS OF AI USE IN MEDICINE

While AI is integrating into medicine, ethical implications have surfaced and shed light on the practicality of the use of AI in healthcare systems. Many ethical considerations are brought up; however, if these considerations can be mitigated, AI could hold great promise for the future. The use of AI in clinical practice has great potential to transform healthcare for the better. Yet, challenges revolve around the informed consent to use, safety and transparency, algorithmic fairness and biases, and data privacy [9].

Informed consent to use has not received attention in the ethical debate category until quite recently. Informed consent to use is the communication between a patient and provider that leads to an agreement or permission to use certain data sets for care, treatment, and medical service. Every patient has the right to know where their data is going and how it is being used. The data collected from the patient is medical data used by bioinformatics, and integrating informed consent to use into AI has had challenges [9]. A couple questions are raised from this sort of ability in the ethical challenges with the informed consent to use. To what extent would clinicians need to inform their patients and under what circumstances? This leads to the ethical debate of using AI and machine learning in healthcare as many AI systems use complex algorithms, which can often result in non-interpretable machine learning techniques that clinicians might not fully understand. This creates a break in knowledge where the provider and patient do not quite understand how a certain diagnosis or measurement is calculated. Health applications for mobile have become popular in integrating AI into their systems to show and monitor various metrics such as sleep, exercise, and diets. This brings forth the same issue of the data breach for informed consent. Many do not understand how that data

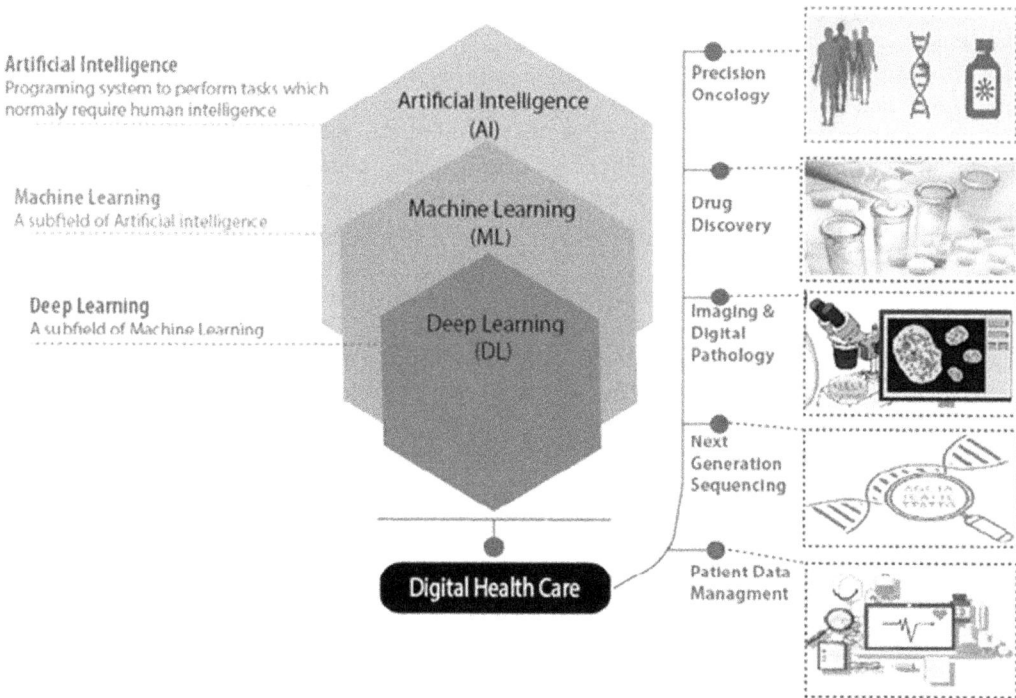

FIGURE 6.2 Applications from AI, ML, and in digital health care and oncology to solve healthcare issues and predict optimal treatment outcomes [1].

collected by wearable technology interprets data and similarly come to a conclusion and then whose responsibility it would be to administer the knowledge of how this technology works. This can then be turned into a legal issue from a patent perspective of infringement and intellectual property.

Next, safety and transparency is another ethical issue regarding AI in healthcare. AI algorithms can use patient data from medical records and cross reference them with other clinical metrics and medical studies currently in the market and provide different avenues to explore treatment options [9]. It has also been known that AI can interpret data using alternative methods and display incorrect and unsafe treatment options that are found verified by physicians. Thus, this solidifies that AI cannot completely transition into healthcare and there must be human factors in the practice of medicine. Transparency relates to informing when AI is being used with patient data and how the AI machine determines the output. In an ideal world, the complex algorithms would open for public use and examination; however, this would bring legal issues mentioned previously. Despite this, there must be a level where AI developers should be sufficiently transparent. This would require developers to inform the public about any shortcoming in the software and the use of data. By allowing this to happen, it can create trust in stakeholders, in specific clinicians and patients, who are the main stakeholders for AI use in healthcare.

Algorithmic fairness and bias pose an ethical dilemma for the integration of AI in healthcare. AI uses computer systems and algorithms along with bioinformatics to create an output for multiple metrics. With this, it also has the ability to improve healthcare in low- and middle-income countries. However, some AI can be biased and create issues from different sets of data. There have been many accounts of algorithms exhibiting bias that result in injustice with regards to ethnicity and skin colour, along with gender [9]. In additions, age and disabilities can be seen as additional places of bias. This sort of bias can result from different data sets used to construct the algorithm and the pitfalls of those data sets. For example, if an AI algorithm has the ability to detect skin cancer and the algorithm was created from data from predominantly Caucasian skin, there would be a lack of data to support conclusions drawn from Black or Asian skin tones for the detection of skin cancer. This type of bias can be resolved by collecting more data and precisely allowing proper integration and computation of the data.

Then lastly, maybe the most important ethical challenge would be data privacy. This revolves around who and what can access patient data, along with what it will be used for. Medical records contain a plethora of data on various topics. The amount of information within medical records can be quantified by millions of gigabytes of data that can be used in an enormous amount of research. With all this data, other studies that are unethical can be conducted and used for unscientific purposes. This can make the patients and clinicians distrust AIs and hinder their integration into healthcare.

AI integration in medicine poses many ethical issues, such as the ability for a machine to establish a proper doctor– patient relationship and make judgements based on empathetic understanding for the patient. The Turing test, proposed by Alan Turing, argued that medical professionals need wisdom rather than intelligence in order to be successful physicians. This is still today true because a holistic approach is needed for any patient for a proper understanding of their condition. Considering these factors, it is important to decipher whether or not a machine can pass the Turing test. To date, no machine has passed the test, which makes us think maybe we need more integration rather than complete takeover of AI in medicine [10]. Nonetheless, we also need to consider the issues of patient privacy, patient preference, and even patient safety. Although incorporating AI into the medical field can provide efficiency, it cannot take into account informed consent, autonomy, and confidentiality. These are key factors for achieving a good patient–physician relationship. AI technology may also be involved in medical education, and while that may be an exciting prospect in some aspects, there are many complexities. One of which is the change in medical education from what it is today to a point where students would be learning how to manage AI machines and not the techniques themselves. This may be a major setback as it would mean that less skilled physicians who are not able to make intuitive decisions would be added into the workforce. A legal issue still remains even if we do start to incorporate more AI in the clinical setting. If an AI machine makes a mistake, there would be medical malpractice issues that would arise along with liability issues that would need to be settled with the manufacturers. Patient reports and imaging would also be under threat as the AI would potentially be seeing private information that is only meant to be seen by the caregivers of the patient. All of these issues create a major conflict of interest for the patient and the physician, along with everyone involved. If there were to be an issue, it would mean that the patient would not only have had ill treatment, but the hospital and AI company manufacturers would be under a potential lawsuit. Most important is patient satisfaction and the ability to treat the patient without doing harm. If that is compromised, then heavy AI integration will not be possible [11]. All things considered, a look into the stakeholders at all levels in AI technology is needed to understand the complexity of issues that would arise if something wrong occurred with the patient.

6.5 LEGAL IMPLICATIONS OF AI IN MEDICINE

As some of the ethical issues present themselves, they can quickly turn into legal challenges. The subject of patient data and medical records is based solely on security; thus, it is tightly monitored and serves legal action if reprimanded. Regulations are often set to hold a certain standard and prevent future changes from affecting the standard and keeping it below a certain threshold. The Food and Drug Administration (FDA) is the entity that typically sets regulations to regulate the safety and efficacy of new and emerging medical technologies. Thus, the incorporation of AI in medicine and the healthcare field must first abide by the FDA regulations. However, a challenge that the FDA faces in regulating AI algorithms is that the relative complexity of black box medical algorithms doesn't concern itself with the medical device [12]. Yet, in many medical devices there are programs that can be analysed, giving legal implications if the code is altered or serves an alternative purpose. In 2016, President Barack Obama signed into law the *21st Century Cures Act* [9]. This act applies to electronic patient health information in hopes of accelerating medical product development and bringing innovations to patients. This also allowed the FDA to regulate medical advisory tools [13]. While the FDA also has software regulations that are monitored closely, many legal actions can be taken when certain regulations are not followed. For example, if a software regulation pertaining to the AI of a certain medical device is not upheld or is faulty, warning letters, seizures, injunctions, and prosecution suits can be implemented by the FDA.

Next, AI raises challenges for liability regimes. AI can affect the liability of patient care. With AI present in healthcare, questions of liability begin to arise. For example, if an AI-based medical device harms a patient, the liability can be spread to several different sources. The physician or healthcare team who recommends the medical device can be liable for providing faulty care, while the manufacturer and developer of the AI algorithm and device can be liable for providing a faulty product and a failed program. Thus, this brings the issue of tort. The law of tort concept is to provide compensation for the wrong done to a person and provide that person relief from the wrongful act of other entities. This ensures that if any wrong is caused, compensation will be provided to the individual for relief by the party that is liable for the wrong caused [13].

Lastly, intellectual property rights become a very important concern that is brought up with the legal implications of AI integration within healthcare. Intellectual property is the work or creation resulting from one's own creativity to produce works such as a manuscript or design that can one day be used to apply for a patent or copyright and trademark filings. Intellectual property laws deal with protecting and enforcing the rights of owners and creators of intellectual creations. A common place where intellectual property rights and laws can apply are the medical device industry. Within the medical device industry, many forms and regulations are placed to protect the property of inventors and owners. The incorporation of AI into the medical device industry along with intellectual property protection rights creates obstacles and challenges for AI black box algorithms and medicine. Due to the vast expenses, resource availability, and expertise needed to develop these algorithms and then train them accordingly, intellectual property rights become very important in

TABLE 6.1

Various Stakeholders Involved with the Use of Artificial Intelligence Technology [4]

Stakeholders	Healthcare professionals	Healthcare industry	Public health officials	Government agencies	Research scientists
Goals	To provide the best possible care for patients	To be cost effective and patient centric to boost efficiency	Protect the rights of the general public and help increase health outcomes	Safely regulate the use AI and ensure the overall safety of citizens through legislature	Work on research for the betterment of human health outcomes
Areas to be vigilant	If AI takes over, then there is chance that medical professionals may not be as competent during medical training	We still do not have enough evidence for the efficacy of AI to be heavily integrated in medicine	Must educate the public so they are informed with the changes in healthcare	Patient privacy and data must be considered as a priority along with patient benefit	Have to ensure that their development is utilised for the right reasons and does no harm to the patient
Scientific challenges	Correct diagnosis along with an individualised treatment plan per patient	AI has no method of providing each patient with an individual approach to care	Must be informed and ready to give widespread messages to the public for their benefit	Must be fully informed and knowledgeable about AI and medicine to pass laws that are beneficial to the patients	Conduct research that has patient outcome benefits that far outweigh the risks
Legal challenges	If patient care is the goal and AI makes a mistake, then physicians can be held responsible	Widespread issues with AI integration can be taxing on healthcare due to an increase in lawsuits along with higher demand for medical services	Have a critical role in informing the public in a timely manner and aiding those who do not understand	If widespread failure ensues, then the legislatures that allow AI can be revoked	At any point if mistakes are made and it concerns a patients life, then those responsible for the manufacturing may be held liable

keeping confidentiality and respective ownership and ownership rights. It is normally expected that intellectual property rights provide some measure of protection, resulting in companies flooding money into developing black box algorithms and in turn black box medicine [12]. In addition, cases revolving around data mining and data analytics concern themselves with intellectual property rights to protect references to or copying of databases and information [9]. Within healthcare, integration of AI presents itself within drug companies to expand drug portfolios, but AI systems can be used for patent examiners to predict innovations and reveal that the patent was actually ineligible for the protection of the patent due to lack or several factors such as novelty and the inventive step [9]. In all, intellectual property rights and laws are often there to protect investments of inventors and owners from having their intellectual property being spoken for by another individual. Integration of AI into the healthcare system for mental and behavioural care brings forth legal challenges that can lead to litigation; thus, improvements in the intellectual property rights can expedite the integration of AI systems in healthcare for mental and behavioural care.

6.6 MENTAL AND BEHAVIOURAL HEALTHCARE

Mental health care is a particular field in the medical world where extra attention needs to be given by the physician to the patient due to the nature of the patients that are being seen. Given these extra limitations, there are still psychotherapies developed which aid in their care. Increasing research has allowed for AI psychotherapists to aid in many aspects of psychotherapy and psychiatry. Applications on the phone have incorporated chat bots that help patients reduce their anxiety and help them learn about ways to cope with their illness. These AI have the ability to analyse visual cues on the face and determine emotions so that they can adequately respond. Animal robots are used for dementia patients for companionship, which is critical for these patients at this stage of disease. Children with autism spectrum disorder (ASD) have special care robots that help them improve their social skills along with other robots that assist with social behaviours such as empathy, imitation skills, and active engagement. AI is seen in robots that help with disorders such as ED or even in victims of sexual violence to assist them during a sensitive period when they are first seen. Many of these advances in mental health care is a major reason for continued research into the field, and there will be a point where we will need to find a way to work cohesively with AI technologies in a manner that ensures the best patient outcomes [14]. AI is also used in support to virtually embody psychotherapeutic devices such as chatbots that work over short message service text messaging like WhatsApp and Facebook Messenger. These applications provide an interactive base for patients and act as virtual psychotherapy with the aim of engaging with patients to recognize emotions and thought patterns that can be recognized by the AI system. In fact, a study found that depression symptoms decreased when they interacted with Woebot, a chatbot AI system, more than groups who relied on electronic book sources [15]. This shows how AI systems integrated into the healthcare system can show promise with mental health care while giving additional benefits to the healthcare system by allowing doctors and practitioners to use their time on other patients that might need additional attention while these patients who interact with these chatbots can get positive outcomes within their mental state from having interaction with machine and AI algorithms.

Robot therapy seems to have much future integration with AI systems for behaviour and mental health care. In addition to chatbots, robot therapy can consist of animal-like robots that have been seen to be in increasing use for patients with dementia. As stated before, children on the autism spectrum have used special robots to help their symptoms and regulate and improve their social skills. In turn, many of the children have seen vast improvements in their behavioural outlook, showing the benefits of AI integration into the healthcare system for a mental health disorder displaying behavioural implications.

For AI to be useful, algorithm developers must develop algorithms that must be trained with large data sets. Figure 6.3 describes an integrated conceptual model for infectious disease using AI. It outlines how AI implementation can be used to track and model specific diseases that can be present from patterns picked up by the AI algorithm and generates possibilities accordingly. This model is used for infectious diseases, yet the same principle applies and can be correlated to be used in mental and behavioural health care.

From Figure 6.3, the breakdown is shown and can provide insight on how AI can be used in the assessment of mental and behaviour disorders. First the AI algorithm needs to be developed and troubleshooted such that there is a mitigation in error within the computation of the algorithm. This would require large data sets and known outcomes from those data sets to train the algorithm as previously mentioned. The more data that can be run through the system, the more advanced it could become while minimising errors generated by the algorithm. Once this has been done, this creates the shared data repository as seen in Figure 6.3. The shared data repository is basically a data warehouse where the algorithm can search and pinpoint commonality within preexisting data and already logged data to come to a conclusion. This is the basis for AI system integration into healthcare made from a pile of data on specific disorders such as symptoms, age, ethnicity, and the predictability used from certain factors such as smoking and alcohol consumption. All these factors

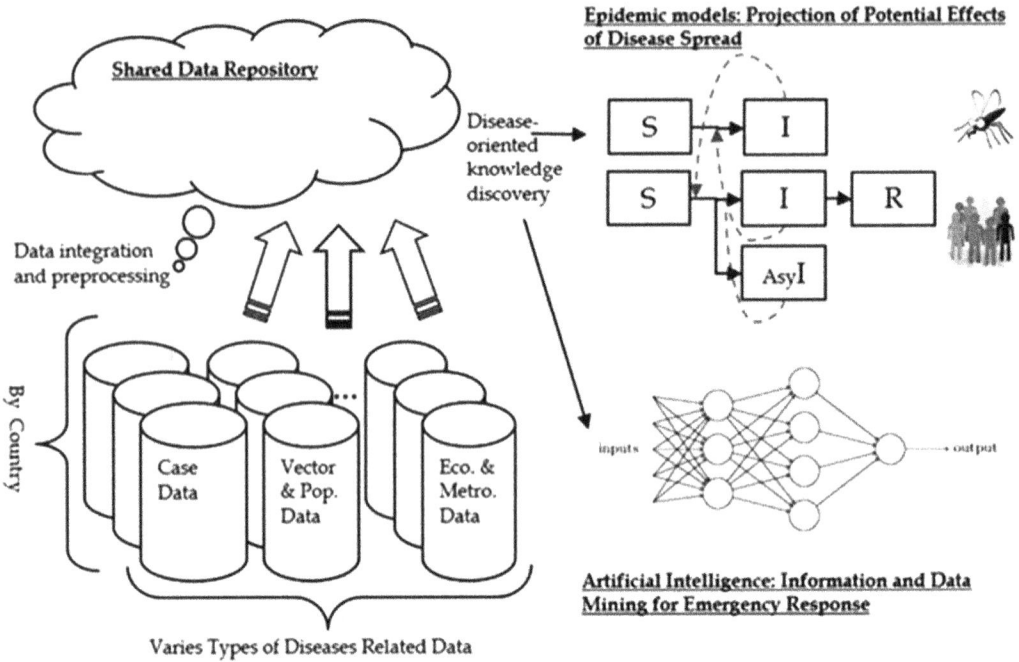

FIGURE 6.3   Integrated conceptual model for infectious diseases using AI [16].

contain data sets that can be stored in this repository for later use in diagnosing and making conclusions drawn by AI. In addition, within AI, there are certain subsets that use the data in the repository to have specific uses. In Figure 6.4, it shows the subsets of each component in AI [16]. Within this diagram, it shows three sets such as deep learning, machine learning, and AI. Obviously, AI is the programmable computation that acts to mimic human intelligence using massive amounts of data interpretation. This is the most basic and most encompassing use for the integration of this technology into healthcare. Within this, machine learning is more specific, where the machine will gather information from previous experience and learn from them to be used in further analysis in more specific measures. This is the second depth as it requires a more specific and fine niche to be able to be used in real-world applications. Lastly, deep learning is the most specific and difficult use of AI in the field of healthcare. This is in the centre as this is the final achievement of AI to produce technology that can mimic human intelligence for better application in healthcare, more specifically mental and behavioural care. Deep learning is associated with technology that can use AI systems and machine learning to create a fully functional machine that can train itself. This becomes a situation where once the machine is up and running, it will be self-sufficient in its ability to interpret data and find exclusions that mimic how human intelligence goes through certain thought processes.

Referring back to Figure 6.3, once a shared data repository is made, using the subset of AI becomes exponentially easier. AI systems can then compute predictions and models of how mental and behavioural disorders can be diagnosed and moreover can create treatment options under various categories. These predictions are based on the processing of algorithms and data integration that can be compiled over a course of inputs and disease-oriented knowledge discovery to formulate the chosen output.

6.7   AI IN PSYCHIATRY

In psychiatry, AI has the potential to revolutionise the diagnosis, prediction, and treatment phases. Psychiatry revolves around understanding the neurobiology that is the underlying cause of most

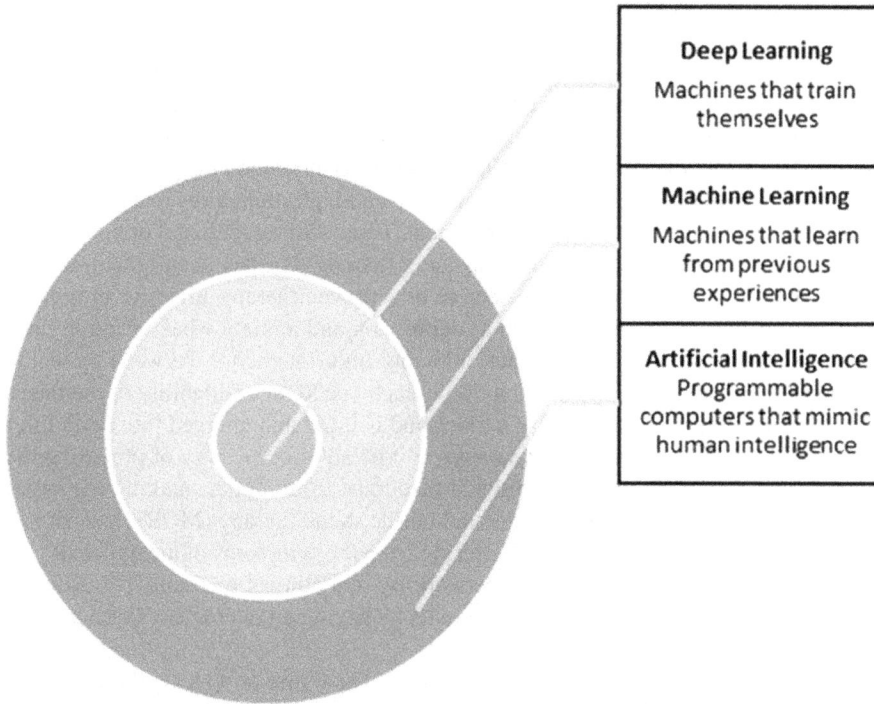

FIGURE 6.4 Diagram illustrating artificial intelligence and main subsets [16].

psychological disorders. Clinicians have found it difficult to diagnose psychiatric disorders due to objective and unreliable clinical measures. Computerised techniques such as latent semantic analysis (LSA), an automated high dimensional tool used to analyse speech, has been successfully implemented to aid clinicians in diagnosing psychiatric disorders [17]. The basis of LSA is that it unravels the relationship between words and passage meaning. Words that are used in similar contexts tend to be more semantically related compared to words that are used in different contexts. By using this analysis, speech transcripts can be gone over and pinpoint patients with schizophrenia and attention deficit hyperactivity disorder (ADHD) by the discrepancy between the meaning of the phrase and how they have chosen specific words to express the meaning. Using AI in this setting could have a huge impact on human clinical ratings in neuropsychological disorders and overall reduce false negatives and false positive diagnoses.

AI used as a prediction tool has become effective with psychiatric symptoms encompassing psychosis, behavioural disorganisation, and catatonia. Along with using the speech transcript analysis for diagnosis of psychiatric disorders, integration of machine learning can hold promise in predicting the development of psychosis in young children, which has been seen to outperform clinical interviews with respect to accuracy as human clinical interviews rely solely on the patient's motivation and accuracy in describing the symptoms and experiences used to formulate a conclusion [17]. In addition, AI integration has been used on mental situations, rather than disorders, such as suicidal individuals. It was seen by a study that an AI system was able to accurately predict suicidal individuals based on their electronic health records which proved to be up to 80% accurate, and it was able to group individuals into the sub-groups: suicidal, mentally ill but not suicidal, and a control group which was "no thought of suicidal intentions and mental illnesss" [17].

Lastly, AI has been assimilating into assisting with therapy options for psychiatric disorders. Oftentimes, individuals that experience symptoms of mental health disorders will enter intervention programs that have poor outcomes due to early dismissal from the program. This opens up a path for AI support in helping the need for ongoing therapy while having no need in restricting intervention programs and clinics in regards to resource availability. Essentially, AI generated programs can continuously give individuals experiencing symptoms of mental health disorders support and therapy without needing the resources typically required or reducing time of the physicians with other patients, which creates long-term changes in the potential outcomes. One example of how AI is integrating into mental health care for psychiatric patients is the use of computer-assisted therapy (CAT). CAT has the potential to deliver aspects of psychotherapy and behavioural treatment from a make-up of programs consisting of videos and questionnaires that are delivered via a computerised platform that is intended to help patients cope with their mental or behavioural health symptoms. Beating the Blues is a computer-assisted therapy that has been recommended by the National Institute for Health and has been seen to be a proven therapy, effective in reducing common mental health disorder symptoms such as depression and anxiety when tested in randomised trials [18]. CAT is delivered via the internet, allowing high interaction between patients and the program while having the maximum amount of outreach due to the availability of the internet. The internet is an essential part of life in today's times and is intricately merged into daily life through all means of communication. Electronic therapies can be an effective way of providing therapy to individuals without having to take up additional resources from clinics, making the system more efficient. Another electronic therapy is moderated online social therapy (MOST), which is an internet-based therapy designed for individuals suffering from the symptoms of mental health disorders; it integrates online peer support and social networking in a clinical medium [17]. Allowing more interaction this space can provide relief in symptoms by having an interaction space monitored and supported by an AI system catered to the use.

Figure 6.5 shows the application of AI in psychiatry. The figure provides a short list of evidence in support of the use of AI systems in diagnosis, prediction, and treatment of psychiatric symptoms.

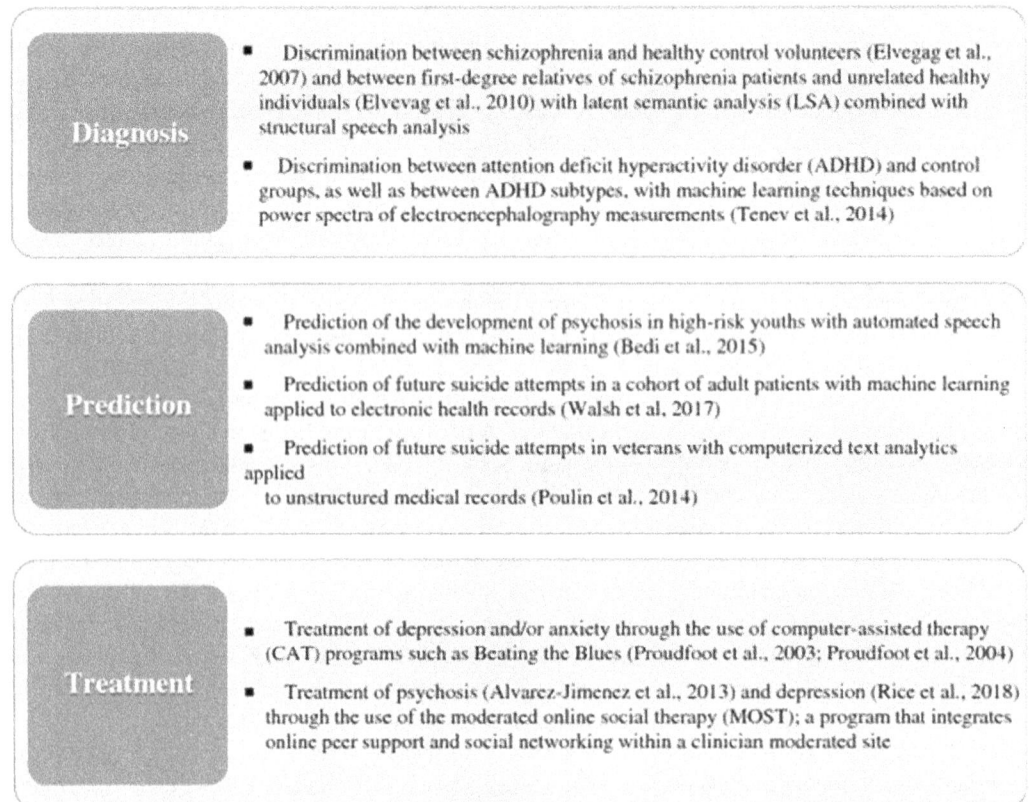

Diagnosis

- Discrimination between schizophrenia and healthy control volunteers (Elvegag et al., 2007) and between first-degree relatives of schizophrenia patients and unrelated healthy individuals (Elvevag et al., 2010) with latent semantic analysis (LSA) combined with structural speech analysis

- Discrimination between attention deficit hyperactivity disorder (ADHD) and control groups, as well as between ADHD subtypes, with machine learning techniques based on power spectra of electroencephalography measurements (Tenev et al., 2014)

Prediction

- Prediction of the development of psychosis in high-risk youths with automated speech analysis combined with machine learning (Bedi et al., 2015)

- Prediction of future suicide attempts in a cohort of adult patients with machine learning applied to electronic health records (Walsh et al, 2017)

- Prediction of future suicide attempts in veterans with computerized text analytics applied to unstructured medical records (Poulin et al., 2014)

Treatment

- Treatment of depression and/or anxiety through the use of computer-assisted therapy (CAT) programs such as Beating the Blues (Proudfoot et al., 2003; Proudfoot et al., 2004)

- Treatment of psychosis (Alvarez-Jimenez et al., 2013) and depression (Rice et al., 2018) through the use of the moderated online social therapy (MOST); a program that integrates online peer support and social networking within a clinician moderated site

FIGURE 6.5   Applications of AI in psychiatry [17].

6.8 BENEFITS OF AI INTEGRATION IN HEALTHCARE

With AI integrating into the healthcare system, an analysis of the potential benefits is conducted. While its legal and ethical implications were analysed in the earlier portions of this chapter, AI still has many benefits that can be looked at in a positive light. A simple advantage of using AI systems in the form of computerised platforms is that they dissipate the need to reveal symptoms to the clinician. It allows patients to find methods of revealing symptoms and receiving therapy from alternative sources rather than their clinician. This frees up clinicians to treat more patients but also spend time on patients who might need additional care. In essence, it allows the healthcare system to become more efficient and treat the most amount of people possible. Another advantage is that AI-based interventions can be cost effective for individuals who lack necessary resources to find other therapies and individuals who are limited in their mobility due to their symptoms [17]. From a human perspective, another advantage would be a virtual non-human means of access. In some cases, individuals might prefer to interact with a virtual environment rather than a nurse or clinician. It can be seen that clinical trial subjects overwhelmingly had a preference for an online virtual platform compared to a human counterpart when conducting the discharge process from the hospital. One reason for this was due to the patients controlling the pace at which information is being presented to them and processed at the same time. This is quite important for individuals that have mental health disorders as it could cause them to have a low literacy rate and require additional time to fully understand the information that is being presented to them. Another factor affecting the presence of a human counterpart compared to a virtual one would be the emotional responses of the patients when they are in the process of gathering and understanding information. One common emotion that might be limited with the implementation of AI systems into healthcare would be the feeling of embarrassment when asking for specific information and services. In addition, it can reduce a feeling of shame when patients do not follow their treatment plan and then must admit that to their physicians, who might display different facial expressions, leading to the patients' altered emotional state. This then could lead to the patient not being completely honest with their clinicians and skew their diagnosis, while AI can eliminate that whole problem as previously mentioned [14].

However, one of the greatest benefits to the integration of AI would be the potential to reach many different populations that have difficulty in receiving treatment in the traditional sense, which is to visit the clinician and be diagnosed by a human individual. For those individuals that are resource insufficient, such as low-income individuals, convenient therapeutic interventions along with chatbots and virtual avatars could be extremely beneficial. Moreover, individuals that may live in remote or rural areas where on-site services are scarce or lacking can see the benefits of accessing therapy or AI applications that overall increase the geographical access to healthcare. It can go the same for individuals who live in high-income countries and areas who do not have access to or cannot afford insurance, along with other individuals who request therapy for their mental or behavioural health disorders that are not covered by their insurance [14]. For all the patient types listed above and others that have not specifically been expressed, AI application and integration into healthcare can exist by complementing existing services and act as an entry point for achieving a better standard of clinical intervention in the future.

6.9 DRAWBACKS OF AI INTEGRATION IN HEALTHCARE

While AI integration into healthcare for mental and behavioural disorders shows many benefits, there have been a number of unorthodox drawbacks that have been expressed in the earlier portions of the chapter. One drawback is the lack of human interaction. AI systems do not have the capabilities to produce human emotion and convey that to the patient. This creates a gap in the prognosis, as treatment plans that rely on computerised techniques do not include a full psychiatric evaluation and convey no emotion awareness or empathic concern like an actual physician would. This can

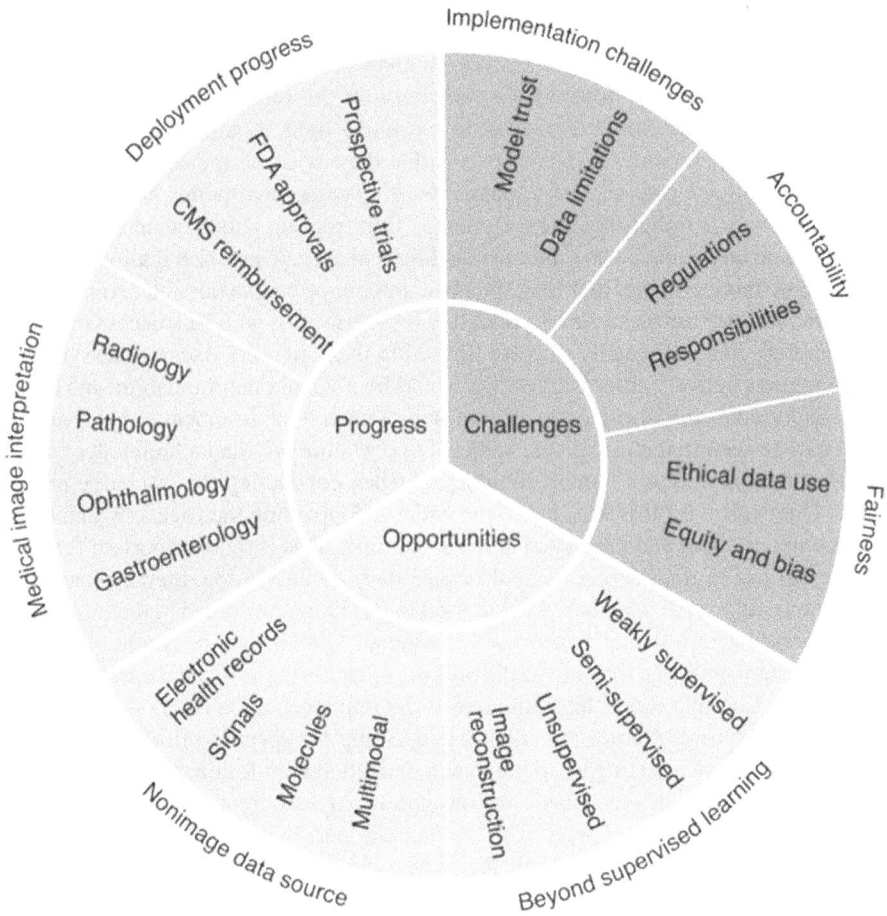

FIGURE 6.6   Summary of the progress, opportunities, and challenges for AI in healthcare [20].

result in patients who do not take the consideration of the AI treatment option and can often become discouraged from the options provided as there is no real empathetic support or reassurance. Along with that, other factors in AI and bioinformatics that deal with public issues come to question the usefulness of those given adaptations. For example, some public issues that present itself with the slow integration of AI in healthcare revolve around bioethics and other ethical concerns, legal implications such as intellectual property rights, privacy, responsibility, accessibility in terms of software and hardware, and lastly development standards [19]. When looking at all these considerations along with how they can affect the incorporation of AI technology into the healthcare system, it poses many questions of if the drawbacks outweigh the benefits. It would be advantageous to be able to retrospectively look into the considerations above and improve on them to be able to integrate the AI technology. The ethical issue presented in this section looks into how the technology could be unethical towards all parties – which includes the patient, clinician, and the developer – interacting with the technology. Minimising these ethical considerations must be done in order to look further into the integration of AI systems. Along with that, these ethical considerations can take a turn into legal implications, causing massive amounts of chaos. For example, ethical issues in AI algorithms can cause harm to the patient by giving misinformed diagnoses or treatment options, worsening their symptoms; they could also cause legal action to be taken, requiring compensation to the wronged party or dismissal of the technology. However, the FDA provides regulation and law that must be followed to aid in the mitigation of wrong doings and other implications.

6.10 CONCLUSION

In conclusion, artificial intelligence technologies are starting to become essential in healthcare. Specifically, there has been momentous integration of AI technology in mental and behavioural healthcare. Not only has this technology slowly been integrating into the healthcare system, it is also starting to be used in bioinformatics. This integration into the healthcare system has shown promise in the potential and anticipated benefits it can bring to the efficiency and accuracy of different factors in healthcare. These potential benefits can be seen as the progress of AI integration use in bioinformatics concerning behavioural and mental healthcare. Figure 6.6 breaks down the progress of AI integration in healthcare along with the challenges it faces while being integrated for use in healthcare, as well as the opportunities that AI can help grow [20]. From the progress, it can be seen that there are two phases, which are medical imaging interpretations and deployment of the technology. AI has had a slow integration but has made some progress. Next, AI has much space and opportunities to grow the healthcare space. For example, Figure 6.6 shows that AI technology can improve nonimage data processing and beyond supervised learning, This is the next movement of AI technology being used to help healthcare. However, while there are opportunities and progress that has been seen with AI technology, it can only be validated if it can overcome some certain challenges. Of all three sections in Figure 6.6, the challenges with AI integration into healthcare have the most categories. The categories seen in the figure are implementation challenges, accountability, and fairness. Despite these categories that are seen in the figure, major ethical and legal issues present itself with this technology. Thus, the integration of AI has many challenges before it can be seen as a clear-cut advantage while being used in behaviour and mental health care. Along with the challenges of incorporating AI technology, a subset of drawbacks also exist. For a smooth integration of AI technology used in bioinformatics concerning behavioural and mental healthcare, these challenges must be worked on and fixed such that it is not a liability to use this technology and that the efficiency and accuracy is at a higher level compared to the human counterpart.

REFERENCES

1. Iqbal, M.J., Javed, Z., Sadia, H. *et al.* Clinical Applications of Artificial Intelligence and Machine Learning in Cancer Diagnosis: Looking into the Future. *Cancer Cell Int* 21(1), 270 (2021). https://doi.org/10.1186/s12935-021-01981-1

2. Amann, J., Blasimme, A., Vayena, E. *et al.* Explainability for Artificial Intelligence in Healthcare: A Multidisciplinary Perspective. *BMC Med Inform Decis Mak* 20(1), 310 (2020). https://doi.org/10.1186/s12911-020-01332-6

3. Laï, M.C., Brian, M., Mamzer, M.F. Perceptions of Artificial Intelligence in Healthcare: Findings from a Qualitative Survey Study among Actors in France. *J Transl Med* 18(1), 14 (2020). https://doi.org/10.1186/s12967-019-02204-y

4. Alexiou, A., Psixa, M., Vlamos, P. Ethical Issues of Artificial Biomedical Applications. In: Iliadis, L., Maglogiannis, I., Papadopoulos, H. (eds), Artificial Intelligence Applications and Innovations. EANN AIAI 2011. *IFIP Advances in Information and Communication Technology*, vol. 364. Springer, Berlin, Heidelberg, 2011. https://doi.org/10.1007/978-3-642-23960-1_36

5. Bickman, L. Improving Mental Health Services: A 50-Year Journey from Randomised Experiments to Artificial Intelligence and Precision Mental Health. *Adm Policy Ment Health* 47(5), 795–843 (2020). https://doi.org/10.1007/s10488-020-01065-8

6. Graham, S., Depp, C., Lee, E.E. *et al.* Artificial Intelligence for Mental Health and Mental Illnesses: An Overview. *Curr Psychiatry Rep* 21(11), 116 (2019). https://doi.org/10.1007/s11920-019-1094-0

7. Loh, E. Medicine and the Rise of the Robots: A Qualitative Review of Recent Advances of Artificial Intelligence in Health. *BMJ Lead* 2(2) (2018). https://doi.org/10.1136/leader-2018-000071

8. Staples, M., Chan, L., Si, D., Johnson, K., Whyte, C., Cao, R. Artificial Intelligence for Bioinformatics: Applications in Protein Folding Prediction. In: *2019 IEEE Technology & Engineering Management Conference (TEMSCON)*, 2019, pp. 1–8. https://doi.org/10.1109/TEMSCON.2019.8813656

9. Gerke, S., Minssen, T., Cohen, G. Chapter 12 - Ethical and Legal Challenges of Artificial Intelligence-Driven Healthcare. In: Bohr, A., Memarzadeh, K. (eds), *Artificial Intelligence in Healthcare*, Academic Press, 2020, pp. 295–336, ISBN 9780128184387. https://doi.org/10.1016/B978-0-12-818438-7.00012-5

10. Powell, J. Trust Me, I'm a Chatbot: How Artificial Intelligence in Health Care Fails the Turing Test. *J Med Internet Res* 21(10), e16222 (2019 Oct 28). https://doi.org/10.2196/16222; PMID: 31661083; PMCID: PMC6914236

11. *AMA J Ethics* 21(2), E121–124 (2019). https://doi.org/10.1001/amajethics.2019.121

12. Price, I.I., Nicholson, W. Artificial Intelligence in HealthCare: Applications and Legal Issues. *Scitech Lawyer* 14, 10 (Nov 28, 2017). U of Michigan Public Law Research Paper No. 599, Available at SSRN: https://ssrn.com/abstract=3078704

13. Hudson, K.L., Collins, F.S. (2017). The 21st Century Cures Act — A view from the NIH. *N Engl J Med* 376(2), 111–113 (2017). https://doi.org/10.1056/nejmp1615745

14. Fiske, A., Henningsen, P., Buyx, A. Your Robot Therapist Will See You Now: Ethical Implications of Embodied Artificial Intelligence in Psychiatry, Psychology, and Psychotherapy. *J Med Internet Res.* 21(5), e13216 (2019 May 9). https://doi.org/10.2196/13216; PMID: 31094356; PMCID: PMC6532335

15. Fitzpatrick, K.K., Darcy, A., Vierhile, M. Delivering Cognitive Behavior Therapy to Young Adults With Symptoms of Depression and Anxiety Using a Fully Automated Conversational Agent (Woebot): A Randomised Controlled Trial. *JMIR Ment Health* 4(2), e19 (2017). https://doi.org/10.2196/mental.7785; PMID: 28588005; PMCID: 5478797

16. Ganasegeran, K., Abdulrahman, S.A. (2020). Artificial Intelligence Applications in Tracking Health Behaviors During Disease Epidemics. In: Hemanth, D. (ed), *Human Behaviour Analysis Using Intelligent Systems. Learning and Analytics in Intelligent Systems*, vol. 6. Springer, Cham. https://doi.org/10.1007/978-3-030-35139-7_7

17. Fakhoury, M. (2019). Artificial Intelligence in Psychiatry. In: Kim, Y.K. (ed), *Frontiers in Psychiatry. Advances in Experimental Medicine and Biology*, vol. 1192. Springer, Singapore. https://doi.org/10.1007/978-981-32-9721-0_6

18. Proudfoot, J., Goldberg, D., Mann, A., Everitt, B., Marks, I., Gray, J.A. Computerized, Interactive, Multimedia Cognitive-Behavioural Program for Anxiety and Depression in General Practice. *Psychol Med* 33(2), 217–27 (2003 Feb). https://doi.org/10.1017/s0033291702007225; PMID: 12622301

19. Kesh, S., Raghupathi, W. Critical Issues in Bioinformatics and Computing. *Perspect Health Inf Manag* 1, 9 (2004 Oct 11). PMID: 18066389; PMCID: PMC2047326

20. Rajpurkar, P., Chen, E., Banerjee, O., Topol, E.J. AI in Health and Medicine. *Nat Med* 28(1), 31–38 (2022). https://doi.org/10.1038/s41591-021-01614-0

7 Practicing Medicine and Ethical Issues Applying Artificial Intelligence

Adithya Kaushal, Divya Kaushal, and Yashwant Pathak

7.1 INTRODUCTION

Societies, specifically people in societies, have been and continue to be inundated with a variety of factors that constitute existence. In particular, people are tasked with learning and the application of said learning in a variety of aspects. Though there are a variety of tasks that result in the application of learned knowledge and skills, the necessity of work/labor and the application of learning to work/labor is repeatedly highlighted.

Work/labor is essential to human existence. Moreover, the ingratiation of work/labor into various aspects of everyday life further emboldens the claim that it is a foundational piece in how we as humans behave. Work/labor not only constitute human behavior but tangentially also result in the production of goods and services that are demanded and consumed by society writ large. While there is an almost innumerable amount of work/labor that could be highlighted, this chapter will focus on healthcare as it is one of the most demanded and consumed services by society [1].

Since the inception of the idea and practice of work/labor, we as humans have striven towards finding ways to increase the efficiency of how we conduct work/labor. Efficiency in work/labor has been promoted and realized through a variety of methods that primarily focus on increasing not only the physical but mental aspects of work/labor. In particular, this form of efficacious work/labor has been a part of the field of automation. However, automation does not perfectly blend the nature of work/labor as we know because work/labor is fundamentally a human experience. While this phenomenon does exist, a concept and field that presents itself as an analog of automation is becoming more prominent because of its ability to recreate that human experience in work/labor: Artificial Intelligence (AI).

The purpose of this chapter will be twofold. First, as previously stated healthcare is one of the most of the most consumed services in the world, therefore it warrants a further understanding of what constitutes the practice of healthcare, or in other terms the practice of medicine. It is critical to elaborate the practice of medicine not only because it is a form of work/labor, but it is almost a necessary form of work/labor that provides society the ability to continue to function. Moreover, just like with any form of work/labor the strive for efficiency in the medical field is a prominent topic of discussion and warrants further examination. This examination will naturally tie in with a conversation about the advent of AI as a tool of efficiency. AI will be explored not only as a singular concept and tool, but as an apparatus in the practice of medicine.

Second, while understanding and describing the practice of medicine as a one of the many forms of work/labor is an important discussion to have, that is not the explicit goal of this chapter. Just as a variety of work/labor contains a multitude of factors that encapsulate the experience of said work/labor, so does the practice of medicine. These variety of factors may result in particular ethical issues that affect how said work/labor is done. Furthermore, it has been demonstrably shown that there are a plethora of ethical issues that arise in the practice of medicine, particularly decision

DOI: 10.1201/9781003353751-7

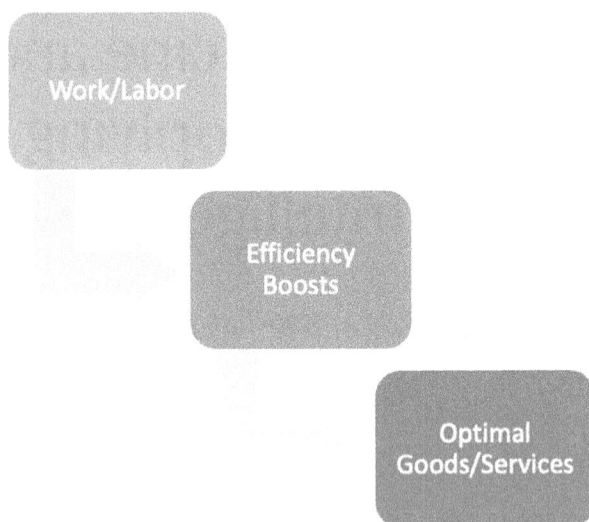

FIGURE 7.1 Proposed model of work/labor output.

making and bias [2]. Therefore, this warrants an examination of how the practice of medicine synthesizes ethical issues of decision making and bias.

Similarly, since work/labor has been framed as a human experience, and AI seeks to replicate or at the very least mimi said human experience, ethical issues will also arise from its application. In conjunction with a discussion of the ethical issues of decision making and bias in the practice of medicine, an elaboration of the same ethical issues that AI presents will also be examined. Lastly, this chapter will then provide a discussion of the ethical issues of decision making and bias when AI has been added to practice of medicine. This discursive amalgamation will serve as this chapter's primary point of accentuation to demonstrate the caution that is needed when striving towards improving the efficiency of work/labor, especially in fields where ethical issues are of paramount concern.

Understanding the need for work/labor and how making said work/labor is an important discussion to have. In particular understanding, one of the most demanded work/labor, the practice of medicine, and the efforts to make that work/labor more efficient, through AI, is an even more important discussion. However, it is also necessary to spotlight that this form of work/labor and efficiency-boosting tool is rife with ethical issues that continue to and may consistently maim them from having a positive impact in society writ large. Therefore, this chapter will examine these ethical issues, particularly decision-making and bias, to further understand the practice of medicine, AI, and the convergence of both. This chapter is in an effort for general elucidation, but hopefully provides a spark for potential solutions to said ethical issues.

7.2 BACKGROUND

7.2.1 PRACTICING MEDICINE

Similar to other forms of work/labor the practice of medicine has an extensive foothold in the annals of history. Some scholars have even presented evidence that the practice of medicine, or at the very least some form of organized healing, existed even in the Prehistoric Era amongst Neanderthals [3]. To summarize, "the practice of medicine includes the diagnosis, treatment, correction, advice, or prescription for any human disease, ailment, injury, infirmity, deformity, pain, or other condition, physical or mental, real or imaginary" [4]. While there are a variety of healthcare professionals, who may administer and provide medical services, such as nurses and physician assistants, the practice of medicine is typically associated with physicians.

Physicians today are viewed as serious practitioners of science and health, however, the origins of physicians and medicine could not be further from modern definitions and conceptions of the profession. The science of medicine was not scientific at all. Specifically, religious and magical practices constituted significant parts of medical practice [5]. Physicians of those times were mostly sorcerers, witches, or other religious/magical figures who emphasized not only the treatment of the body but the soul as well [6]. However, there is documented evidence that the first practitioner of medicine, as we think of contemporarily, was in ancient Egypt [7]. Scholarly evidence has shown that in the year 3000 BCE Imhotep, the chief minister to King Djoser, was a physician [8]. Additionally, through the British discovery of the Ebers and Edwin Smith papyri, there is documented and written evidence of surgical and medical practices that align with what we conceive of as modern medicine [9]. While academic work has shown the prevalence of modern medical practices and the presence of physicians in ancient Egypt, it must be noted that this observation was not unique only to Egypt. India, China, Japan, Italy, and other various parts of Europe, have long-documented histories of medical practices and physicians that abide by modern conceptions of the medical field and profession [5].

While ancient societies do provide an insight into the practice of medicine, a more prescient examination can be found in Greek history. Hippocrates, the famed philosopher, mathematician, and general scholar, has been commonly regarded as the father of modern medicine [10]. Hippocrates came from a family of physicians who derived their genealogy from the Greek god of medicine, Asclepius [10]. Hippocrates is deemed the father of modern medicine due to a seemingly all-encompassing approach to medical practice. More specifically, he is lauded as introducing "ethical rules of conduct, close observation of clinical symptoms, an open mind for any ideas, and willingness to explain the cause of diseases." [10]. For example, it has been noted that Hippocrates defined one of the many tenets of a physician by expressly describing that "the physician had to examine a patient, observe symptoms carefully, make a diagnosis, and then treat the patient." [10]. This guideline of practice for physicians is and continues to remain the primary method of medical practice. While there is extensive documentation of Hippocrates' various medical discoveries and practices, perhaps his most famous and lasting contribution and continuing staple of modern medicine is to the creation of the Hippocratic Oath. The Hippocratic Oath is one of the most commonly known and read medical texts that emphasize that a new physician must abide by a code of ethical principles [11]. While Hippocrates did not write the Hippocratic Oath himself, he repeatedly stated similar ethical guidelines that physicians must follow such as, tending to the sick, refraining from promiscuous and unjust behavior, and secrecy [10]. While it is colloquially believed that contemporary medical schools require physicians to recite the Hippocratic Oath, this is not the case. Many medical schools and physicians have adopted different texts and phrases to recite as an honorary code of ethics [11].

In modern times the practice of medicine has transitioned from attachments to religion and mysticism to a serious profession. Medical care and physicians are grounded by a scientific foundation. Moreover, there is an intense level of dedication required to practice medicine as a physician. Physicians must undergo rigorous undergraduate education, followed by years at medical school, subsequent residencies, and fellowships before they are considered independent physicians [4]. Present-day physicians are not eclectic healers or barbers, they are highly trained professionals. However, their core modus operandi has not wavered significantly from Hippocrates' time. They are still charged with diagnosing and treating patients while upholding the highest of ethical and professional standards. While many focus on the potential lucrative nature of becoming a physician, including some physicians themselves, most are driven by the innate desire to truly help and care for the sick. The job of a physician may seem mundane or simple to understand: treat the sick, but it is a vital profession for the continuation of society. The ongoing COVID-19 pandemic has only reified that position.

7.2.2 ARTIFICIAL INTELLIGENCE (AI)

Artificial Intelligence, commonly known as AI, is an advanced form of computer technology that can mimic or replicate the cognitive functions of humans, such as problem-solving, decision-making, and learning [12]. AI is an umbrella term that encompasses a variety of technologies and research areas, such as natural language processing, machine learning, computer vision, robotics, and more [13]. AI has evolved over the years and is increasingly used in healthcare, transportation, finance, and education. AI has the potential to revolutionize virtually any industry and it is essential for staying competitive in today's fast-paced economy [14].

The history of AI goes back to the 1940s when Alan Turing proposed his famous Turing Test [15]. This test proposed a criterion to judge the intelligence of a machine. He proposed that if a machine could have a conversation indistinguishable from that of a human, then it could be considered an intelligent machine [15]. This notion has led to the development of many different AI technologies, such as speech recognition, natural language processing and robotics.

Over the years, AI has become more and more complex, with the development of game playing systems, expert systems and natural language understanding [16]. In the 1950s, researchers developed the first game playing programs. These programs were able to beat the game of checkers, and the first game playing system to beat a professional player was developed in the early 1960s. In the 1970s, expert systems were developed with the ability to interact with data and make decisions based on the data [17]. In the 1980s, natural language understanding, which allows machines to understand human language, was developed [17].

Since the 1990s, AI has continued to be developed and advanced, with the development of machine learning and deep learning [18]. Machine learning is a branch of AI that focuses on the development of programs that can learn from data and improve their performance over time [19]. Machine learning has been used to create self-driving cars, medical diagnostics systems, search engines and more. Deep learning is a branch of machine learning that uses artificial neural networks to solve complex problems [20]. Deep learning has been used to make advances in fields such as computer vision, natural language processing and healthcare.

AI has become a major part of modern society, and its applications are only growing. For example, AI can automate many mundane and repetitive tasks that humans are often stuck doing, freeing them up to focus on more creative and higher-value tasks [21]. AI can also help to enhance existing processes and operations, making them more efficient, accurate, and effective. In addition, AI can make predictions based on data and pattern recognition, which can be invaluable for businesses and other organizations [21].

AI is also being used to improve customer experience and engagement [21]. For example, many companies use AI-powered chatbots to provide customer service, answer questions, and provide personalized recommendations. AI can also analyze customer data and provide insights into customer behavior and preferences, enabling companies to target their marketing efforts better and tailor their products and services [21].

AI is also used to augment human capabilities in the medical field [22]. For example, AI-powered medical devices, such as robotic surgical systems and diagnostic tools, are becoming increasingly available to hospitals and medical centers. AI-powered medical devices can assist physicians in diagnosis, surgery, and other medical procedures [23]. AI is also used to develop personalized treatments and medications tailored to a patient's medical needs.

In addition, AI is being used to create more efficient and accurate systems for decision-making. AI-powered decision support systems can help to identify patterns in large datasets and provide recommendations that can improve decision-making accuracy and speed [23]. AI-powered systems can also provide insights into customer trends and behaviors, enabling businesses to target their products and services to customers better.

Finally, AI is also used to power intelligent systems, such as robots and virtual assistants. AI-powered robots can automate tedious and repetitive tasks, freeing humans to focus on higher-value tasks [24]. AI-powered virtual assistants can answer questions, provide customer service, and provide personalized recommendations. AI is revolutionizing how we live and work, and its applications are only growing [24]. AI has the potential to improve efficiency, accuracy, and safety in many areas, and its use is becoming increasingly widespread.

7.2.3 Practicing Medicine with AI

The medical field has become increasingly reliant on the use of technology. Artificial intelligence (AI) is the field of computer science that creates intelligent machines that can reason, think, and learn [25]. Within recent years, AI has made an extraordinary impact in the medical profession, aiding research and allowing physicians to diagnose and treat patients accurately. The implications of practicing medicine with AI and its potential to revolutionize the healthcare sector will be discussed.

One of the primary benefits of practicing medicine with AI is increased accuracy [26]. AI-based technologies can read and interpret medical images quickly and accurately and identify abnormalities that humans may overlook. AI can also help speed up the diagnosis process and reduce the risk of mistakes [27]. This could lead to earlier intervention, potentially resulting in better patient outcomes. Additionally, AI can help automate cumbersome tasks, such as collecting data from medical records or entering information into patient databases [28]. By automating these processes, AI allows healthcare providers to work quicker and more efficiently, delivering better patient care.

An AI-driven approach to medicine can also lead to more personalized medical care. AI systems like IBM Watson can analyze a patient's medical history, symptoms, genetic profile, and lifestyle to develop a unique "health portrait." [29] This personalized portrait can give healthcare providers more insight into the individual's health, allowing them to make a more informed diagnosis and treatment decisions.

AI can be used to detect and prevent diseases in potentially vulnerable populations [30]. AI-driven models can be designed to screen large populations for disease risk factors and provide early warning alerts for individuals likely to develop health issues [30]. AI can also be used to monitor patients' conditions over time. For instance, AI-enabled sensors on medical devices can capture real-time data that could be used to detect signs of disease or adverse reactions before they become a problem.

Despite these potential benefits, several critical issues need to be addressed before AI can be used safely and effectively in the medical field [31]. First, many AI systems lack transparency and cannot fully explain the results [32]. This can lead to decision-making that is based on incomplete information, as well as unfavorable outcomes for patients. Additionally, ethical concerns are associated with using AI to make decisions about patient care. If a physician bases their diagnosis or treatment on the results of an AI system, there are potential risks for the patient, such as misdiagnosis or incorrect treatments [32].

AI technology has been widely adopted in the medical field but has some drawbacks. AI can potentially reduce some of the physicians' workloads, but it can also lead to a lack of human interaction and empathy when diagnosing patients. AI systems can also not identify subtle nuances that a human doctor may be able to detect, such as facial expressions or body language [33]. AI systems can also lead to bias, as they are only as good as the data they are trained on, which may be incomplete or inaccurate [34]. Additionally, AI systems can be expensive and require much effort to maintain, making it difficult for smaller medical practices to afford them. Finally, AI systems can be vulnerable to hacking, which can put patient information at risk [35].

Implementing AI in healthcare requires significant investment. AI-enabled technologies, such as the IBM Watson AI platform, can be quite expensive and cost-prohibitive for some healthcare providers [35]. Additionally, there is still much work to be done in terms of training and educating healthcare providers to use AI-based technologies safely and effectively.

AI has the potential to revolutionize the healthcare sector, providing more accurate diagnoses and treatments while allowing physicians to spend more time and energy on providing personalized care to their patients. However, it is important to recognize that AI systems still need to be refined, and many issues, such as ethical concerns and the cost of implementation, must be addressed before AI can be used safely and effectively in the medical field.

7.3 USE OF AI IN HEALTHCARE

AI is increasingly becoming a part of healthcare, from diagnosis to treatment to post-care follow-ups. AI is being used to help healthcare professionals diagnose, predict, and manage diseases more accurately and efficiently. AI systems can quickly analyze data from medical records and images, such as MRI and CT scans, to identify patterns that aid in diagnosis. AI also provides personalized treatments for patients, leveraging patient data to determine the best treatment plans based on individual needs. Additionally, AI is being used to improve post-care follow-ups, helping healthcare providers monitor patient health and alert them of potential issues. AI's ability to quickly and accurately analyze large amounts of data is helping healthcare providers make more informed decisions, leading to better patient outcomes [36].

7.3.1 Current Use of AI in Healthcare

The potential of artificial intelligence (AI) to revolutionize healthcare is immense. AI can be defined as the science and engineering of creating intelligent machines, which refers to a machine's ability to emulate human intelligence and reasoning [37]. AI is increasingly being used in healthcare as a powerful tool to improve the quality of healthcare services, reduce costs, and unlock the potential of data. AI can be used in various ways, such as in diagnostics, treatment, monitoring and surveillance, drug development, and operational efficiency [37]. AI is already being used in healthcare to improve patient care, reduce costs, and increase efficiency.

AI has a wide range of potential applications in healthcare, ranging from diagnosis to treatment and monitoring [38]. In diagnostics, AI can be used to interpret medical imaging, such as X-rays and CT scans. AI can be used to identify patterns in medical images, detect anomalies, and provide an accurate diagnosis [38]. AI can also be used to automate the interpretation of lab results and patient records, which can help reduce the time and cost associated with diagnosis. AI can also be used in treatment and monitoring, to provide personalized treatments based on a patient's individual characteristics, and to monitor a patient's vital signs and other health parameters in real-time [38].

AI can also be used in drug development [39]. AI can be used to analyze large amounts of data to identify patterns and correlations between drug compounds and disease states. This can help speed up the drug development process and reduce the cost of drug development [39]. AI can also be used to monitor drug safety and efficacy and to identify potential side effects.

AI is also being used to improve operational efficiency in healthcare [40]. AI can automate administrative tasks, such as scheduling appointments, managing patient records, and billing [40]. AI can also be used to optimize patient flow and reduce wait times. AI can also be used to automate patient care, such as by providing personalized care plans and reminders [40].

AI has the potential to revolutionize healthcare, but there are some challenges associated with the use of AI [41]. Privacy and security are major concerns, as AI systems can collect and store large amounts of sensitive patient data [41]. AI systems must also be trained and validated to ensure accuracy and reliability. In addition, the use of AI in healthcare raises ethical and legal concerns, as AI systems can make decisions that may have ethical and legal implications [42].

Despite these challenges, AI has the potential to revolutionize healthcare and improve patient outcomes [43]. AI can be used to improve diagnosis, treatment, and monitoring, reduce costs, and improve operational efficiency. AI can also be used to improve drug development and speed up the process of discovery [43]. AI can revolutionize healthcare, but it is important to understand the potential implications and ensure that appropriate safeguards are in place [44].

AI is increasingly being used in healthcare to improve patient care, reduce costs, and increase operational efficiency [45]. AI has a wide range of potential applications in healthcare, ranging from diagnosis to treatment and monitoring. AI has the potential to revolutionize healthcare, but there are some challenges associated with the use of AI, such as privacy, security, and ethical and legal concerns [46].

7.3.2   Advantages of AI in Healthcare

In recent years, Artificial Intelligence (AI) has become a driving force in a variety of industries, including healthcare. AI is a computer-based technology that can be used to make decisions and automate processes, enabling healthcare organizations to reduce costs, improve efficiency, boost effectiveness, and even save lives. In this paper, we will discuss the various advantages of AI in healthcare, including its ability to automate processes, reduce costs, improve patient outcomes, and increase access to care.

7.3.2.1   Automation

AI can be utilized to automate many of the manual processes that have traditionally been handled by healthcare personnel [47]. Automation can improve efficiency, reduce errors, and free up healthcare workers to focus on more important tasks. For example, AI can be used to automate administrative tasks such as patient scheduling, billing, and documentation [47]. AI can also be used to automate medical device operations and monitor patient data for early detection of diseases and other health problems. AI can also be used to automate the drug delivery process, which can reduce medication errors and improve patient safety [47].

7.3.2.2   Cost Reduction

The use of AI in healthcare can help reduce costs by streamlining processes and eliminating manual labor [48]. AI can be used to automate data entry and analysis, streamline workflow processes, and reduce paperwork. AI can also help to reduce costs by identifying and eliminating redundant or unnecessary processes in a healthcare organization [48]. Additionally, AI can be used to identify cost-saving opportunities, such as reducing the use of expensive medications or procedures [48].

7.3.2.3   Improved Patient Outcomes

The use of AI in healthcare can help improve patient outcomes by providing early detection of disease and reducing the risk of misdiagnosis [49]. AI can be used to identify patterns in patient data and develop diagnostic models to predict the likelihood of a certain outcome. AI can also be used to automate the monitoring of patient vital signs and provide real-time feedback to clinicians [49]. This can help to identify potential health issues before they become serious, thus improving patient outcomes.

7.3.2.4   Increased Access to Care

AI can be used to increase access to care by providing more efficient and cost-effective care. AI can be used to automate the process of scheduling appointments, reducing wait times and improving access to care [50]. AI can also be used to automate the process of collecting data from patients and analyzing it for early detection of diseases [50]. This can help to reduce the need for costly tests and procedures, thus increasing access to care.

AI can be used to improve the efficiency, effectiveness, and cost-effectiveness of healthcare organizations [51]. AI can be used to automate processes, reduce costs, improve patient outcomes, and increase access to care. As AI advances, it is likely that healthcare organizations will benefit even more from its use.

7.3.3   Disadvantages of AI in Healthcare

The use of Artificial Intelligence (AI) in healthcare is revolutionizing how healthcare providers and patients interact. However, a few potential drawbacks are associated with implementing AI into healthcare systems. This paper discusses the potential disadvantages of AI in healthcare, including potential errors, privacy issues, and lack of efficiency.

7.3.3.1   Errors

One potential disadvantage of using AI in healthcare is the potential for errors [52]. For example, AI can be programmed to use algorithms to diagnose and treat patients. However, there is a possibility that an algorithm may not be able to interpret the data to reach a correct diagnosis accurately. AI can also be used to interpret medical images, such as x-rays and CT scans; however, if the AI is not adequately trained on the range of images or is given incomplete or corrupted instruction data, it could lead to incorrect diagnosis due to misinterpretation of the images [52]. AI can also be used to make decisions concerning treatment, but it may not be able to differentiate between symptomatic and asymptomatic factors contributing to the patient's condition. Consequently, AI may miss or misinterpret important data or provide incorrect treatments, leading to negative patient outcomes.

7.3.3.2   Privacy

Another potential disadvantage of AI in healthcare is the potential for privacy issues [53]. AI requires data to train its algorithms; however, the data collected may contain sensitive information about patients, such as medical history, genetic information, and personal data collected for billing purposes. In order to use this data, healthcare systems must ensure that the data is kept secure. Unfortunately, as AI systems become more widespread, there is a growing risk of data breaches or misuse of patient data [53]. Additionally, AI systems may have access to patient data from multiple sources, such as medical records and insurance companies. These data sources may be located in different jurisdictions, which could create conflicts with privacy laws and regulations [53].

7.3.3.3   Lack of Efficiency

AI in healthcare can be very expensive and may require significant amounts of time and resources to implement [54]. Healthcare providers must invest in both hardware and software, as well as training and installation processes. This can be a heavy financial burden for many healthcare providers, particularly smaller ones. Moreover, AI can be slow to respond to changes in the healthcare environment. This can be especially challenging in the case of technology-driven treatments or diagnostic processes, which may require AI to be re-trained and re-coded regularly. This can be very costly and time-consuming, leading to a decreased level of efficiency [54].

While AI has the potential to improve the quality and efficiency of healthcare, it is important to consider the potential disadvantages of using AI in healthcare. The potential for errors and privacy issues, as well as the associated costs and lack of efficiency, suggest that more research must be conducted to ensure that AI is able to accurately and securely meet healthcare needs.

7.3.4   Ethical Concerns of AI in Healthcare

Artificial intelligence (AI) is a rapidly emerging technology that has already made its presence felt in the healthcare sector. AI has enabled the healthcare industry to achieve accuracy and efficiency levels that were previously impossible [55]. AI has enabled medical professionals to make more informed decisions quickly and accurately. However, with the introduction of AI into the healthcare sector, there has been an increased focus on the ethical concerns associated with using this technology. This section will discuss the ethical concerns of AI in healthcare, with a particular focus on the implications for patient privacy, patient autonomy, and the potential for deception.

7.3.4.1 Privacy and Security

One of the main ethical concerns posed by the use of AI in healthcare is the potential for the violation of patient privacy [56]. AI systems are designed to capture and use a vast amount of patient data, and this data could be used to make decisions about patient care that may not always be in the best interests of the patient. Furthermore, if the data collected is not securely stored, it can be accessed and used for malicious purposes, such as identity theft [56]. To ensure that patient data is securely stored and not accessed by unauthorized individuals, healthcare organizations must implement comprehensive data security measures, such as encryption and strict access control measures.

7.3.4.2 Patient Autonomy

Another ethical issue posed by the use of AI in healthcare is the potential for AI to override patient autonomy [57]. AI systems are designed to make automated decisions based on the data they collect, and this can lead to decisions being made without consulting the patient. This could potentially conflict with the patient's right to make the final decision about their own care. To avoid this potential conflict, healthcare organizations must ensure that patients are always consulted before decisions are made and given the opportunity to voice their opinion.

7.3.4.3 Potential for Deception

Finally, there is a potential for AI systems to deceive both healthcare professionals and patients. AI systems are designed to make decisions quickly and efficiently, and this can lead to decisions that may not always be in the best interest of the patient [58]. Furthermore, AI systems can be easily manipulated by malicious individuals, allowing the system to make decisions that are not based on correct information. To avoid these issues, healthcare organizations must ensure that they have strict protocols in place to verify the accuracy of information and that they have mechanisms to detect and mitigate any potential malicious activity [58].

The ethical concerns raised by the use of AI in healthcare are significant. Healthcare organizations must take all necessary steps to ensure that patient privacy is protected, that patient autonomy is respected, and that potential for deception is minimized [59]. This includes implementing comprehensive data security measures, ensuring that the patient is consulted before decisions are made, and verifying the accuracy of the information [60]. Only by taking such measures can healthcare organizations ensure that AI is used in a responsible and ethical manner.

7.4 ETHICAL ISSUES SURROUNDING AI

With the introduction of AI into healthcare and the practice of medicine, it should be noted that various challenges are presented. There are clinical, social, and ethical risks presented when incorporating AI into the practice of medicine, but for the scope of this section, the focus will be on ethical issues. [61] While using AI in the clinical practice of medicine can be transformative, it raises ethical challenges that can negatively affect the practice.

7.4.1 Decision Making

Decision-making in ethical issues surrounding AI is becoming increasingly important due to the rapidly changing nature of the technology. AI is being used in various industries such as healthcare, finance, and transportation, and it is essential to consider ethical issues when making decisions about its use. This section will examine some of the ethical issues surrounding AI, the challenges of making ethical decisions, and how these decisions can be made in a responsible and ethical manner.

One of the main ethical issues surrounding AI is the potential for it to be used in a biased and discriminatory manner [62]. AI algorithms are often trained on datasets that are biased or lack representation from certain groups, which can lead to AI systems making decisions that are discriminatory or biased against certain groups. This can be especially problematic in areas such as healthcare, where AI systems can be used to make decisions about treatments or diagnoses that can have serious consequences for individuals [62]. As such, it is important for decision-makers to ensure that AI systems are trained on unbiased data and that appropriate measures are taken to prevent bias or discrimination in using AI.

Another ethical issue related to AI is privacy [63]. AI systems often collect and store large amounts of data, which can be used to make decisions and predictions about individuals. This raises the issue of how this data should be used and how individuals should be protected from potential misuse of their data [63]. Decision makers need to consider the ethical implications of collecting, storing, and using data when making decisions about the use of AI.

Decision makers need to consider the potential for AI to be used in unethical ways [64]. AI systems may be used to carry out activities that are unethical or illegal, such as targeted surveillance or manipulation of data. As such, it is important for decision makers to consider the potential for AI to be used in unethical ways when making decisions about its use.

Making ethical decisions about the use of AI can be challenging, as it requires considering a wide range of ethical issues and potential consequences. It is important for decision makers to carefully consider the potential risks and ethical implications when making decisions about AI. This includes considering the potential for bias, discrimination, and misuse of data, as well as the potential for AI to be used in unethical ways [65].

In order to ensure ethical decision making in the use of AI, it is important for decision makers to have a clear understanding of the ethical implications of AI, as well as the potential risks and consequences [66]. It is also important to ensure that AI systems are trained on unbiased data and that appropriate measures are taken to prevent bias or discrimination in the use of AI. In addition, decision makers should ensure that appropriate measures are taken to protect individuals' privacy and that AI systems are not used in unethical ways [67].

Decision making in ethical issues surrounding AI can be challenging, but it is essential to ensure that AI is used in a responsible and ethical manner [68]. By understanding the ethical issues, considering the potential risks and consequences, and taking appropriate measures to protect privacy and prevent bias and discrimination, decision makers can ensure that AI is used ethically and responsibly.

Decision making in ethical issues surrounding AI in healthcare is an important area of research. According to Sengupta, Prakash, and Sengupta (2020), AI in healthcare has the potential to revolutionize medical care, but it also carries many ethical issues [69]. These ethical issues include data privacy, data security, and algorithmic bias. To ensure ethical decision making in the application of AI in healthcare, Sengupta et al. (2020) suggest the use of ethical frameworks and principles such as the Hippocratic Oath, the Belmont Report, the Nuremberg Code, and the Declaration of Helsinki. Additionally, guidelines proposed by the International Medical Informatics Association (IMIA) have been developed to provide ethical guidance for the use of AI in healthcare [70]. These guidelines address issues such as data privacy, data security, and algorithmic bias. Finally, Sengupta et al. (2020) suggest that decision makers should consider the ethical implications of AI in healthcare when making decisions and should ensure that the decisions are made transparently, with input from all stakeholders.

7.4.2 BIAS

Bias is one of the most prominent ethical issues surrounding artificial intelligence (AI). AI systems are designed to automate decision-making processes, but they are also prone to incorporating human biases that can lead to poor or even dangerous outcomes [71]. Bias in AI can manifest in a variety of ways, from discriminatory outcomes due to the data used in training to algorithmic decisions that are based on flawed assumptions [71]. As AI systems become increasingly pervasive, it is essential to understand the potential risks of bias and to develop systems that are designed to be fair, transparent, and accountable [72].

One way in which bias can be introduced into AI systems is through the data used in training [73]. If the data used to train the system is not balanced or if it contains stereotypical or prejudiced information, then the AI system could produce biased results [74]. For example, if a system is trained using facial recognition data that is predominantly male, then the system may produce gender-biased results. Similarly, if an AI system is trained using income data that is skewed toward higher-income households, then it could produce results that are biased against lower-income households [75].

Another way in which bias can manifest in AI systems is through algorithmic decision-making [76]. Algorithms are designed by humans and are subject to human biases. If an algorithm is based on flawed assumptions, then it could produce decisions that are unfair or discriminatory. For example, if an algorithm is based on the assumption that all people of a certain gender or race are more likely to commit a crime, then it could produce decisions that are biased against certain groups [77].

Bias can also be introduced through human intervention in the decision-making process. If a human is given too much control over the decision, then they could introduce their own biases into the system [78]. This could lead to outcomes that are not in line with the values of the society in which the AI system operates.

In order to reduce the potential for bias in AI systems, it is important to ensure that the data used in training is balanced, that algorithms are designed to be fair and transparent, and that human intervention is minimized [78]. By doing so, we can help to ensure that AI systems are designed with fairness, equity, and transparency in mind [79].

Bias in AI systems can be a significant ethical issue in the healthcare sector. AI systems can be trained on data that contains historical biases, leading to healthcare applications that can perpetuate those biases. For example, an AI system trained on a dataset of medical records could make inaccurate predictions when evaluating people from traditionally marginalized communities [80]. Additionally, AI systems may be used to replace human decision-making in healthcare, which could lead to a decrease in patient autonomy and a lack of physician accountability [81].

7.4.3 Data Privacy and Surveillance

Data privacy and surveillance in ethical issues surrounding AI is an important and timely topic within the world of artificial intelligence [82]. As AI technology progresses and becomes more pervasive in everyday life, it is important to consider the ethical implications of data privacy and surveillance, particularly in regards to the protection of individuals' rights and freedoms [82].

Data privacy is a fundamental right that all individuals should have, and its importance is only growing as AI technology continues to advance [83]. As AI is embedded into an ever-increasing number of devices, systems, and services, the potential to collect and store personal data increases. The collection of this data can be used for a range of purposes, from targeted marketing to predictive analytics, but individuals must be made aware of exactly what data is being collected and how it will be used [84]. Without knowledge of the data being collected and how it will be used, individuals are at risk of having their data misused or exploited, which can have a detrimental effect on their privacy and security [84].

Surveillance is another ethical concern when it comes to AI and data privacy [85]. As AI technology becomes more integrated into everyday life, there is a risk that individuals' activities and movements can be tracked and monitored without their knowledge or consent. This not only violates individuals' right to privacy and security, but can also lead to a lack of trust in the technology, which can hinder its adoption and use [86]. As such, it is important to ensure that all AI systems are designed to allow individuals to opt-in or opt-out of data collection and surveillance, and that any data collected is done so in a transparent and secure manner [87].

Overall, data privacy and surveillance are ethical issues that must be addressed when considering the development and use of AI technology [88]. It is essential that individuals' rights to privacy and security are respected, and that any data collected is done so in a transparent and secure manner. By taking measures to ensure that data is collected and used ethically and responsibly, AI technology can be utilized in ways that benefit individuals and society as a whole [89].

Data privacy and surveillance are two of the major ethical issues surrounding AI in healthcare [90]. Data privacy is a major concern because AI systems are often used to process sensitive personal data such as health records, which can be used to identify individuals and make decisions about their care [90]. Surveillance is a concern because AI systems can be used to monitor and track individuals, raising issues of autonomy and privacy.

Data privacy is a major ethical concern in AI in healthcare, as AI systems are often used to process sensitive personal data such as health records [91]. This data can be used to identify individuals and make decisions about their care, which can raise ethical issues around consent, accuracy, and fairness. AI systems can also be used to track and monitor individuals, which can raise privacy and autonomy issues [91].

Surveillance is another ethical concern in AI in healthcare. AI systems can be used to monitor and track individuals, raising issues of privacy and autonomy [92]. This can be especially concerning in contexts where AI systems are used to make decisions about care, as individuals may not be aware of the extent to which they are being monitored.

Overall, data privacy and surveillance are two of the major ethical issues surrounding AI in healthcare [93]. These issues are important to consider when developing and deploying AI systems, as they have the potential to have major implications for individuals' data privacy and autonomy.

7.5   TRENDS OF AI IN THE PRACTICE OF MEDICINE

AI has the potential to revolutionize the practice of medicine and healthcare delivery in the near future. AI has already been used to help diagnose diseases, improve treatments, and provide personalized healthcare services. AI technologies can be used to automate mundane tasks, improve accuracy and reduce mistakes, reduce costs, and enhance the patient experience [94]. AI can also be used to provide insights into patient data and to identify previously unknown correlations between symptoms and treatments [95]. AI can also be used to provide real-time feedback to clinicians and to detect subtle changes in patient conditions [96]. AI is already being used in a variety of ways in the healthcare field, and it is likely that the use of AI in medicine will continue to grow as more sophisticated applications are developed (Figure 7.2).

7.5.1   CURRENT TRENDS

AI is becoming increasingly prevalent in the practice of medicine, with many healthcare providers now relying on AI to make faster, more accurate decisions and to provide more personalized medical care. AI is being used to diagnose and treat diseases, monitor patient health, and provide virtual care [98]. AI also has the potential to revolutionize the delivery of healthcare, improving efficiency and safety while providing more personalized care [99].

The use of AI in healthcare is becoming increasingly commonplace, with AI-based systems being used to diagnose illnesses, monitor patient data, and provide personalized care [100]. AI can be used to detect and diagnose diseases by analyzing medical images, such as X-rays, CT scans, and MRIs. AI-based systems are also being used to analyze large data sets, such as electronic health records, to identify patterns and trends in patient health and to predict future health outcomes [101]. AI can also be used to monitor patient health and provide alerts when a patient's health is at risk.

AI is also being used to provide virtual care, with some healthcare providers now offering virtual visits with AI-based systems [102]. AI is also being used to improve the efficiency and accuracy of healthcare. AI-based systems can be used to automate tasks such as data entry and analysis, freeing

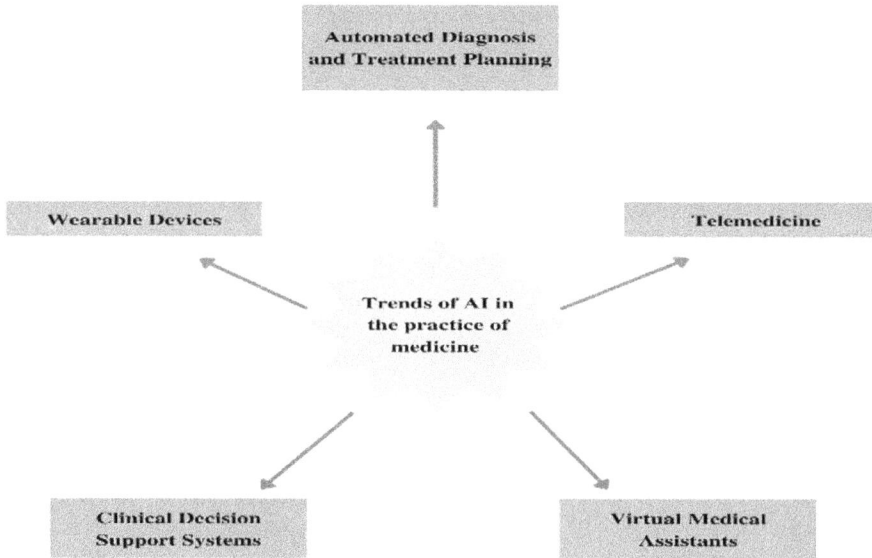

FIGURE 7.2   The 5 major trends of AI in the practice of medicine.

up healthcare providers to focus on providing personalized care [103]. AI can also be used to ana-lyze large data sets to identify patterns and trends in patient health, helping healthcare providers to make more informed decisions (Table 7.1). AI can also be used to provide personalized care, with AI-based systems being used to provide tailored treatments and advice.

AI is being used to provide more personalized care, with AI-based systems being used to pro-vide tailored treatments and advice [104]. AI-based systems can analyze patient data to provide

TABLE 7.1
Trends of AI in the Practice of Medicine

Trend	Definition
Clinical Decision Support Systems	Clinical decision support systems (CDSS) are being developed to provide physicians with real-time guidance and recommendations for diagnosis and treatment. CDSS are expected to become more sophisticated and personalized as AI technology advances.
Automated Diagnosis and Treatment Planning	AI can be used to develop automated systems for diagnosis, treatment planning, and management of chronic diseases. These systems can be used to provide personalized recommendations for diagnosis and treatment based on a patient's medical history and current condition.
Wearable Devices	Wearable devices are being developed to monitor the health of patients in real time and provide physicians with data to help inform treatment decisions. AI can be used to analyze the data from these devices and provide more accurate diagnoses and more tailored treatments.
Virtual Medical Assistants	Virtual medical assistants powered by AI are being developed to help physicians with administrative tasks such as scheduling appointments and managing medical records. These systems can also provide real-time advice for diagnosis and treatment decisions.
Telemedicine	AI-powered telemedicine systems are being developed to provide remote care to patients. These systems can be used to provide advice to patients in real time, as well as to monitor their health and provide follow-up care.

Sources: [97]

personalized care plans, providing tailored advice and treatments for each patient. AI-based systems can also be used to provide personalized reminders, helping patients to stay on track with their health goals [105].

In addition, AI is being used to improve safety in healthcare settings. AI-based systems can be used to monitor medical equipment and alert healthcare providers if a device is malfunctioning or if a patient is at risk [106]. AI-based systems can also be used to detect potential errors in medical records and alert healthcare providers if a mistake has been made.

Overall, AI is becoming increasingly prevalent in the practice of medicine and is being used to diagnose and treat diseases, monitor patient health, provide virtual care, improve efficiency and accuracy, and provide more personalized care. AI has the potential to revolutionize healthcare, improving safety, efficiency, and patient outcomes.

7.5.2 FUTURE POTENTIAL TRENDS

The practice of medicine is rapidly changing due to the introduction of artificial intelligence (AI). AI is a broad term that includes a range of technologies such as machine learning, natural language processing, and computer vision [107]. As AI technology advances, its potential applications in healthcare are becoming increasingly relevant and promising [108].

AI has the potential to revolutionize the practice of medicine in the near future, and it is likely to have a significant impact on clinical decision-making, diagnosis, and patient outcomes [109]. AI has already begun to transform the practice of medicine by providing physicians with a more accurate and timely diagnosis [110]. AI systems can analyze large amounts of data and identify patterns that may not be apparent to a human observer. This can result in more precise diagnoses and more accurate treatment plans. This can also lead to a reduction in medical errors and improved patient safety [111]. AI has also been used to develop new treatments and to identify potential drug targets [112]. This has the potential to reduce the cost of healthcare and to improve the effectiveness of treatments.

In the future, AI is likely to become even more integrated into clinical practice [112]. AI technology is expected to be used more widely in medical imaging, patient monitoring, and drug discovery. AI can be used to diagnose diseases more quickly and accurately, allowing physicians to make decisions more quickly [113]. AI can also be used to automate some of the more mundane tasks involved in the practice of medicine, such as data entry and paperwork. AI can also be used to provide personalized treatments tailored to individual patients, resulting in improved outcomes.

AI is also likely to have a significant impact on the delivery of healthcare. AI-enabled systems can be used to automate administrative tasks such as scheduling, patient registration, and billing [114]. AI can also be used to analyze patient data and identify potential health risks. This has the potential to improve the efficiency of healthcare delivery and to reduce costs. AI can be used to assist physicians in making decisions. AI-enabled systems can provide doctors with real-time access to the latest medical research and clinical guidelines. This can help physicians make more informed decisions and provide better care for their patients.

AI is rapidly transforming the practice of medicine and has the potential to revolutionize healthcare. AI-enabled systems can provide more accurate diagnoses, improve the efficiency of healthcare delivery, and automate administrative tasks. AI can also assist physicians in making decisions and provide personalized treatments tailored to individual patients. AI is likely to become even more integrated into the practice of medicine in the near future, and its potential applications are numerous and exciting.

7.6   CONCLUSION

The prospect of advancement in any field or profession is typically met with acclaim. The idea that novel theories, techniques, technology, etc. were developed and subsequently implemented is an idea that thrills many. The field of medicine is experiencing said excitement. Rapid advancement in medicine is not a novel concept. From the creation of new surgical techniques, diagnostic testing, and expeditious vaccine development, the field of medicine continues to evolve and continually push the boundaries of scientific discovery and application. One novel tool in the field of medicine that is being explored and used is AI. Through the use of AI the medical field is poised for a significant transformation of how people are treated. The implementation of AI in medicine presents the opportunity for medical care to become efficient and effective. AI in medicine could potentially open a new frontier in how physicians practice medicine allowing for more lives to be saved and possibly from illness even occurring in the first place. While the prospect of the positive impacts from AI in medicine seem alluring and initially advantageous, there are a variety of issues that stem from the use of AI in medicine. In particular, the ethicality of AI in medicine is something of paramount concern. Biases, decision making, privacy, and many other similar aspects are already notable as issues of concern as it pertains to physicians treating patients. These issues are also prevalent in the application of AI in medicine.

Practicing medicine with AI is becoming increasingly more common as technology advances. AI is being used to improve patient care, streamline operations, and reduce costs. Current use of AI in healthcare includes diagnostics, drug discovery, robotic surgery, and medical imaging. The advantages of using AI in healthcare are improved accuracy, increased efficiency, and the ability to detect and diagnose diseases earlier. However, there are also some disadvantages of AI in healthcare including lack of understanding of the algorithms, privacy and data security concerns, and potential bias. Ethical concerns related to AI in healthcare include decision-making, privacy and surveillance, and bias.

When it comes to decision-making, AI can be used to help make decisions about diagnosis, treatments, and patient outcomes. This can be beneficial, but it can also create ethical dilemmas if decisions are being made without input from the patient or healthcare provider. Additionally, AI algorithms have the potential to be biased, which could lead to unequal access to healthcare and incorrect or unfair diagnoses. Privacy and surveillance is another ethical concern when it comes to AI, as the technology is able to collect large amounts of data and create detailed profiles of individuals.

In terms of current trends, AI is being used in healthcare to improve patient care, streamline operations, and reduce costs. AI is being used for diagnostics, drug discovery, robotic surgery, and medical imaging. In the future, AI may be used to personalize treatments and predict outcomes. However, ethical considerations will need to be taken into account when implementing AI in the practice of medicine.

While there are considerable ethical concerns about the use of AI, it does not mean that AI should be tossed aside and rendered unusable. The advent of AI and in particular AI in medicine should be met with optimism. The field of medicine has the potential to become not only more efficient, but accurate. Physicians will be better equipped with an incredibly useful tool that will allow diagnoses, treatments, and overall care to be improved substantially. The prospect of that should drive the conversation and efforts to rectify the serious ethical concerns AI presents when applied to the field of medicine. It may even result in a self-reflection and correction of the ethical issues that physicians face themselves.

The expanding of implementary horizons in the field of medicine is something that should be met with applause. Continuous medical development is important as society and our environment continue to evolve. However, while there is applause it is also vital for healthy skepticism to be displayed to ensure tools, techniques, and treatments are used in an efficacious and ethical manner.

REFERENCES

1. Babalola, O. (2017). Consumers and their demand for healthcare. *Journal of Health & Medical Economics*, 3(1), 6.

2. Featherston, R., Downie, L. E., Vogel, A. P., & Galvin, K. L. (2020). Decision making biases in the allied health professions: A systematic scoping review. *PLOS One*, 15(10), e0240716. https://doi.org/10.1371/journal.pone.0240716

3. Spikins, P., Needham, A., Tilley, L., & Hitchens, G. (2018). Calculated or caring? Neanderthal healthcare in social context. *World Archaeology*, 50(3), 384–403. https://doi.org/10.1080/00438243.2018.1433060

4. Doctor of Medicine Profession (MD). (n.d.). https://medlineplus.gov/ency/article/001936.htm

5. Underwood, A. E. (1998b, July 20). History of medicine | History & facts. Encyclopedia britannica. https://www.britannica.com/science/history-of-medicine

6. Rhodes, P. (1998, July 20). History of medicine | History & facts. Encyclopedia britannica. https://www.britannica.com/science/history-of-medicine

7. Mikić, Z. (2008). *Medicinski Pregled*, 61(9–10), 533–538.

8. Richardson, R. G. (1998, July 20). History of medicine | History & facts. Encyclopedia britannica. https://www.britannica.com/science/history-of-medicine

9. Guthrie, D. J. (1998, July 20). History of medicine | History & facts. Encyclopedia britannica. https://www.britannica.com/science/history-of-medicine

10. Siqueira, A. A. F., & da Silva, V. A. (2009). The hypoglycemic effect of Lippia alba (Millettia alba) in streptozotocin-induced diabetic rats. *In Vivo*, 23(4), 507–513. https://doi.org/10.2174/092986709787750789

11. Greek Medicine - The Hippocratic Oath. (n.d.). https://www.nlm.nih.gov/hmd/greek/greek_oath.html

12. Xu, Y., Liu, X., Cao, X., Huang, C., Liu, E., Qian, S., Liu, X., Wu, Y., Dong, F., Qiu, C.-W., Qiu, J., Hua, K., Su, W., Wu, J., Xu, H., Han, Y., Fu, C., Yin, Z., Liu, M., … Zhang, J. (2021). Artificial Intelligence: A powerful paradigm for scientific research. *Innovation*, 2(4), 100179. https://doi.org/10.1016/j.xinn.2021.100179

13. Booth, S., & Wigington, P. (2020). Artificial Intelligence (AI): Definition, types, and examples. The balance small business. https://www.thebalancesmb.com/what-is-artificial-intelligence-ai-4159141

14. Garvie, C., & Bellovin, S. M. (2018). Artificial intelligence for the real world. *IEEE Security and Privacy*, 16(5), 66–69. https://doi.org/10.1109/MSP.2018.2828094

15. Rapaport, W. (2006). Turing test: Encyclopedia of language &Amp. *Linguistics*, 151–159. https://doi.org/10.1016/b0-08-044854-2/00933-0

16. Lohr, S. (2018). How artificial intelligence is changing medicine. *The New York Times*. https://www.nytimes.com/2018/07/06/business/artificial-intelligence-medicine.html

17. Rosen, D. (2018). What is artificial intelligence (AI)? Definition, types, and examples. *Forbes*. https://www.forbes.com/sites/davidrosen/2018/11/15/what-is-artificial-intelligence-ai-definition-types-and-examples/#3a8a3dfa6fbe

18. Bennett, K. P., & Matwin, S. (2017). AI overview: A guided tour of artificial intelligence. *IEEE Intelligent Systems*, 32(1), 22–27.

19. Gandhi, B. (2018). How artificial intelligence is changing the world. LinkedIn. Retrieved from https://www.linkedin.com/pulse/how-artificial-intelligence-changing-world-budhendra-gandhi/

20. Peddlekar, M. (2019). The implications of artificial intelligence in healthcare. Healthcare IT news. Retrieved from https://www.healthcareitnews.com/news/implications-artificial-intelligence-healthcare

21. Tai, M. T. (2020). The impact of artificial intelligence on human society and bioethics. *Tzu Chi Medical Journal*, 32(4), 339. https://doi.org/10.4103/tcmj.tcmj_71_20

22. Davenport, T., & Kalakota, R. (2019). The potential for artificial intelligence in healthcare. *Future Healthcare Journal*, 6(2), 94–98. https://doi.org/10.7861/futurehosp.6-2-94

23. Parikh, R. B., & Helmchen, L. A. (2022). Paying for artificial intelligence in medicine. *Npj Digital Medicine*, 5(1). https://doi.org/10.1038/s41746-022-00609-6

24. Manickam, P., Mariappan, S. A., Murugesan, S. M., Hansda, S., Kaushik, A., Shinde, R., & Thipperudraswamy, S. P. (2022). Artificial intelligence (AI) and Internet of medical things (IoMT) assisted biomedical systems for intelligent healthcare. *Biosensors*, 12(8), 562. https://doi.org/10.3390/bios12080562

25. Topol, E. J. (2019). High-performance medicine: The convergence of human and artificial intelligence. *Nature Medicine*, 25(1), 44–56.

26. Ahuja, A. S. (2019). The impact of artificial intelligence in medicine on the future role of the physician. *PeerJ*, 7, e7702. https://doi.org/10.7717/peerj.7702

27. Buehler, C., & Landman, B. A. (2020). Artificial intelligence in radiology: A review of applications and implications. *Radiology*, 296(3), 536–552.

28. Huang, Y., & Li, Y. (2020). Artificial intelligence for medical imaging: A comprehensive survey. *IEEE Transactions on Medical Imaging*, 39(6), 1645–1663.

29. Lee, K. Y., & Kim, J. (2016). Artificial intelligence technology trends and IBM Watson references in the medical field. *Korean Medical Education Review*, 18(2), 51–57. https://doi.org/10.17496/kmer.2016.18.2.51

30. Prasad, A., & Jain, A. (2017). Artificial intelligence in healthcare: Past, present, and future. *Healthcare*, 5(2), 63.

31. Topol, E. J., & Krumholz, H. M. (2020). Artificial intelligence in health care: Promises, perils, and prospects. *Science*, 368(6490), 527–532.

32. Daneshjou, R., Smith, M. P., Sun, M. D., Rotemberg, V., & Zou, J. (2021). Lack of transparency and potential bias in artificial intelligence data sets and algorithms: A scoping review. *JAMA Dermatology*, 157(11), 1362. https://doi.org/10.1001/jamadermatol.2021.3129

33. Frank, M. R., Wang, D., Cebrian, M., & Rahwan, I. (2019). The evolution of citation graphs in artificial intelligence research. *Nature Machine Intelligence*, 1(2), 79–85. https://doi.org/10.1038/s42256-019-0024-5

34. Kleinberg, J., Ludwig, J., Mullainathan, S., & Rambachan, A. (2018). Algorithmic fairness. *AEA Papers and Proceedings*, 108, 22–27. https://doi.org/10.1257/pandp.20181018

35. Meyer, E. T., Shankar, K., Willis, M., Sharma, S., & Sawyer, S. (2019). The social informatics of knowledge. *Journal of the Association for Information Science and Technology*, 70(4), 307–312. https://doi.org/10.1002/asi.24205

36. Shaban-Nejad, A., Michalowski, M., & Buckeridge, D. L. (2018). Health intelligence: How artificial intelligence transforms population and personalized health. *Npj Digital Medicine*, 1(1). https://doi.org/10.1038/s41746-018-0058-9

37. Al-Hail, A., Al-Ansari, M., & Al-Jumah, M. (2020). Artificial intelligence in healthcare: Applications and potential benefits. *Healthcare*, 8(4), 124. https://doi.org/10.3390/healthcare8040124

38. Ghadimi, A., Mirzaei, A., & Ghadimi, M. (2020). Artificial intelligence in healthcare: A comprehensive review. *Computers in Biology and Medicine*, 124, 103783. https://doi.org/10.1016/j.compbiomed.2020.103783

39. Oberoi, J. S., & Liu, Y. (2020). Artificial intelligence in drug development: Current progress and future opportunities. *Trends in Pharmacological Sciences*, 41(8), 681–696. https://doi.org/10.1016/j.tips.2020.05.004

40. McDonald, C. M. (2019). Using AI to improve operational efficiency in healthcare. Retrieved November 8, 2020, from https://www.forbes.com/sites/christinamcdonald/2019/05/17/using-ai-to-improve-operational-efficiency-in-healthcare/#2f8d8bdd78d9

41. Pfeifer, M. (2019). AI and the future of healthcare. *Harvard Business Review*. https://hbr.org/2018/05/ai-and-the-future-of-healthcare

42. Gao, Y., & Yu, K. (2020). Artificial intelligence in healthcare: A comprehensive survey. *IEEE Journal of Biomedical and Health Informatics*, 24(3), 1143–1158. https://doi.org/10.1109/JBHI.2019.2932063

43. Kawamoto, K., & Chino, Y. (2020). The potential of artificial intelligence for improving healthcare and patient outcomes. *International Journal of Environmental Research and Public Health*, 17(14), 5046. https://doi.org/10.3390/ijerph17145046

44. Kutac, M., Rogers, D., & Balas, E. (2020). Artificial intelligence for healthcare: Opportunities and challenges. *Journal of the American Medical Informatics Association JAMIA*, 27(7), 992–997. https://doi.org/10.1093/jamia/ocaa037

45. Perez-Marin, D., & Kotsiantis, S. B. (2020). Artificial intelligence in healthcare: A survey. *IEEE Access*, 8, 83596–83614. https://doi.org/10.1109/ACCESS.2020.2987420

46. Rastegari, M., & Ghafoorian, M. (2020). The ethical, legal and social implications of artificial intelligence in healthcare. *Artificial Intelligence in Medicine*, 99, 101749. https://doi.org/10.1016/j.artmed.2020.101749

47. Chen, H., & Liao, Y. (2018). The potential of artificial intelligence in healthcare. *International Journal of Medical Informatics*, 113, 1–9.

48. Fortuna, R. J., & Chiang, M. F. (2020). Artificial intelligence in healthcare: A review of its current applications. *Journal of Medical Internet Research*, 22(3), e20804.

49. Koppel, R., & Chen, J. (2018). Artificial intelligence in healthcare: Current applications and implications for the future. *Mayo Clinic Proceedings*, 93(4), 488–501.

50. Peng, J., & Wang, Y. (2018). Artificial intelligence in healthcare: Current status and future perspectives. *IEEE Reviews in Biomedical Engineering*, 11, 195–202.

51. Shah, P., & O'Connell, M. (2017). Artificial intelligence in healthcare: Applications and value. *Managed Healthcare Executive*, 27(3), 42–48.

52. Razavian, S., & Shaya, M. K. (2018). The potentials of artificial intelligence in healthcare: A review. *Data in Brief*, 20, 944–956.

53. Ho, Y. W., Shome, M., Lau, B. H., & Zhang, Q. (2019). Artificial intelligence in healthcare: A review of the literature. *Journal of Medical Internet Research*, 21(1), e13291.

54. Zhang, J., Lu, Z., & He, Q. (2019). A survey of deep learning in healthcare: Achievements and trends. *Artificial Intelligence in Medicine*, 95, 1–14.

55. Kant, I. (1781). *Lectures on ethics*. Cambridge: Cambridge University Press.

56. Ahooei, Z., Norris, J. L., & Garg, A. (2019). Artificial intelligence in healthcare: Privacy and security considerations. *International Journal of Healthcare Information Systems and Informatics*, 14(2), 15–29.

57. Hannun, A., et al. (2017). Deep learning in healthcare: The promise and the challenges. *Nature Biotechnology*, 35(5), 434–443.

58. Kilov, D., Ari, G., & Wootten, R. (2017). The challenges and opportunities of artificial intelligence infused healthcare. *International Journal of Healthcare Delivery Reform Initiatives*, 8(1), 12–18.

59. Kouki, P., & Zoghlami, T. (2019). Ethical impact of artificial intelligence deployment in healthcare: A seven-dimensional analysis. *Healthcare*, 7(3), 104.

60. Pani, M., et al. (2020). Artificial intelligence applications in healthcare: A review of ethical, legal, and regulation challenges. *Electronic Physician*, 12(3), 586–593.

61. Lekadir, K., Quaglio, G., Garmendia, A. T., & Gallin, C. (2022). Artificial intelligence in healthcare: Applications, risks, and ethical and societal impacts. EPRS (European Parliamentary Research Service), Scientific Foresight Unit (STOA) PE 729.512 - June 2022. https://www.europarl.europa.eu/stoa/en/document/EPRS_STU(2022)729512.

62. Buehler, S., Wallach, W., & Danks, D. (2018). Ethical decision-making in AI: A checklist for responsible innovation. *AI Magazine*, 39(4), 70–80. https://doi.org/10.1609/aimag.v39i4.2792

63. Herath, H. M. K. K. M. B., & Miital, M. (2022). Adoption of artificial intelligence in smart cities: A comprehensive review. *International Journal of Information Management Data Insights*, 2(1), 100076.

64. Sanders, A. (2017). Ethical decision-making in the age of artificial intelligence. *Journal of Business Ethics*, 143(3), 459–473. https://doi.org/10.1007/s10551-015-2790-7

65. Wagner, C., & Floridi, L. (Eds.) (2019). *AI, ethics, and society: The quest for good AI*. Springer Nature. ECIS 2022 Research Papers. 60.

66. Hughes, J. E. (2017). The ethics of artificial intelligence. In *The Oxford handbook of ethical theory*, Markus Dirk Dubber, Frank Pasquale, and Sunit Das (eds.), (pp. 725–746). Oxford University Press.

67. Groth, P., & Mittelstadt, B. (Eds.) (2019). *The ethics of artificial intelligence and robotics*. Oxford University Press. 4. AI Now Institute. (2018). The AI Now 2018 Report. Retrieved from https://ainowinstitute.org/AI_Now_2018_Report.pdf 5

68. Vermeulen, E. (2020). *Artificial intelligence and ethics: A guide for decision-makers*. Springer Nature. ECIS 2022 Research Papers. 60.

69. Siau, K., & Wang, W. (2020). Artificial intelligence (AI) ethics: Ethics of AI and ethical AI. *Journal of Database Management*, 31, 74–87.

70. International Medical Informatics Association (IMIA). (2020). IMIA statement on ethical principles for artificial intelligence and machine learning in health and healthcare [Press release]. Retrieved from https://www.imia-medinfo.org/wp-content/uploads/2020/03/IMIA_AI_Statement_Final.pdf

71. Angwin, J., & Larson, J. (2020). *Automating inequality: How high-tech tools profile, police, and punish the poor*. New York: St. Martin's Press.

72. Binns, A. (2019). *AI ethics: How to avoid bias in artificial intelligence*. New York: McGraw-Hill Education.

73. Hancock, K. (2020). AI and bias: What you need to know. *Harvard Business Review*. Retrieved from https://hbr.org/2020/02/AI-and-bias-what-you-need-to-know

74. Rosenfeld, P. (2020). Reducing bias in artificial intelligence. *Nature Machine Intelligence*, 2(12), 785–793. https://doi.org/10.1038/s42256-020-00276-y

75. Drexler, M. (2020). Bias in AI: What it is, why it matters, and what we can do about it. Retrieved from https://medium.com/future-crunch/bias-in-ai-what-it-is-why-it-matters-and-what-we-can-do-about-it-e12bdc1f14d3

76. Gutkin, T. (2019). AI ethics: What are the ethical implications of AI? Retrieved from https://www.techopedia.com/definition/34676/ai-ethics-ethical-implications-of-ai

77. Cappelli, A. (2020). AI ethics: Ethical challenges and approaches to responsible AI development. Retrieved from https://academy.microsoft.com/en-us/professional-program/tracks/ai-ethics/

78. Mittelstadt, B. D., Allo, P., Taddeo, M., Wachter, S., & Floridi, L. (2019). The ethics of artificial intelligence. *Nature Machine Intelligence*, 1(9), 469–474. https://doi.org/10.1038/s42256-019-0086-3

79. Gellman, R. (2020). AI is everywhere, and so is the risk of unfair bias. Here's how to fight it. Retrieved from https://www.nytimes.com/2020/02/10/technology/ai-bias-discrimination-risk.html

80. Nguyen, T. T., Hashemian, M., D'Souza, A., & Alimohamed, S. (2020). A framework for understanding bias in health care artificial intelligence. *JMIR Medical Informatics*, 8(2), e16373.
81. Kugelman, M. A., Pang, J. S., & Batz, M. J. (2020). Ethical considerations in artificial intelligence in healthcare. *International Journal of Medical Robotics and Computer Assisted Surgery*, 16(3), e2045.
82. Dietz, T., & Custers, B. (2020). Data protection in the age of AI: An introduction to the ethical and legal challenges. *Information and Communications Technology Law*, 29(2), 190–207.
83. Goguen, A. (2019). Data privacy and surveillance in the age of AI: A legal and ethical framework. *AI and Society*, 34(3), 545–561.
84. Šedo, J., & Žilvinas, B. (2019). Ethical challenges in AI and big data: Mapping the field. *AI and Society*, 34(2), 211–228.
85. Denning, D. E. (2018). Big data and privacy: A technological perspective. *Communications of the ACM*, 61(4), 24–26.
86. Denning, D. E. (2018). Big data and privacy: A technological perspective. *Communications of the ACM*, 61(4), 24–26.
87. Koppel, J., & Bozdag, E. (2019). Algorithmic accountability: The implications of surveillance technologies for privacy and civil liberties. *Surveillance and Society*, 17(3), 337–354
88. Datta, A., Datta, A., & Telang, R. (2015). Understanding privacy in the age of big data and surveillance. *Communications of the ACM*, 58(2), 36–41.
89. Newman, M. (2018). Surveillance and privacy in the age of big data. *Communications of the ACM*, 61(4), 27–30.
90. Eggers, W. D., & Tucker, C. E. (2017). Artificial intelligence and privacy: Challenges and opportunities. *Communications of the ACM*, 60(2), 37–39.
91. Chen, S. X., Shilton, K., & Datta, A. (2018). Privacy, AI, and machine learning: Challenges and opportunities. *Communications of the ACM*, 61(6), 26–29.
92. Palmore, J., & Yu, H. (2018). Ethical issues in artificial intelligence. *Communications of the ACM*, 61(6), 30–35.
93. Nguyen, M., & Nguyen, P. (2020). Ethical issues surrounding AI in healthcare: Data privacy and surveillance. In C. D. Dwork & D. O'Connor (Eds.) *Artificial intelligence for medicine and healthcare* (pp. 5–15). Boston, MA: Springer.
94. Liang, C., & Zhou, K. (2019). Artificial intelligence in healthcare: Current applications and future opportunities. *International Journal of Medical Informatics*, 124, 10–20.
95. Chan, T. (2020). Artificial intelligence in healthcare: Applications, challenges, and future directions. *Journal of Medical Internet Research*, 22(7), e16804
96. Bruno, V., Pecoraro, F., Civitarese, C., & Giannotti, F. (2020). Artificial intelligence in healthcare: State-of-the-art and challenges. *Applied Sciences*, 10(8), 3030.
97. Bajwa, J., Munir, U., Nori, A., & Williams, B. (2021). Artificial intelligence in healthcare: Transforming the practice of medicine. *Future Healthcare Journal*, 8(2), e188–e194.
98. Weber, J., Böhmer, M., & Reichel, J. (2020). Artificial intelligence in medical practice: Current applications and future perspectives. *Clinical Chemistry and Laboratory Medicine (CCLM)*, 58(9), 1417–1425. https://doi.org/10.1515/cclm-2019-0904
99. Armbrust, M. et al. (2010). Above the clouds: A Berkeley view of cloud computing. Technical Report No. UCB/EECS-2009-28 http://www.eecs.berkeley.edu/Pubs/TechRpts/2009/EECS-2009-28.html.
100. Chen, Y. et al. (2019). Deep learning for medical image analysis. *Annual Review of Biomedical Engineering*, 21, 221–248.
101. Kumar, S. et al. (2020). Artificial intelligence in healthcare: A comprehensive review. *Journal of Medical Internet Research*, 22(1), e13702.
102. Liu, X. et al. (2019). Artificial intelligence in healthcare: A concise review. *IEEE Reviews in Biomedical Engineering*, 12, 285–301.
103. Sachdeva, S. et al. (2019). Artificial intelligence in healthcare: A review of applications and challenges. *Journal of the American Medical Informatics Association JAMIA*, 26(1), 17–27.
104. Alsaadi, M. E., Chaudhry, S. A., & Alansari, M. (2020). Artificial intelligence in medicine: Current applications and future opportunities. *Clinical Medicine Insights: Circulatory, Respiratory and Pulmonary Medicine*, 14, 1179546820902517. https://doi.org/10.1177/1179546820902517
105. Chen, S., & Gong, E. (2020). Artificial intelligence in medical practice: Current status and future prospects. *Frontiers in Medicine*, 7, 597. https://doi.org/10.3389/fmed.2020.00597
106. Maruthappan, V. V., Miller, J. E., & Liu, Y. (2020). Artificial intelligence in medicine: Current applications, opportunities, and challenges. *Current Opinion in Endocrinology, Diabetes and Obesity*, 27(6), 464–471. https://doi.org/10.1097/MED.0000000000000609

107. Ahmad, S., & Khan, M. F. (2021). Artificial intelligence and its applications in medical practice. *Journal of Advanced Research in Medical Education*, 7(1), 1–4.

108. Meng, Y., & Wang, Y. (2020). Artificial intelligence in medicine: A review and future direction. *Frontiers in Medicine*, 7, 596.

109. Tao, A., & Duan, L. (2020). Artificial intelligence in healthcare: Applications, challenges and future directions. *International Journal of Medical Informatics*, 135, 103900.

110. Amini, A., Reiter, S., & Topol, E. J. (2019). The future of artificial intelligence in medicine: Promises and challenges. *JAMA*, 322(17), 1678–1686. https://doi.org/10.1001/jama.2019.15393

111. O'Connor, M. J., & Beaulieu-Jones, B. K. (2020). Artificial intelligence in medicine: Current applications and future potential. *JAMA*, 324(1), 90–101. https://doi.org/10.1001/jama.2020.19095

112. Ghassemi, M., Bhardwaj, A., Badri, M., & Faisal, A. (2019). Artificial intelligence in healthcare: past, present and future. *Nature Reviews Endocrinology*, 15(9), 559–572. https://doi.org/10.1038/s41574-019-0276-6

113. Khullar, D. (2020). Artificial intelligence in medicine: A primer and perspective. *JAMA*, 324(1), 68–69. https://doi.org/10.1001/jama.2020.19815

114. Topol, E. J., & Collins, F. S. (2020). High-performance medicine: The convergence of human and artificial intelligence. *Science*, 368(6491), 539–542. https://doi.org/10.1126/science.abb7015.

8 Cybersecurity and Intraoperative Artificial Intelligence Decision Support
Challenges and Solutions Related to Ethical Issues

Surovi Saikia, Vishnu Prabhu Athilingam, and Firdush Ahmed

8.1 INTRODUCTION

All dimensions of healthcare have been profoundly affected by digitization of care and delivery, and cybersecurity is a critical concern in this regard. An increased reliability on technology has given rise to other risk factors in healthcare which has not been able to make them immune to cyber-attacks, which also affect other industries. Remarkable operational disruptions such as delayed surgeries, ambulance divergence, and disabling entire hospital networks were caused by cybersecurity attacks [1]. Seventy percent of the hospitals have reported an increase in cyber-attacks for significant recent victim for security incident [2]. Several factors account for the vulnerability of hospital systems to cyber-attacks, for example, they have old, limited IT resources in hospitals, EHRs (electronic health records), and other thousands of endpoints as many of the electronic devices are connected to the internet, which has only increased the probability of cyber-attacks in the future [3].

Technical mishaps are often blamed for operative complications, but many errors may occur due to human-erratum in visual perception resulting in judgmental error that drives actions and behaviors resulting in adverse events [4]. An ongoing process for interpreting the field of surgery which allows a physician to make a smart decision can be offered by expert intraoperative performances in nature, as it has been observed that expert surgeons have the ability to visualize and get the safe and dangerous zones in dissection either in an unknown anatomy or hostile environment [5]. The risk of unintentional injuries can be minimized to a large extent by this type of mental mode, and guidance on a real-time interoperative here has the potential to improve surgeon's performance during surgery by improving the mental state.

Significant advances were made in recent times in medical AI, including surgical applications. A number of opportunities at the current stage of this field have clinical significance with some methodological shortcomings. Healthcare at the present time is experiencing 2–3 times in average cyber-attacks as in different industries which may reach to a thousand in a month, and it has been an international problem [6]. Different forms of cyber-attacks exist, such as ransomware attacks, which disable EMR databases and workstations. Sometimes the attackers offer the restoration in exchange for payment, and this has not been the case always but many have paid the criminals. If the infection affects the ransomware and encrypts it, it can very well be expected that the data most probably is stolen, opening the door for abuse with the PHI (patient health information). Many medical devices are connected to send and receive data to the hospital network, and therefore are highly prone to cyber-attacks. Recent such ransomware cyber-attacks crippled the online services

DOI: 10.1201/9781003353751-8

of AIIMS (All India Institute of Medical Sciences), New-Delhi on November 30, 2022, after which around 6,000 hacking attempts were made about a week later on the ICMR (Indian Council of Medical Research) website [7]. Online services for a day was suspended in November, at Safdarjung Hospital, New-Delhi, on December 3, 2022; another cyber-attack was done to a Tirupur-based hospital in Tamil Nadu, India, where personal details of 1.5 lakh patients were put on sale by the attackers via Telegram channels and other cybercrime forums; examples of the details include date of birth, address, guardian's name, and doctor's details of the patients [8].

Both personal and financial information are loaded in PHI, which makes them a goldmine for cyber criminals as they can be sold on the dark web for a lucrative amount as compared to a simple credit card. "Social engineering" strategy as mentioned by Kevin Mitnick, one of the successful hackers from 1960s–1995, continues to be the basic for many hackers even today [9]. Destroying the potential success of a cyber-attack is the duty of the IS (Information Systems) department, and one major strategy in destroying this success is to consider the architecture of hardware and software to create the defense in depth that complicates the navigation for the attacker, thereby limiting the malware spread.

8.2 INTRAOPERATIVE COMPLICATIONS DUE TO HUMAN ERRORS

"A skillfully performed operation is about 75% decision making and 25% dexterity" as mentioned by Dr. Frank Spencer, a cardiovascular surgeon. AI is the study of algorithms that are fed into machines which gain the ability to reason and perform cognitive functions. It plays a wide role in medical fields for surgery, diagnosis, and treatment applications. Recent advancements in the medical field were developed with the use of robotics and artificial intelligence, which are incorporated into the equipment and instrumentations used. Human errors and mishandling are some of the major drawbacks in surgery, which leads to serious consequences later. Intraoperative decision-making includes three processes: analyzing the situation accordingly, taking action relevant to the situation, and evaluating the action made.

8.2.1 APPLICATIONS OF INTRAOPERATIVE DECISION MAKING

In parathyroid abnormal patients with the growth of noncancerous adenoma, growth is seen where non-invasive surgery is preferred by most of the patients as compared to open surgeries. In some patients, multiple adenomas may result in hyperparathyroidism, which, if not removed preoperatively, PTH (parathyroid hormone) helps surgeons to determine if all the adenoma responsible for the PT is removed. By considering the PTH level as mathematical data and feeding the results into the final stage of the logistic regression model, the ratio of cure can be predicted. Testing the model on unscreened population of 100 patients, none of them had hyperparathyroidism; short- and long-term follow-up found that out of 81, 78 were with single adenoma cured later, and out of 19, 17 patients were with multiple adenomas [10].

During laparoscopic hysterectomy surgery, to visualize internal organs, instruments are inserted through small incisions via the abdomen or vagina; this is carried out under general sedation. During this surgery, there is minimal physical response from the electronic device to the user, and the instrument, which uses this minimal tactile feedback, uses vibration audible click and examination of one's body through palpation. Hence it is difficult to find the anatomic structures. Due to this improper diagnosis, AI is considered; ANN (Artificial Neural Network) (a modified Google Net architecture which is a convolution neural network that has 22 layers) has been created to distinguish the uterine artery from the ureter during laparoscopic hysterectomy. Using this AI, over 94.2% accuracy resulted from 2,500 images obtained from 35 videos of patients. Of post data augmentation of 8,000 images, 2,000 images were tested [11]. Based on the intraoperative warnings and recommendations, aid in automated case retrieval and provide help using real-time surgical video [12].

8.3 CLINICAL DECISION-MAKING USING AI

Implementation of EHRs has paved the way for maintenance and growth of personalised and automated data essential for immediate patient care as made possible by AI and its subfields like ML (Machine Learning) and deep learning. Systems for keeping EHRs have proliferated throughout healthcare systems, and large, diverse datasets produced by the resulting EHR databases present new prospects for designing and implementing better healthcare systems. Manual data entry is reduced, bringing objectivity where hypothetical-deductive reasoning offers precise forecasts and classifications with customized treatment schedule. Several significant barriers to clinical adoption still exist, despite the tremendous potential of AI techniques to improve patient care. Cardiovascular disease, a major one in avoidable death and morbidity, can be screened for and diagnosed using ECG (electrocardiography). Heart failure is one of the most challenging cardiovascular disorders to diagnose because of the diversity of underlying pathology and clinical presentations. Grun et al., 2021 show that using ordinary 12-lead ECGs, AI is capable of properly predicting heart failure, showing the potential for AI to support early diagnosis and treatment utilizing common clinical data [13]. Large volumes of patient data are measured in audiology; however, they are generally dispersed over local clinical databases with distinctive architecture, data elements, and variable names, which makes it difficult for external validation and multi-centre-based studies [14]. These studies use distinct lasso regression, elastic net and random forest methods, which have similar, good predictive performance to demonstrate the viability of autonomous prediction, typically audiological functional characteristics. The trained model serves as the prototype foundation for a widely applicable clinical decision-support system for audiology that also works well across regional clinical databases in spite of their distinct organizational structures, data items, and variable names. ICUs provide care for seriously ill patients who need advanced organ support or near-constant monitoring. As gastrointestinal, hepatic, and renal dysfunction frequently affect critically sick patients, it might be difficult to provide medication because these conditions impair medication absorption, metabolism, and excretion. Many data-driven pharmaceutical dosing models have been put forth, but their capacity to recognize inter-individual variations and calculate customized doses is restricted. Deep reinforcement learning was used by Eghbali et al., 2021 to construct sedation management agents that perform better in the ICU when managing blood pressure as compared to physicians [15]. The author's methodology has the potential to automate the dosage of other drugs often used in ICUs.

The use of ML in the creation of clinical decision-support systems for audiology has the potential to increase the objectivity and accuracy of diagnostic choices made by clinical specialists. However, such a tool must be accurate to be accepted and trusted by doctors for clinical application.

Giordano et al., 2020 studied the intervention of artificial intelligence in clinical decision-making [16]. Clinicians will be able to identify high-risk patients and optimise resource utilisation and perioperative decisions with the aid of ML models that can risk-stratify patients prior to surgery. Clinical staff, patients, and their families can benefit from the fast processing of all available data by ML and AI to produce well-informed, evidence-based recommendations and take part in shared decision-making to determine the best course of action. ML algorithms can be used in a variety of scenarios across the care spectrum, such as perioperative circumstances or the management of diseases [17].

In healthcare, risk-prediction algorithms have been used to detect high-risk patients and to make the right treatment decisions afterward. In this era of value-based care, appropriate risk classification should lead to proper resource use. Historically, statistical regression models have served as the foundation for most risk-prediction systems. Examples include the National Surgical Quality Improvement Program, the Framingham risk score, and QRISK3 (for coronary heart disease, ischemic stroke, and transient ischemic attack) (NSQIP). Unfortunately, a lot of these risk stratification techniques either lack specificity and accuracy at the patient level or call for specially educated clinicians to analyse the records and determine the risk. Healthcare systems are increasingly looking to ML to help with risk stratification, and these models may perform better in calibration and discrimination than statistical models. Enhancing preoperative support for high-risk patients and high-cost populations is a rising national initiative [18, 19].

The clinical decision-making process and the subsequent patient treatment depend heavily on optimization for each or all of the possible patient outcomes. Usually, healthcare professionals meet with family members to decide the necessary optimal steps, their timing, and the best order. Despite having the best of intentions, these choices may result in subpar care because of the complexity of patient care, the growing amount of responsibility placed on healthcare professionals, or just because of human mistakes. Clinical decision-making frequently adheres carefully to set standards, protocols, and guidelines that meet accountability and safety standards. It may be advantageous for the patient to adjust therapies for a more individualized regimen, though, when there is a complex care environment.

Particularly in acute care settings, ML techniques can be useful tools in such dynamic situations for optimizing patient care outcomes in a data-driven manner. Modern deep-learning approaches and ML often maximize an objective function (such as medicine dose) using complicated and multidimensional data (such as patient medical histories taken from EHRs). ML tools have been employed in a variety of settings to improve care outcomes, including critical care to improve sepsis treatment [20], management of chronic illnesses [21], and improving surgical results [22]. Utilizing sequential decision-making methods with inspiration from allied disciplines like operation research is another more sophisticated strategy.

As an example, deep reinforcement learning models [23] are based on widely used ideas like Q-learning and the Markov decision process (MDP) [24] applied to neural networks. Models that use reinforcement learning discover the best policies based on a reward function. The best policy is identified by defining the policies as a series of behaviours that lead to the highest reward. Recently, reinforcement learning and deep reinforcement learning have been applied in a variety of clinical settings, such as the best timing for interventions, the best drug dosage, and the best individual target laboratory values. Prasad et al. [25] employed reinforcement learning to wean patients from mechanical breathing in the intensive care unit, and Nemati et al. [26] used deep reinforcement learning to optimize drug dosing. Although these tools have huge potential to improve the patient care process, responsibility and safety come first. Due to the black-box nature of contemporary deep-learning techniques, this might become more difficult. The resulting policies may be flexible and tailored, but it may be difficult to understand and articulate their reasoning. Additionally, using reinforcement learning in therapeutic contexts is significantly more difficult than it is in conventional simulation and gaming scenarios.

The widespread use of EHRs in healthcare systems around the world has produced enormous collections of personal data sets that are ideal for AI to analyse, build, and make predictions. Deep learning networks and ML as subfields have demonstrated progress in addressing the healthcare issues of risk stratification and patient outcome optimization. As this technology is progressively incorporated into complex healthcare systems, its use will grow dramatically. By enabling multiple outcome optimizations of outcomes that are too challenging to perceive and traverse on an individual and isolated basis, AI capabilities will help physicians assess competing for healthcare goals and numerous hazards. It will be required of healthcare professionals to easily operate within this new AI frontier and relate it to their patients.

Additionally, doctors must be able to deconstruct the models from which these AI systems get their predictions and analyse them. Additionally, doctors will need to understand the training cohort that was used to develop the model in order to evaluate possible bias and the proper patient population applicability. To achieve this understanding, all new students currently enrolled in medical school will need to have their curriculum completely revised as well as get an additional medical education and professional development [16].

8.4 LAWS, POLICIES, AND PRINCIPLES FOR AI REGULATION IN HEALTHCARE

Limited laws are formulated for managing and regulating the use of AI specifically in the health sector. Although numerous guidelines and principles are developed for the use of AI in both the public and private sector and research facilities, no consensus can be found in its definition [27]. Other frameworks, laws and norms are applied to AI use, such as data protection laws, bioethical policies and laws, human right obligations, and regulatory standards, which are described below.

8.4.1 DATA PROTECTION LAWS

These laws are "right-based approaches" which provide standards for regulating data processing that protects individual rights and establishing boundaries for data controllers and processors. It also ensures that people don't have the right to take decisions solely based on automated process. GDPR (General Data Protection Regulation) is one such well-known data protection law of the EU. Use of personal data for AI is also regulated by some standard guidelines such as Ibero-American Data Protection Network, consisting of 22 data protection authorities in Spain, Portugal, Mexico, and Central and South American countries [28]. Their guidance includes recommendations for using and processing personal data in AI and specific guidelines for compliance with rights and principles which protect personal data when in use with AI projects [29].

8.4.2 BIOETHICAL POLICIES AND LAWS

These policies and laws help in regulating AI use, and recent years have seen an increase in the revision of these bioethical laws to include the growing use of AI in healthcare, science, and medicine. Fot instance, the recent revision of the French government's national bioethical law [30] makes standards to account for the rapid growth of digital technologies in healthcare systems. It includes the standards to be followed for human supervision that requires evaluation done by patients and clinicians at essential points of R&D of AI and the provisions to include informed free consent forms for using data, thereby creating a national platform to collect and process of health data.

8.4.3 HUMAN RIGHTS OBLIGATIONS AND AI

Generally, most of the international instruments prepare a baseline for the protection and promotion of human dignity worldwide from the human rights listed there which are enforced through national legislation. The Office of High Commissioner for Human Rights has given several insights for the use of AI towards the realization of human rights. In the March 2020 guideline, the Office noted that big data and AI have the potential to improve the human right to health when contemporary technologies are designed in an accountable fashion [31]. It also mentioned that such novel technologies could very well dehumanize care, looking down the independence and autonomy of an aged population, which may pose a serious contrary to human right for health [31]. A number of concerns were raised by the United Nations during the speech to the Human Rights Council on February 2021 related to the human rights associated with the ever-increasing collection and use of COVID-19 pandemic data, and the governments were asked to place human rights at the center of legislation and regulatory frameworks during the use and development of digital technologies [32].

Whenever essential interpretations were made by Human Rights Organizations and adapted, the existing human rights standards and laws for assessment of AI and scrutiny in the advent of challenges and advantages associated with AI are also made. Even with such robust standards of human rights, it is recognized by institutions and organizations that better definitions are required for the relation of human rights and AI. The Council of Europe has prepared new legal guidance and in 2019–2020, it established an Ad-hoc Committee on AI to make broad multi-collaborator consultations for determining practicability and potential problems for a legal framework to design and apply AI as per the Europe's standard of democracy, rule of the land and human rights, and the guideline on AI and data protection in 2019 [33]. Also, the ethical charter of European Commission contains five principles with respect to the use of AI in health in the Efficiency of Justice [34].

8.4.4 REGULATORY STANDARDS FOR USING AI IN HEALTH

Health regulatory authorities are responsible for giving the warrant in safety, efficacy, and proper use of AI in healthcare and therapeutic development. Experts from WHO are on the bench preparing the considerations on how AI should be used in healthcare to be considered by stakeholders, regulators, and developers for exploring new AI technologies. This exploration includes transparency, documentation, data quality, life-cycle approach, risk management, collaboration, engagement, clinical and analytical validation, and data and privacy protection. There exist many ethical issues and challenges which have to be addressed by the regulatory agencies.

8.5 ETHICAL PRINCIPLES FOR AI IN HEALTH

The ethical principles of AI are based on the inherent worth of humanity and dignity so as to guide developers, regulators, and users in overseeing and improving such technologies. These aspects include the following while considering the ethical aspects in the application of AI in healthcare.

8.5.1 FOSTERING ACCOUNTABILITY AND RESPONSIBILITY

During the application of AI, some things may go astray for which there should be accountability. There should be a proper mechanism to address the questioning, and a redressal mechanism should exist for groups and individuals that are negatively affected with algorithmically informed conclusions which should include quick, effective redressal from companies and governments that distribute the AI technologies. Factors such as compensation, restitutions, rehabilitation, sanctions, and wherever possible a promise for non-repetition should be included in the redressal process.

In the development and disbursement of AI technologies, the factor of responsibility can be assured by using "human warranty," which comes as an evaluation by clinicians and patients. Regulatory principles are applied upstream and downstream of the algorithm by having critical points of human supervision which are identified by patients, professionals, and designers. The aim here is to keep the algorithm on a development path of machine learning which is medically effective, ethically responsible, and can also be interrogated with active involvement of public and patient [35]. AI should be used by health professionals responsibly as it is the responsibility of humanity to ensure that AI technologies can perform at their optimum under appropriate conditions.

8.5.2 AUTONOMY PROTECTION

Adopting AI shifts the decision-making power to machines; the principle of autonomy comes so that any extension of machine autonomy doesn't downgrade human autonomy. When the context of healthcare arises, it should imply that humans should be in full control of medical decisions and healthcare systems. The design of AI should be such that they conform to principles and human rights so that they can assist humans—either patients or medical practitioners—in making informed decisions. Duties for protecting confidentiality and privacy and to provide informed valid consent within the existing legal framework of data protection falls under the domain of respect for autonomy. If in any circumstances an individual withholds consent, they should not be offered other incentives by either the government or private partners, nor should any essential service be denied for the same. Data protection laws are for safeguarding individual rights and also act as means to lay down obligations on data processors and controllers along with protection of patient's data privacy and confidentiality, including the establishment of the patient's control over their own data. These protection laws should also pave the way for easy access to patient's own data and to shift and share those data as they wish.

8.5.3 Promoting Sustainable and Responsive AI

AI technologies should only be introduced if they can be fully integrated and sustained in health-care systems. Particularly in health systems that are under-resourced, where new technologies are not used, repaired, or updated, there is a wastage of very scare data which could otherwise have been inventions for proven interventions. Also, the AI systems should be designed in such a way that their ecological footprints are minimized and energy efficiency is increased for the consistent use of AI with the efforts of the society to reduce the human impact on earth's ecosystem, climate, and environment. Sustainability also needs the help of companies and government to convey the predicted disruptions to the workplace, training to healthcare workers to adopt AI, and the possible job losses when automated systems will take over in healthcare routine and administrative works.

The responsiveness of AI technologies can only be achieved when developers, designers, and users continuously in a systematic, transparent way scrutinize for an accurate, appropriate response by the AI and to validate if the response received is at par with the requirement and expectations and in the purpose for which it was used. Responsive duty also calls for the termination of AI technologies when it is ineffective and causes dissatisfaction.

8.5.4 Securing Equity and Inclusiveness

Accidental biases which may arise with AI should be avoided at the earliest and mitigated as soon as possible. Developers are expected to be aware of the possible biases while developing their design, its implementation, and the possible harm these biases may create in future to individuals and society as a whole. For example, a diagnostic system developed to detect cancerous skin lesions trained on a certain skin colour may not generate the same accurate results when tested on a different skin colour. Developers should be careful that the training data used for AI development does not include sampling bias and are diverse, complete, and accurate. Monitoring and evaluation must be done on the effect from the use of AI technologies and must also include the disproportioned effects on specific individual groups when they highlight the existing bias and discriminations; appropriate provisions should be made to readdress such biases.

Inclusiveness in AI used in healthcare is encouraged to include equitable use and access, widest possible access and use despite gender, age, income, and other aspects. Employees should be hired by institutions from different backgrounds, disciplines, and cultures to develop, monitor, and employ AI technologies. Active participation of the potential users of the system, including patients and providers who should be diverse sufficiently so that it can be amplified by using open-source software or by making the source code available publicly.

AI technologies should be shared to the maximum, and they should not only be available to high-income countries but also to low- and middle-income countries. The diversity of languages and forms of communications that exist around the world should be considered by AI developers; this will avoid barriers in use. Governments and industries should ensure to minimize the gap of "digital divide" between and within the countries so that equitable access to AI is made.

8.5.5 Encouraging Human Safety, Human Well-Being, and Public Interest

Human safety ensures that AI use doesn't cause any physical and mental harm; a careful and balanced analysis has to be made against any "duty to warn" when an AI system warns an individual against a certain ailment that he is unable to address due to lack of access to affordable and appropriate healthcare systems, thus preventing an individual from discrimination due to health concerns.

AI technologies should fulfill the regulatory requirements for efficacy, safety, and accuracy prior to deployment, and quality control and improvement should be essentially fulfilled. Thus, a continuous responsibility lies on the shoulders of developers, funders, and users to monitor and measure the performance of the AI and ensure it has no detrimental influence on people.

8.5.6 Ensuring Explainability, Transparency, and Intelligibility

AI technologies should be explainable for those to whom the explanation is directed, and data protection laws have specified certain boundaries for explainability for automatic decision-making. As many AI technologies are complex, this complexity may sometimes confuse the person receiving the explanation and the explainer.

When sufficient information is published before the design and distribution of the AI technology—which also supplies meaningful public debate (and discourse is the base of transparency)—it improves the quality of the system and provides protection to public and patient health. Transparency is required by system evaluators to identify errors, and proper effective oversight can be conducted by the government based on transparency. Accurate information about expectations and limitations of AI technology, methods for data collection, labelling, and processing, protocols of operation and development procedure of the algorithm should be included in the transparency data.

8.6 BUILDING ETHICAL APPROACHES, LIABILITY, AND GOVERNANCE FRAMEWORK

There is an approach to integrate human rights and ethics in the "Design for values," based on the values of freedom, human dignity, solidarity, and equality, decoding them as non-functional requirements [36]. Several questions are raised when AI is used to support clinical decision-making. Liability rules may help clinicians to rely on AI to inform and validate their clinical judgement if an algorithm concludes an unpredicted outcome, but this will discourage the fullest use of AI [37].

Governments should have clear data-protection laws with independent data protection authorities armed with adequate powers and resources to monitor and enforce data protection laws.

8.7 EMERGING PROBLEMS IN AI FOR HEALTHCARE

The introduction of AI in healthcare has brought many challenges to doctors when it comes to making complex clinical decisions [38, 39]. AI has the ability to increase biased findings despite specific clinical results, as for instance, patients with pneumonia alone were classified as high-risk by AI as compared to patients with pneumonia and asthma as low-risk accompanied by comorbid asthma complications [40]. It has been considered that participants cannot assume moral responsibility when they can no longer predict the outcome future behavior of the machine, which increases the contradiction between doctors and patients, which is not achievable in healthcare [41]. Studies have shown that the original expectations in healthcare related data that was expected originally from privacy protection laws such as the HIPAA Act of the US Congress in 1996 [42].

Security remains the most important factor in AI application in the medical industry and needs a continuous review. Most of the AI products currently use electronic devices such as mobile phones, bracelets, and computers. Three critical aspects regarding the security of these hardware may be looked into, such as, firstly, factors such as temperature, cost, and electromagnetic interference will affect them; secondly, the professionalism and complexity of information technology and medical knowledge makes it inaccessible to engineers and doctors for using multi-technology integrated AI. Medical workflow and data leakage may occur if engineers are allowed to access and process medical data, while doctors on the other hand may have less knowledge about the working and principles of AI technologies. Thirdly, the issue of network security exists, which may result in a global downfall if complex nodes of transmissions are bombarded by cyber-attacks [43].

8.8 CHALLENGES FOR THE ETHICAL USE OF AI IN HEALTHCARE

The health services sector is quickly being dominated by AI. In order to manage patients and medical resources, it converts the manual health system into an automatic one, where humans still do routine activities. Digital health services present new technological issues for developers, and tasks are carried out by artificial intelligence. AI has the ability to significantly enhance patient care while lowering healthcare expenses. The need for health services is anticipated to increase as the population grows. The healthcare industry requires creative ideas to figure out how to be more effective and efficient without spending too much money [44].

Technology can provide remedies in this case. Rapid technological advancements, particularly in the disciplines of AI and robotics, can support the expansion of the healthcare sector. Robotics and AI in healthcare are developing swiftly, especially when used for early detection and diagnostic purposes [45]. At the same time, AI is becoming more powerful. It makes it possible for them to execute tasks that humans do, frequently more quickly, easily, and affordably. There are several hazards and difficulties, such as the risk of patient harm from system faults, the risk of patient privacy while obtaining data, and using artificial intelligence to make conclusions and more [46]. The ability of AI preventative care to help people maintain their health is crucial for public health. This cutting-edge technology is utilized to administer medical care more effectively. There are benefits to the procedure; therefore, the future of AI in healthcare is not entirely bright. There are several concerns about how AI will exercise doctor's rights and obligations and protect privacy, and the relevant law is not yet fully prepared for this advancement. The fact that AI is used throughout the global healthcare system shows that the laws in place today support it. It has been demonstrated that medical care can be provided according to the rules established for the development of technology and health technology goods [47]. This study attempted to determine the potential for AI and its risks in the healthcare industry.

8.8.1 METHODS USED IN AI

Early detection and diagnosis both heavily rely on AI. It is utilized in several ways to more accurately, reliably, and promptly diagnose diseases like cancer. It compares data from certain patients, including images of a sizable amount of data from other patients, to compare data from other patients. A diagnosis is proposed by the independent learning system after an association is found. For instance, Google's Deep Mind Health Technology mixes ML with a neuroscience system to model the human brain using AI and offers support for diagnosis and decision-making to medical practitioners. PRISMA flowchart 7 employs the PICO approach of writing a literature review. It serves as the framework for creating literature search plans and for applying quantitative research techniques. The following search method was used: Data were gathered using a combination of word searches on the terms "implement artificial intelligence in improving healthcare in hospitals or primary healthcare," "artificial intelligence in health services" and other related terms from three databases, including Web of Science, Google Scholar, and EBSCOhost. Paper journals can be accessed by visiting the Science Direct website. Only original articles from the papers from 2010 to 2020 in English were chosen for study. In a 10-year span, 259 papers were obtained. To find the right paper, some of the 209 papers that were more specialized were chosen. The papers chosen have the titles that are most appropriate for this literature evaluation. There were 11 papers that were selected. Additionally, the writers examined the chosen studies' abstracts and conclusions before analysing the papers. The best path for the development of the health industry, notably the health service, is through the application of AI. AI applications will, however, be impacted by things like system errors, ethical problems, and legal restrictions. Moderate photos, text from patient reports on medical conditions, diagnosis and care, and cost reimbursement codes may all be included in the context of health, but they are not the only ones. For AI models, inaccurate and under-representative training data sets can lead to bias, false predictions, unfavourable outcomes, and even widespread prejudice, as was previously discussed [48].

8.8.2 APPLICATION OF AI IN HEALTH SERVICES

The use of technology is crucial in this day and age. Technology has a place and works to make human labour easier and more productive. Additionally, technology helps to reduce mistakes brought on by human incompetence in the health field. For instance, if technology is not used in surgical procedures carried out by doctors, the procedure may be risky and unsuccessful. In other terms, AI refers to the emulation of human-made intelligence in robots that have been designed to think similarly to humans. Clinical decision-making, diagnostics, prevention, and therapy can be done with the help of AI [49].

Although the deep learning model has demonstrated highly promising results in the analysis of medical images and clinical risk prediction, this model is also challenging to comprehend and explain. In the field of medicine, where transparency and the capacity to defend clinical judgments are crucial, the challenge creates certain issues as discussed earlier in this chapter. In the case of AI, doctors must consider the causes and effects of medical issues as well as the methods and models employed to aid in the decision-making process. In addition to the issues, the potential for autonomous functions of AI applications and the potential for malicious or unintentional tampering with these applications to produce unsafe results may present a significant barrier for doctors to adopt AI in their medical practices [48]. The health industry is not yet developed to its full potential. The application of AI in the public health sector is a promising one. Due to the potential cost reductions at the core of healthcare delivery, AI is being used more and more, despite its delayed implementation. In many ways, it has been widely accepted that using AI to reinvent the health industry is feasible. The study of in-depth knowledge processing for health management systems, electronic health records, and the active monitoring of clinicians in their medical judgments are all covered under the virtual branch of AI.

Recent research has focused in particular on the application of AI systems to help clinicians with patient diagnosis. In the future, when updated with more advanced technology and given access to more full data, this AI will be able to detect a variety of additional ailments. Technology in healthcare has obvious advantages, such as improving patient treatment options and results, as well as potential secondary advantages including fewer referrals, lower costs, and more efficient use of time. Additionally, it can encourage rural recruiting and retention efforts and lessen professional isolation [50]. This can thus help create a more equitable global healthcare system in low-resource situations in both high- and low-income nations.

8.9 CONCLUSIONS

Doctors will never be replaced by AI, and its application is not a threat to doctors; rather, it will help in the reshaping the role of a doctor in healthcare. Besides concentrating on the sensitivity and accuracy of the reports, AI research should also focus on the pathogenesis and etiology [51]. The time of AI has arrived, and it has been intervening in all walks of life including healthcare, and research on the relevant laws, ethics, and supervision is currently needed.

CONFLICT OF INTEREST

The authors declare no conflict of interest, financial or otherwise.

ACKNOWLEDGEMENTS

We thank the Vice Chancellor, Bharathiar University, Coimbatore-641046, Tamil Nadu for providing the necessary facilities. We also thank UGC-New Delhi for Dr. D S Kothari Fellowship ((No. F-2/ 2006 (BSR) / BL / 20-21 / 0396)), TANSCHE (RGP/ 2019-20/BU /HECP-0005) No.C3/ CRTD/ 995/ 2021).

REFERENCES

1. Skahill E, West D. Why hospitals and healthcare organizations need to take cybersecurity more seriously. 2021. https://www.brookings.edu/blog/techtank/2021/08/09/why-hospitals-and-healthcare-organizations-need-to-take-cybersecuritymore-seriously/.
2. HIMSS. 2020   HIMSS cybersecurity survey. 2020. https://www.himss.org/sites/hde/files/ media/file /2020/11/16/2020_himss_cybersecurity_survey_final.pdf.

3. Gordon WJ, Fairhall A, Landman A. Threats to information security-public health implications. *N Engl J Med* 2017; 377(8), 707–709.

4. Guru V, Tu JV, Etchells E, et al. Relationship between preventability of death after coronary artery bypass graft surgery and all-cause risk-adjusted mortality rates. *Circulation* 2008; 117(23), 2969–2976.

5. Madani A, Vassiliou MC, Watanabe Y, et al. What are the principles that guide behaviors in the operating room? *Ann Surg* 2017; 265(2), 255–267.

6. Suliburk JW, Buck QM, Pirko CJ, Massarweh NN, Barshes NR,Singh H, Rosengart TK. Analysis of human performance deficiencies associated with surgical adverse events. *JAMA Netw Open* 2019; 2(7), e198067.

7. After cyberattack on AIIMS, ICMR website faces 6,000 hacking (http://timesofindia.indiatimes.com/articleshow/96031036.cms).

8. Tamil Nadu: Data of 1.5 lakh patients put up for sale online (http://timesofindia.indiatimes.com/article-show/95970041.cms).

9. Flin R, Youngson G, Yule S. How do surgeons make intraoperative decisions? *Qual Saf Health Care* 2007; 16(3), 235–239.

10. Udelsman R, Åkerström G, Biagini C, Duh QY, Miccoli P, Niederle B, Tonelli F. The surgical management of asymptomatic primary hyperparathyroidism: Proceedings of the fourth international workshop. *J Clin Endocrinol Metab* 2014; 99(10), 3595–3606.

11. Harangi B, Hajdu A, Lampe R, Torok P. Recognizing ureter anduterine artery in endoscopic images using a convolutional neural network. In 2017 IEEE 30th International Symposium on Computer-Based Medical Systems (CBMS). 2017, 726–727.

12. Quellec G, Lamard M, Cazuguel G, Droueche Z, Roux C, Cochenern B. Real-time retrieval of similar videos with application tocomputer-aided retinal surgery. *Conf Proc IEEE Eng Med Biolsoc* 2011, 4465–4468.

13. Grün D, Rudolph F, Gumpfer N, et al. Identifying heart failure in ECG data with artificial intelligence-a meta-analysis. *Front Digit Health* 2021; 2, 584555.

14. Saak SK, Hildebrandt A, Kollmeier B, Buhl M. Predicting Common Audiological Functional Parameters (CAFPAs) as interpretable intermediate representation in a clinical decision-support system for audiology. *Front Digit Health* 2020; 2, 596433.

15. Froutan R, Eghbali M, Hoseini SH, Mazloom SR, Yekaninejad MS, Boostani R. The effect of music therapy on physiological parameters of patients with traumatic brain practice. *Randomized Controlled Trial* 2020; 40, 101216.

16. Giordano C, Brennan M, Mohamed B, Rashidi P, Modave F, Tighe P. Accessing artificial intelligence for clinical decision-making. *Front Digit Health* 2021; 3, 65.

17. Debnath S, Barnaby DP, Coppa K, et al. Machine learning to assist clinical decision-making during the COVID-19 pandemic. *Bioelectron Med* 2020; 6, 14.

18. Corey KM, Kashyap S, Lorenzi E, et al. Development and validation of machine learning models to identify high-risk surgical patients using automatically curated electronic health record data (Pythia): A retrospective, single-site study. *PLOS Med* 2018; 15(11), e1002701.

19. Tsilimigras DI, Sahara K, Hyer JM, et al. Trends and outcomes of simultaneous versus staged resection of synchronous colorectal cancer and colorectal liver metastases. *Surgery* 2021; 170(1), 160–166.

20. Costa C, Tsatsakis A, Mamoulakis C, et al. Current evidence on the effect of dietary polyphenols intake on chronic diseases. *Food Chem Toxicol* 2017; 110, 286–299.

21. Ko WY, Siontis KC, Attia ZI, et al. Detection of hypertrophic cardiomyopathy using a convolutional neural network-enabled electrocardiogram. *J Am Coll Cardiol* 2020; 75(7), 722–733.

22. Chang K, Bai HX, Zhou H, et al. Residual convolutional neural network for the determination of IDH status in low-and high-grade Gliomas from MR imagingneural network for determination of IDH Status in Gliomas. *Clin Cancer Res* 2018; 24(5), 1073–1081.

23. François-Lavet V, Henderson P, Islam R, Bellemare MG, Pineau J. An introduction to deep reinforcement learning. arXiv 2018. doi: 10.1561/9781680835397.

24. Bellman R. A Markovian decision process. *J Math Mech* 1957; 6(4), 679–684.

25. Prasad N, Cheng L-F, Chivers C, Draugelis M, Engelhardt BE. A reinforcement learning approach to weaning of mechanical ventilation in intensive care units. arXiv 2017. arXiv:170406300.

26. Nemati S, Ghassemi MM, Clifford GD. Optimal medication dosing from suboptimal clinical examples: A deep reinforcement learning approach. *Annu Int Conf IEEE Eng Med Biol Soc* 2016, 2978–2981.

27. Jobin A, Ienca M, Vayena E. The global landscape of AI ethics guidelines. *Nat Mach Intell* 2019; 1(9), 389–399.

28. General recommendations for the processing of personal data in artificial intelligence. Brussels: Red Ibero America de Proteccion de Datos, European Union; 2019 (https://www.redipd.org/sites/default/files/2020 -02/guide-general-recommendations-processing-personal-data-ai.pdf, accessed 27 December 2020).

29. Specific guidelines for compliance with the principles and rights that govern the protection of personal data in artificial intelligence projects. Brussels: Red IberoAmerica de Proteccion de Datos, European Union; 2019 (https://www.redipd.org/sites/default/files/2020-02/guide-specificguidelines-ai-projects .pdf, accessed 27 December 2020).

30. French bioethics law: An original participatory approach for the national bioethics consultation. Paris: Institut Pasteur, 2 September 2019 (https://www.pasteur.fr/en/home/research-journal/reports/french -bioethics-law-original participatory-approach-national-bioethics-consultation, accessed 16 November 2022).

31. Question of the realization of economic, social and cultural rights in all countries: the role of new technologies for the realization of economic, social and cultural rights: Report of the secretary general. 2020. Geneva: Office of the High Commissioner for Human Rights (https://www.ohchr.org/EN/HRBodies/ HRC/RegularSessions/Session43/Documents/A_HRC_43_29.pdf, accessed 9 January 2021).

32. Secretary-General Guterres calls for a global reset to recover better, guided by human rights. Geneva: United Nations Human Rights Council; 2021 (https://www.ohchr.org/EN/HRBodies/HRC/Pages/ NewsDetail.aspx?NewsID=26769&LangID=E, accessed 30 March 2021).

33. Guidelines on artificial intelligence and data protection. Strasbourg: Council of Europe; 2019 (https://rm .coe.int/guidelines-on-artificial-intelligence-and-data-protection/168091f9d8, accessed 13 April 2020).

34. European ethical charter on the use of artificial intelligence in judicial systems and their environment. Strasbourg: Council of Europe; 2018 (https://rm.coe.int/ethical-charter-en-forpublication-4-december -2018/16808f699c, accessed 26 April 2020).

35. Public debate. Strasbourg: Council of Europe; 2021 (https://www.coe.int/en/web/bioethics/publicdebate, accessed 17 October 2022).

36. van den Hoven J, Vermaas PE, van de Poel I, editors. Handbook of ethics, values, and technological design: Sources, theories, values, and application domains. Cham: Springer; 2015 (https://www.springer .com/gp/book/9789400769694#aboutAuthors, accessed April 2022).

37. Price WN, Gerke S, Cohen IG. Potential liability for physicians using artificial intelligence. *JAMA* 2019; 322(18), 1765–1766.

38. Tobia K, Nielsen A, Stremitzer A. When does physician use of AI increase liability? *J Nucl Med* 2021; 62(1), 17–21.

39. Gelhaus P. Robot decisions: On the importance of virtuous judgment in clinical decision making. *J Eval Clin Pract* 2011; 17(5), 883–887.

40. Nabi J. How bioethics can shape artificial intelligence and machine learning. *Hastings Cent Rep* 2018; 48(5), 10–13.

41. Basu T, Engel-Wolf S, Menzer O. The ethics of machine learning in medicalsciences: Where do we stand today? *Indian J Dermatol* 2020; 65(5), 358–364.

42. Golbus JR, Price WN 2nd, Nallamothu BK. Privacy gaps for digital cardiology data: Big problems with big data. *Circulation* 2020; 141(8), 613–615.

43. Sohn JH, Chillakuru YR, Lee S, et al. An open-source, vender agnostic hardware and software pipeline for integration of artificial intelligence in radiology workflow. *J Digit Imaging* 2020; 33(4), 1041–1046.

44. Pee LG, Pan SL, Cui L. Artificial intelligence in healthcare robots: A social informatics study of knowledge embodiment. *J Assoc Inf Sci Technol* 2019; 70(4), 351–369.

45. Coeckelbergh M. Health care, capabilities, and AI assistive technologies. *Eth Theor Moral Pract* 2010; 13(2), 181–190.

46. Hamid S. The opportunities and risks of artificial intelligence in medicine and healthcare. SAGE Communications Summer; 2016.

47. Indonesia R. Law of the Republic of Indonesia number 36 of 2009 concerning health. Jakarta Republic of Indonesia; 2009.

48. Reddy S, Allan S, Coghlan S, Cooper P. A governance model for the application of AI in health care. *J Am Med Inform Assoc* 2020; 27(3), 491–497.

49. Triantafyllidis AK, Tsanas A. Applications of machine learning in real-life digital health interventions: Review of the literature. *J Med Internet Res* 2019; 21(4), 12286.

50. Romiti S, Vinciguerra M, Saade W, Anso Cortajarena I, Greco E. Artificial intelligence (AI) and cardiovascular diseases: An unexpected alliance. *Cardiol Res Pract* 2020; 27, 4972346.

51. Ghosh A, Kandasamy D. Interpretable artificial intelligence: Why and when. *AJR Am J Roentgenol* 2020; 214(5), 1137–1138.

9 Artificial Intelligence and Robots in Individuals' Lives

How to Align Technological Possibilities and Ethical Issues

Partha Pratim Kalita, Manash Pratim Sarma,
and Ankur Pan Saikia

9.1 INTRODUCTION

Artificial intelligence is a real science that uses a lot of strategies, methods, and procedures to solve real-world problems. Because of its relentless ascent into the future, there are additionally a few conversations about its morals and wellbeing. The ability to create a setting in which both humans and robots can find a context where they share values by creating an AI-friendly environment for humans. Therefore, the aim of this chapter is to investigate the ethical dilemmas posed by ethical algorithms, pre-existing or acquired values, and AI-related ethical issues. The essential focal point of the chapter will likewise be on computer-based intelligence security. In general, the chapter will make readers aware of AI safety as a related area of research and provide a brief analysis of the concerns as well as potential solutions to the ethical issues that are presented.

9.1.1 ETHICAL ISSUES

9.1.1.1 Privacy

The Cambridge Analytica scandal, in which data from 87 million Facebook profiles was stolen and used to influence the US election and the Brexit campaign, has brought the importance of privacy into the spotlight. A fundamental human right, privacy ought to be protected from abuse.

DOI: 10.1201/9781003353751-9

9.1.1.2 Misinformation and Fake News

The spread of false information and our inherent inability to perceive reality based on evidence—a phenomenon known as confirmation bias—threaten an informed democracy. Russian hackers' efforts to influence US elections, the Brexit campaign, and the crisis in Catalonia are just a few examples of how social media can spread misinformation and fake news rapidly. Recent advances in computer vision have made it possible to completely fabricate a President Obama video. It is unknown how institutions will respond to this threat.

9.1.1.3 Cybersecurity

Cyber security is of particular concern to governments and businesses, particularly banks. In 2015, the crypto currency exchange CoinCheck reported a $500 million theft, and banks in Russia, Europe, and China reported a $1 billion robbery. AI can reduce these flaws, but hackers can also use AI to come up with novel and sophisticated ways to harm institutions.

Like any other technological system, AI systems can make mistakes. A common misconception is that robots are error-free and infinitely precise.

9.1.1.4 The Algorithm

We must put in a lot of effort to avoid bias and discrimination when developing AI algorithms. For example, face recognition with Haar Fountains has a lower location rate for individuals with brown complexion than for individuals with fair complexion. This is because the algorithm was developed to locate a double T pattern in a grayscale image of the person's face that corresponds to the nose, mouth, and eyebrows of that person. This pattern is harder to spot in people with dark skin.

> However, many people may be insulted. How can an algorithm be racist if Haar Cascades are not racist? When programming these algorithms, we need to be aware of their limitations, be transparent with users by explaining how they work, or use a more effective approach with people with dark skin.

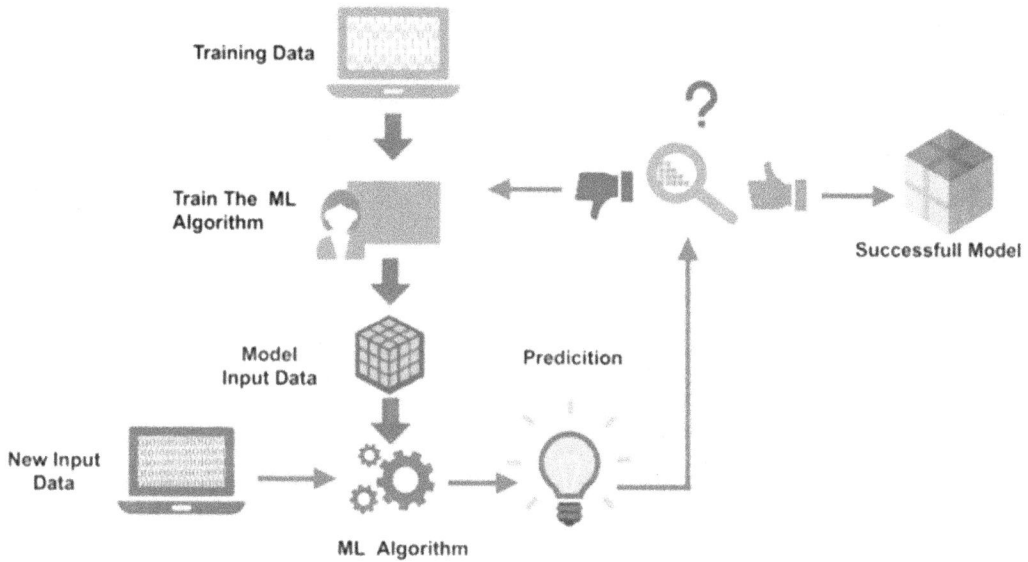

FIGURE 9.1   The Flow Chart of ML Algorithm. (Image Courtesy: http://brainstormingbox.org)

9.1.1.5   Regulation

Microsoft's Chief Legal Officer, Brad Smith, suggested that governments "strike a balance between innovation support and the need to ensure consumer safety by holding the makers of AI systems responsible for harm caused by unreasonable practices" even though AI-based products and services are subject to regulation because there were no laws written with AI in mind. Professionals, researchers, and policymakers ought to collaborate to ensure that AI benefits humanity.

9.1.1.6   Job Loss

The scientific revolution of the 18th century and the industrial revolution of the 19th century brought about complete social change. Monetary development was practically non-existent for millennia before it. In the 19th and 20th centuries, social development was remarkable. To oppose the textile industry's automation in the 19th century, a group in the United Kingdom known as the Luddites destroyed machinery.

Since then, the concern that automation and technological advancements will result in widespread unemployment has persisted. It is undeniable that there has been challenging work relocation, despite the fact that that forecast has proven to be incorrect. According to PwC, around 30% of jobs will be automated by 2030. Governments and businesses should give workers the tools they need to adapt to these changes in these circumstances by supporting education and moving jobs.

FIGURE 9.2 A visual depiction of how AI could potentially replace human beings in certain jobs. (Image Courtesy: https://www.forbes.com/)

9.1.2 Can't See How AI Will Affect Ethical Issues? Any Solution!

A project between Bank of America and Harvard University aims to make the intricate technologies that underpin artificial intelligence (AI) and machine learning (ML) more universally beneficial and accessible to the global workforce. The Council on the Responsible Use of Artificial Intelligence aims to address ethical and policy issues to help businesses and their leaders adapt to the interface between humans and machines, which is rapidly changing.

The council aims to capitalize on the speed and convenience that AI and machine learning can offer a variety of industries, including finance, healthcare, retail, manufacturing, and others, by developing best practices and maintaining a transparent dialogue

Cathy Bessant's (Chief Operations and Technology Officer of Bank of America, one of the founding members of the council) concerns are supported by the fact that a lot of people see AI as a threat to their livelihoods, which inspires scepticism and, in extreme cases, fear. The council was set up to look into the most recent developments in this new technology, educate the public about its legal and moral implications, and come up with better, more useful, and responsible ways to use a gift that is still misunderstood in a lot of people's minds in order to help change the current ideology. Despite the fact that AI has enormous potential to simplify our lives, this is the case.

Bessant made the observation that "Machines cannot by definition be better than humans."

They were made by people; they were designed by humans. Therefore, a human being is the driving force behind every aspect of a machine's operation, regardless of whether it can function independently after being programmed. The quality of the cycles and the people who train them is determined.

9.1.3 What "Rights" Does Artificial Intelligence Possess, and Can a Conscience Be Embedded?

"Well, if droids could think, there'd be none of us here, would there?"

- Obi-W-Kenobi

Despite the fact that fully autonomous robots with human-like capabilities are still largely a fantasy, lawmakers, legal experts, and manufacturers are already debating their legal status, or "legal personality," as well as the ethical issues associated with their production and use: finally, whether these machines or people should be held accountable for their actions.

FIGURE 9.3   By definition, machines cannot be superior to humans. (Image Courtesy: https://pedanco.com)

Debate has erupted regarding the capability of self-learning machines to independently determine moral equivalence in relation to ethical decisions that have traditionally been the responsibility of humans. Is it justifiable, for example, for a machine to kill an enemy combatant it has identified without relying on human decision-making? Or is the robot morally equivalent to a weapon without brains?

Is there a fundamental difference in ethics between a "sexbot" and the typical brainless sex toy?

9.2   TRANSFORMATION OF AN INDIVIDUAL'S LIFE

Numerous technological shifts are taking place in our society. In just a few decades, it will be very different from what it is now. The quick development of the computerized reasoning and advanced mechanics industry is one critical variable that affects and changes various parts of daily existence.

Scientists, practitioners in the field, and everyday people discuss the potential outcomes of active AI and robotics development from a variety of perspectives. In addition, while some people are convinced that smart technologies have unending power and enormous benefits, others are terrified of the "rise of the machines" and human extinction.

A few examples of the kinds of jobs that can be effectively completed by robots include assembly line work, software testing, financial reports, and other documents compiled based on data. In addition, they effectively take the place of humans when it comes to hard manual labour or dangerous work, such as mining or the substance industry.

Cognitive tasks can be performed by AI systems in place of employees like doctors, accountants, and financial experts. Consequently, artificial intelligence can greatly benefit medical and healthcare processes like drug formulation, clinical trials, and diagnostics. Despite this, the majority of medical professionals will remain employed due to ethical concerns and the machines' lack of communication skills.

Robots will have the greatest impact on people management, social interactions, and creativity in education and natural science. Most of the time, these jobs are hard to automate, and even if they were, doing so would be expensive, which doesn't appeal to businesses. Specialist occupations like electricians and plumbers, for instance, are subject to uncertainty.

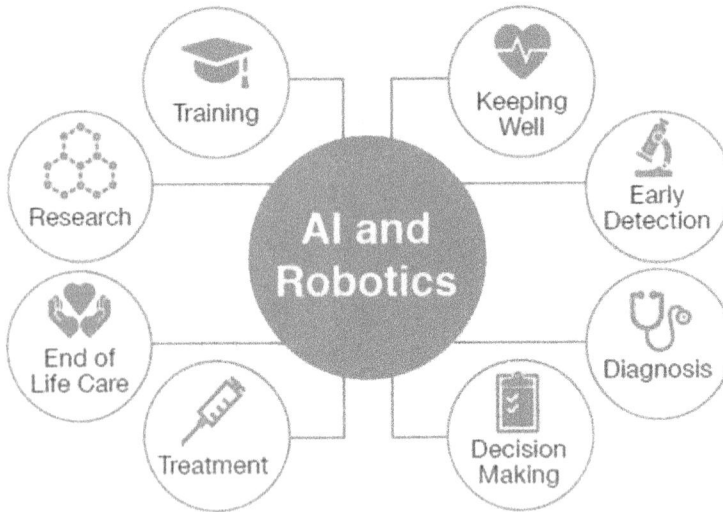

FIGURE 9.4 Various approaches or subfields within the field of AI/ML.

9.2.1 THE INDIAN SCENARIO

Artificial intelligence refers to the future technology that is capable of imitating natural features in a manner that is comparable to the human mind. It would be extremely beneficial if utilized in the desired manner. The only question is whether humans will make ethical use of this technology. As a result, this kind of technology should only be implemented with extreme caution because it will be hard to undo once it is implemented. If we use AI without adhering to ethical standards, we must consider the following outcomes:

Inequality: We already have a society with a lot of inequality, and this technology would only make it worse. The poor would suffer, while the wealthy would benefit more because of a lack of resources.

Communication barriers: Our generation has already become so diverse due to advancements in science and technology that it is challenging to maintain a family. As artificial intelligence was implemented, this gap would get even wider.

Cultural unrest: The Indian culture is well-known around the world. Machines won't be able to comprehend our culture's tolerance of one another, so our values and culture will suffer.

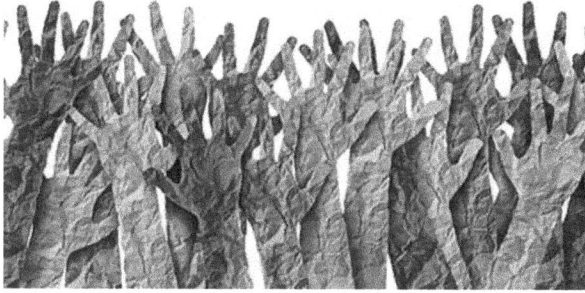

Artificial intelligence will cut down on the need for human labour, bring down operating costs, and boost yield. It is being used to drive automobiles, examine banking applications, and extract hazardous chemicals, among other tasks that were once considered blue collar jobs. The accompanying moral issues will emerge because of the far and wide reception of computer-based intelligence in all fields:

Depression: If employment opportunities cease to exist, the struggling lower middle class may find resilience in the face of poverty.

Social crisis: There is a social crisis like a variety of forms of social and domestic violence.

Emotionless: Because AI is built on algorithms, it cannot tell the difference between good and bad when confronted or pursued by humans.

Governments in developing nations like India must take the necessary steps to strike a balance between the demand for AI and human employment as a result of widespread poverty and unemployment.

9.2.2 THE INDUSTRY 4.0

With the advent of the fourth industrial revolution, AI will emerge as the new domain of the future in the upcoming years. However, there are a few ethical questions that might have something to do with the problems caused by this new workforce.

The following are examples of these worries:

- Loss of employment as a result of an increase in the number of robots—industries will undoubtedly experience a labour shortage;
- Social mayhem—with the deficiency of occupations, social inconsistencies among the unfortunate workforce will be an issue.
- Developed nations will dominate all global industries and produce results that are quicker and better.
- Humans are able to make quick decisions, but robots aren't. This could be a problem in difficult situations.
- Humanoid war will undoubtedly have a negative impact on humanity if it gets out of hand.

9.3 ETHICAL ISSUES IN ARTIFICIAL INTELLIGENCE

The first use of artificial intelligence in healthcare was first reported in 1970s where researchers developed an AI program called MYCIN used for identifying blood clotting disease as well as

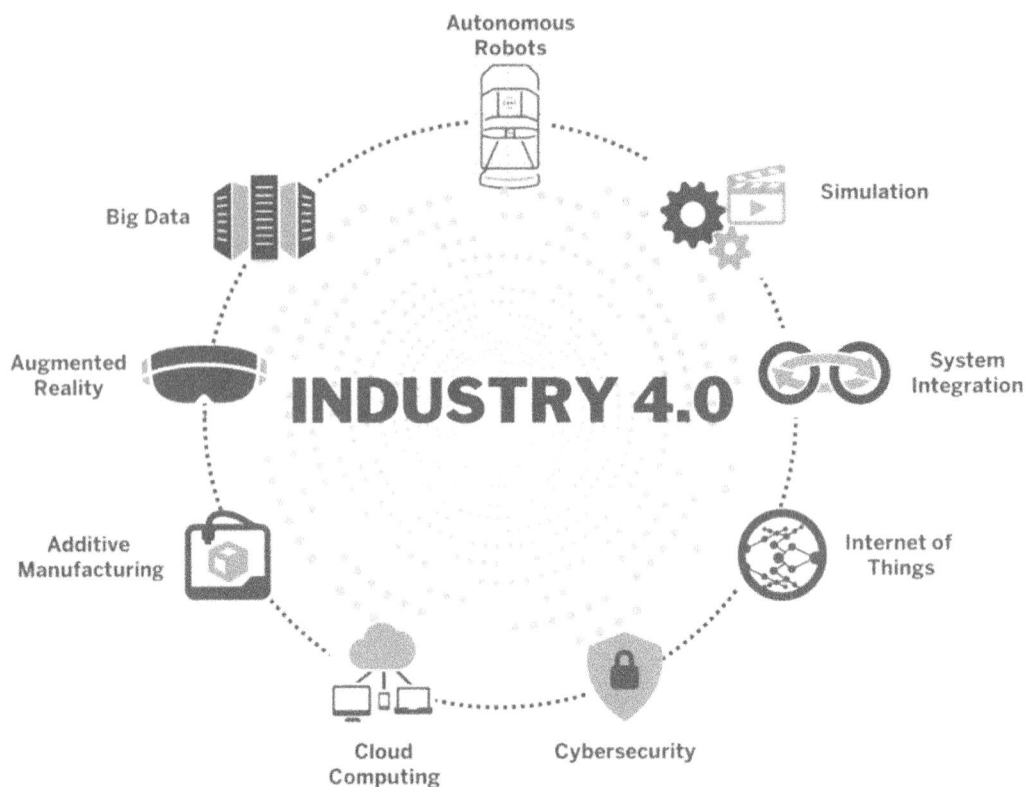

FIGURE 9.5   Life Cycle of Industry 4.0. (Image Courtesy: https://www.calsoft.com/)

bacterial infection which was actually a computer-based consultation system to help the physicians. The development of innovative AI systems during the 1980s and 1990s contributed to the important medical advancements like collecting and processing data in a faster way, helping in different surgical procedures as well as the implementation of electronic health records of patients. With the advancement in the technologies in recent years, AI is also evolved in the area of health sciences, which has importance in radiology, psychiatry, primary care, disease diagnosis, telemedicine, etc. Neural networks are also used to diagnose stroke.

Even though AI technologies are gaining a significant amount of attention in medical research, practical implementation still faces several challenges. The restrictions are the first hindrance. There are currently no guidelines for assessing the effectiveness and safety of AI systems. The US FDA made its first attempt to offer recommendations for evaluating AI systems in order to get over the challenge. The first suggestion categorises AI systems as "basic wellness products," that are minimally regulated if they are intended exclusively for general wellness and pose no danger to users. The second piece of instruction justifies relying on real data to evaluate how well AI systems work. Last but not least, the instruction explicitly states the principles for the adaptive design in clinical trials, which will be extensively employed to evaluate the operational capacity of AI systems. Soon after these recommendations were made public, Arterys' medical imaging platform, the first FDA-approved deep learning clinical platform that can aid cardiologists in the diagnosis of cardiac diseases, became available. Data exchange becomes the next problem in AI systems. AI systems must be trained using data from clinical research in order to perform well. However, after an AI system has been successfully trained using historical data and deployed, maintaining the source data becomes a critical challenge for the system's further development and improvement.

The current healthcare industry does not offer incentives for system data sharing. However, a healthcare revolution is underway in the USA to boost data exchange. Concerns about privacy will continue to be a big deal. The industry's stakeholders must be vigilant and keep a close eye out for different challenges as more information becomes public, even though new computer technology can easily delete names and other personal information from records being transferred into big databases. The healthcare industry must also learn something from previous data-driven revolutions. Companies have indeed too frequently abused data availability by self-enhancement that only benefits them, and this may also happen in the healthcare industry. For instance, MRI equipment owners may decide to use big data exclusively to target underserved people and diseased regions in order to spread out fixed expenditures among a greater number of patients.

> It's necessary to programme AI with a defined good conduct, such as "a certain activity (x) could have effects that threaten humans; therefore, (x) is wrong," This is accomplished by extrapolating acts' long-term implications. Building machine learning-based AI that may be used in situations where actions have multiple effects would be safe in this sense. Since creating a supportive environment for citizens to prosper is the major goal of the justice system, a predictability element in AI is crucial for legal challenges. Another crucial idea is that AI requires strong safeguards against manipulation and unlawful reprogramming.

It is universally believed that present AI systems lack morality. Computer programmes may be adjusted in any way we desire, at least as far as the algorithms themselves are concerned. The moral obligations we have in our interactions with modern AI systems are all based on our commitments to other beings, such our fellow citizens, and not on any obligations to the systems themselves.

AI systems are prone to sudden, severe failures when the setting or context changes. In a split second, AI can change from being incredibly smart to being completely dumb. Even if AI bias is addressed, all AI systems will have their limitations. The system's capabilities must be stated clearly to the human decision-maker, and the system's requirements must be taken into consideration when designing it. Medical professionals may become complacent and neglect to maintain their skills or enjoy their work when a diagnostic and treatment system is largely accurate. It will happen again in other fields, like criminal justice, when judges have changed their judgements based on risk estimates that were later found to be unreliable.

> The adoption of new technology creates issues about the potential for it to evolve as a brand-new cause of inaccuracy and data breach. Mistakes can have significant effects for the patient who is the victim of the error in the high-risk healthcare sector. This is important to keep in mind since patients interact with physicians at times while they are most vulnerable.

There are different limitations in the use of AI in healthcare, but in near future the use of AI will increase for better decision making by clinicians. Development of appropriate algorithms based on un-biased, real-time data may be helpful in making any decision by the AI system in a very short period of time.

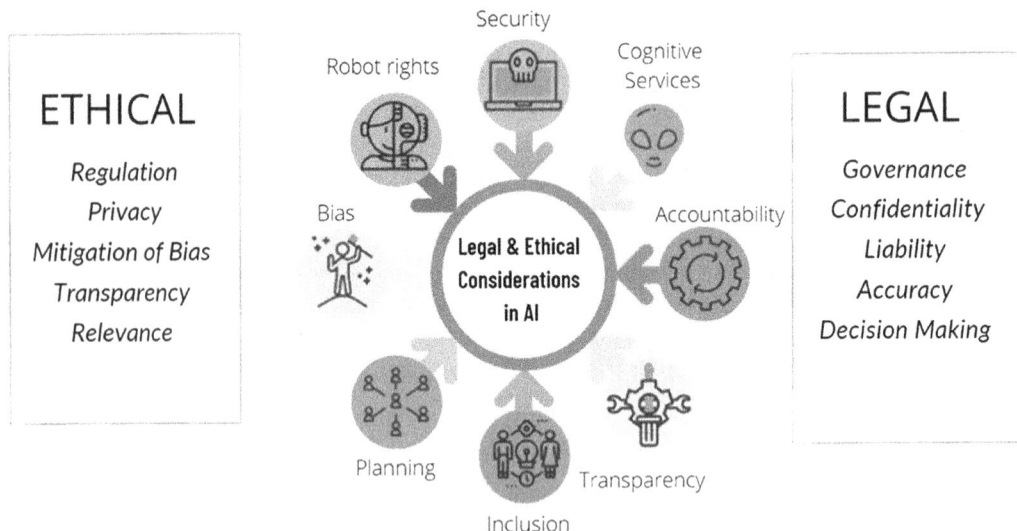

FIGURE 9.6 Fairness, transparency, and interpretability of AI systems. (Image Courtesy: https://www.frontiersin.org/)

9.4 TRUST AND BIAS IN ROBOTS

Imagine that you are approaching an intersection as you walk on a sidewalk. You are allowed to cross the street if a sign flashes. To cross safely, you must rely on drivers to follow the rules of the road. But if a self-driving vehicle comes into contact with pedestrians or other people, what should happen? As pedestrians, should we have faith that this car, which does not have a driver, will come to a stop when we enter a crosswalk? Are we prepared to put our lives at risk for it?

Self-driving cars and emergency evacuation robots, for example, are examples of robots that are influencing human decision-making processes on the road as well as in buildings. However, it is difficult to determine whether or not to trust the technology when it has the potential to either save or kill someone. For instance, do the sensors of a self-driving car have sufficient representative data to use as training material to determine whether an adult or a child walks between two cars into the street?

There are numerous varieties of robots utilized in healthcare. Patient care is also beginning to incorporate telepresence robots. Socially assistive robots are becoming more and more popular as a treatment option to help people with disabilities or impairments get better after rehabilitation. Humanoid robots that can interact with others have been used, for instance, to help children with cerebral palsy participate in interactive play-based physical activities. The success of these kinds of interventions depends on the creation of robot behaviours that increase a patient's engagement and, as a result, their compliance with therapeutic goals.

The healthcare sector has the potential to highlight the relationship between trust and bias, despite the fact that robotic technology offers patients significant advantages. Some patients may be reluctant to question a doctor's authority when they meet them; similar to this, patients may be reluctant to question their assistive robot's direction. These populations may be particularly susceptible to aspects of over-trust in social bonding situations with robots because they may expect robots to produce viable solutions that enhance their quality of life.

IMPACT OF AI IN HEALTHCARE

Areas of Impact for Artificial Intelligence in Healthcare

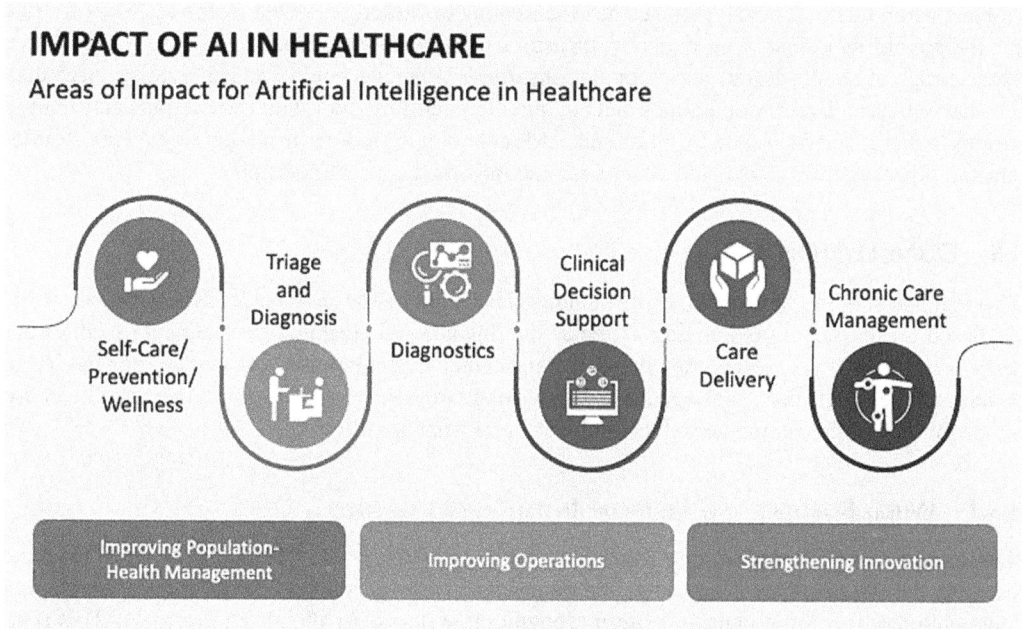

FIGURE 9.7 Trust factors in the adoption of robots, and designing robots.

When participants interact with robot therapy partners rather than a human therapist, our research demonstrates that trust and compliance rise. We demonstrate that participants are not only compliant with the exercise guidance provided by their robot therapy partners in these therapy exercise scenarios, but also that there is a trend toward increased self-reported trust. Adults and healthcare professionals frequently overly trust robotic systems. When deciding whether or not to continue a rehabilitation exercise, for instance, even experts may rely on a robot because humans typically rely on computers.

When developing the algorithms that are embedded in robots, the data that has been learned from human experts, such as a group of physical therapists, is frequently utilized. A robot's decision-making can be supported by data that can be used to classify a child's emotional state or make suggestions based on a person's gender. However, the data set may be misinterpreted or not accurately reflect reality. For instance, if a rehabilitative robot is interacting with a female child, its training data set might reveal that female patients do not enjoy video games, indicating that this type of therapy would not be utilized. The robot's decision is especially problematic because it could lead to an unintended and potentially harmful outcome if it followed instructions based on historical biases.

Because a patient resembles other patients in some way, a robot may perpetuate a false perception of that patient. Because some studies have shown that girls respond better to emotional stimuli from robots than boys do, the training data might conclude that robots shouldn't be emotionally engaging with boys. The robot could be programmed to evaluate which emotional behaviours or video games to deliver rather than making a blanket statement that video game therapy or emotional stimuli should not be tried with a particular gender.

The quality of the data that robots rely on could be compromised by a number of things. Assume that the training data set consists of radiologists' decisions regarding how to read an MRI; because radiologists are human beings who, like the rest of us, have biases, the relevant diagnoses may be influenced by bias. Confirmation bias, for instance, can make it difficult to determine whether a patient has a brain tumour if a condition is assumed to be present prior to examining the patient's MRI.

The overall quality of each patient's healthcare may be directly affected by the tendency to trust and the possibility of bias. Consequently, measures must be taken to resolve the issue; to guarantee the accuracy of results across these populations, these strategies could include creating large data sets that represent diverse populations and having AI systems monitor their own outputs and analyse various "what-if scenarios." In order to make AI technologies more transparent to patients, regulators and other organizations could also revisit the informed consent principles.

9.5 CONCLUSION

A wide range of brand-new applications in industrial robotics and robotic process automation (RPA) are based on the rapidly expanding artificial intelligence (AI) technology sector. AI's ethics are being scrutinized more closely than those of transformative technologies before it, which has led to policies and regulations restricting its use. What safeguards are in place to guarantee the ethical use of one of today's most empowering technologies, and what are the risks?

9.5.1 WHERE ROBOTICS AND ARTIFICIAL INTELLIGENCE (AI) MEET

What exactly is AI, and how does it relate to technologies for robotics? The broad fields of machine learning, natural language processing, and data science now encompass the concept of employing computational, reasoning-capable pattern recognition systems. Artificial intelligence (AI) is typically used in industrial robotics to safely automate operations (like a robotic arm on an automotive assembly line) by sensing and interpreting the environment in which a physical machine operates. Industrial robots rarely pose a threat because they are typically restricted to performing routine tasks that do not require ongoing training.

However, AI encompasses significantly more than industrial machine sensors and actuators; it is also the driving force behind the creation of cutting-edge software for modelling, decision support, predictive analytics, and a wide range of other intelligent applications that are able to generate autonomous output that raises questions. As a result, AI applications that implement ongoing learning and ingest massive amounts of data in real time and require real-time interpretation and analysis are more vulnerable to risk than machine robotics.

9.5.2 WHICH RISKS ARE MORALLY QUESTIONABLE AND WHICH ARE NOT?

Because the technology's scope is still being defined and the environment in which it operates can be hazy, it can be difficult to ethically realize the benefits of AI. Regardless, ongoing debate surrounds AI's potential drawbacks, with a focus on the following key issues:

Security and privacy: AI is based on data because all data collection and storage are now digital and networked. However, individuals and businesses may be harmed by cybersecurity vulnerabilities.

Transparency and opaqueness: What use and analyses are made of the data? It's possible that the analytical decision output is not accurately represented by the patterns that AI systems recognize. How is the quality of the selected data determined? In order to guarantee that AI-derived output complies with ethical standards and is impartial and free of bias, the system needs to incorporate transparency, community engagement, and "algorithmic accountability."

Biases: There are a variety of biases that can affect algorithms, such as statistical bias, which is the use of flawed data or datasets that have nothing to do with the problem at hand; bias that is unconsciously attributed to the subject of the analysis, also known as unconscious bias; as well as confirmation bias, which is the tendency to interpret the data in a manner that supports one's preconceived notions.

> **Ethical considerations**: The development of laws and regulations is catching up to this enormous growth and opportunity, just as it did with other disruptive technologies that came before AI. Despite the fact that AI systems are still in their infancy, significant technical efforts are being made to identify and eliminate bias. Additionally, technological solutions are limited by the difficulty of mathematically comprehending fairness.

There have been some notable beginnings, despite the fact that very little actual policy has been developed. The Centre for Data Innovation proposed in a 2019 EU policy document that "trustworthy AI" should be legal, ethical, and technically robust, and it outlined the requirements necessary to achieve these goals: Additionally, the EU framework is being codified at the moment. "A first-of-its-kind policy that outlines how companies and governments can use a technology seen as one of the most significant, but ethically fraught, scientific breakthroughs in recent memory," the *New York Times* wrote in April 2021.

9.5.3 A Framework for Evaluating AI Applications

While AI can be beneficial to businesses that are able to successfully harness its power, it can also harm a business's reputation and future performance if it is not implemented ethically. However, it is challenging to develop standards or legislation, in contrast to Bluetooth and other recently introduced technologies. This is because AI includes a lot of ambiguous concepts, like robots on the battlefield and automated legal assistants used to review contracts. Due to the fact that almost everything related to data science and machine learning is now considered AI, it is not a good candidate for industry standards. Regardless, something needs to be done, and it will almost certainly happen.

A proposed framework to guarantee ethical implementations are emerging as we proceed, building on previous work. There are four main pillars to the framework: Trust, fairness, privacy, and transparency; the first and most important requirement is the capacity to demonstrate an AI

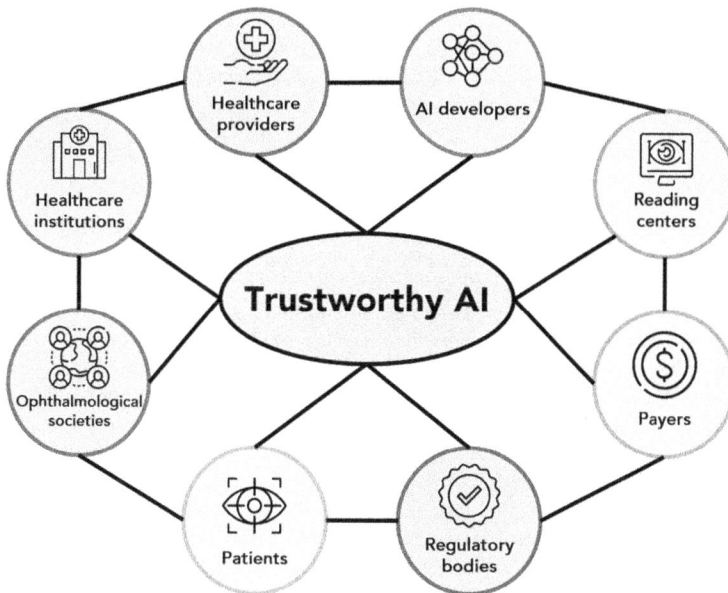

FIGURE 9.8 Transparency and explain-ability components of trustworthy AI.

application's trustworthiness. The AI applications that people use must have been developed under supervision that is accountable and believable, and they must come from a trustworthy source.

Transparency: Openness about how AI is being used and the benefits it offers; in particular, use-case scenarios will help lessen concerns and increase adoption.

Fairness: Developers must demonstrate that AI deployment is impartial and fair. Algorithms need to be fine-tuned to remove biases because AI does not have the ability to use judgment and instead focuses primarily on pattern recognition. Additionally, procedures should be put into place to avoid biases caused by human experiences.

Privacy: When utilizing AI, developers ought to take into consideration any personally identifiable information (PII) that is embedded in the data that is being processed. While AI processing alleviates some privacy concerns by preventing humans from interacting with sensitive data, it raises others, such as information use scope, storage location, and access rights.

In conclusion, efforts to limit AI's power will undoubtedly continue. In order to figure out how to incorporate their applications into the framework that was mentioned earlier, industry players are employing proactive thinking. Utilizing a peer review system to promote trust and transparency is one option. An AI Community, which would operate similarly to previous open-source environments, would receive use case submissions from AI developers.

The situation eventually improved despite initial opposition to open source. Global chief financial officers in technology companies were initially terrified by the concept of providing free source code. It became abundantly clear over the course of time that the transparency brought about by the distribution of open-source software, in which a community monitored, updated, and provided feedback, resulted in significant advancements and efficiency enhancements. Open-source code is used by every software company in the world today.

The creation of an ad hoc organization where businesses would commit their projects and applications to a centralized AI registry is another method for clarifying AI applications. The registry would be a self-reporting body for gathering feedback, advice, and affirmation, in contrast to a formal standards body, which is typically dominated by large organizations that control the agenda. In the end, incorporating ethics into computer science curriculums from the beginning is the most effective strategy for ensuring that AI applications are implemented ethically.

BIBLIOGRAPHY

Amnesty International and Access Now. 2018. "The Toronto declaration: Protecting the right to equality and non-discrimination in machine learning systems." https://www.accessnow.org/cms/assets/uploads/2018/08/The-Toronto-Declaration_ENG_08-2018.pdf.

Andersen, K. E., S. Köslich, B. Pedersen, B. Weigelin, and L. C. Jensen. 2017. "Do we blindly trust self-driving cars?" In *Proceedings of the Companion of the 2017 ACM/IEEE International Conference on Human-Robot Interaction*, pp. 67–68. New York: Association of Computing Machinery.

Buolamwini J., and T. Gebru. 2018. Gender shades: Intersectional accuracy disparities in commercial gender classification." *Proceedings of Machine Learning Research* 81:1–15.

European Commission. 2018. "A European approach on artificial intelligence." Accessed April 25, 2018. http://europa.eu/rapid/press-release_MEMO-18-3363_en.htm

Goddard, K., A. Roudsari, and J. C. Wyatt. 2012. "Automation bias: A systematic review of frequency, effect mediators, and mitigators." *Journal of the American Medical Informatics Association* 19:121–127.

Howard, A., C. Zhang, and E. Horvitz. 2017. "Addressing bias in machine learning algorithms: A pilot study on emotion recognition for intelligent systems." In *IEEE International Workshop on Advanced Robotics and its Social Impacts*. doi:10.1109/ARSO.2017.8025197.

Institute of Electrical and Electronics Engineers. 2017. "Ethically aligned design: A vision for prioritizing human wellbeing with artificial intelligence and autonomous systems." Piscataway, NJ: IEEE Press. http://standards.ieee.org/develop/indconn/ec/ead_brochure.pdf

Robinette, P., W. Li, R. Allen, A. Howard, and A. Wagner. 2016. "Overtrust of robots in emergency evacuation scenarios." In *The Eleventh ACM/IEEE International Conference on Human Robot Interaction*, pp. 101–108. Piscataway, NJ: IEEE Press.

Wagner, A, J. Borenstein, and A. Howard. 2018. "Overtrust in the robotic age: A contemporary ethical challenge." *Communications of the ACM* 61.9: 22–24.

Van Melle, William. 1978. "MYCIN: A knowledge-based consultation program for infectious disease diagnosis." *international Journal of Man-Machine Studies* 10.3: 313–322.

Kaul, Vivek, Sarah Enslin, and Seth A. Gross. 2020. "History of artificial intelligence in medicine." *Gastrointestinal Endoscopy* 92.4: 807–812.

Jiang, Fei, et al. 2017. "Artificial intelligence in healthcare: Past, present and future." *Stroke and Vascular Neurology* 2.4.

https://www.fda.gov/medical-devices/software-medical-device-samd/artificial-intelligence-and-machine -learning-aiml-enabled-medical-devices

https://www.forbes.com/sites/theapothecary/2016/08/19/artificial-intelligence-machine-learning-and-the-fda/ ?sh=15a09fff1aa1

Kayyali, B., D. Knott, and S. V. Kuiken. 2013. "The big-data revolution in US healthcare: Accelerating value and innovation." http://www. mckinsey.com/ industries/ healthcare- systems- and- services/ our-insights/ the-big-data- revolution- in- us- health- care

Yudkowsky, Eliezer, and Nick Bostrom. 2011. "The ethics of artificial intelligence." In *Cambridge Handbook of Artificial Intelligence*. Cambridge: Cambridge University Press.

Mannes, Aaron. 2020. "Governance, risk, and artificial intelligence." *AI Magazine* 41.1: 61–69.

Smith, Helen. 2021. "Clinical AI: Opacity, accountability, responsibility and liability." *AI & SOCIETY* 36.2: 535–545.

The Coming Technological Singularity: How to Survive in the Post-Human Era Archived 2007-01-01 at the Wayback Machine, by VernorVinge, Department of Mathematical Sciences, San Diego State University, (c) 1993 by VernorVinge.

Article at Asimovlaws.com Archived 2012-05-24, at the Wayback Machine, July 2004, accessed 7/27/09.

Winfield, A. F., K. Michael J. Pitt, and V. Evers March 2019. "Machine ethics: The design and governance of ethical AI and autonomous systems [scanning the issue]." In *Proceedings of the IEEE*. 107.3: 509–517. doi:10.1109/JPROC.2019.2900622. ISSN 1558-2256. S2CID 77393713. Archived from the original on 2020-11-02. Retrieved 2020-11-21.

Al-Rodhan, Nayef. December 7, 2015. "The moral code." Archived from the original on 2017-03-05.

Wallach, Wendell, and Colin Allen. November 2008. *Moral Machines: Teaching Robots Right from Wrong.* Oxford University Press. ISBN 978-0-19-537404-9.

Bostrom, Nick, and Eliezer Yudkowsky. 2011. "The ethics of artificial intelligence (PDF)." Cambridge Handbook of Artificial Intelligence. Cambridge Press. Archived (PDF) from the original on 2016-03-04.

Santos-Lang, Chris. 2002. "Ethics for artificial intelligences." Archived from the original on 2014-12-25.

Howard, Ayanna. "The regulation of AI – Should organizations be worried? I Ayanna Howard." *MIT Sloan Management Review*. Archived from the original on 2019-08-14.

Jobin, Anna, Marcello Ienca, and Effy Vayena. September 2, 2020. "The global landscape of AI ethics guidelines." *Nature* 1.9: 389–399. arXiv:1906.11668. doi:10.1038/s42256-019-0088-2. S2CID 201827642.

Floridi, Luciano, and Josh Cowls. July 2, 2019. "A unified framework of five principles for AI in society." *Harvard Data Science Review* 1. doi:10.1162/99608f92.8cd550d1. S2CID 198775713.

Open Source AI. Archived 2016-03-04 at the Wayback machine bill Hibbard. 2008 *Proceedings of the First Conference on Artificial General Intelligence*, eds. Pei Wang, Ben Goertzel, and Stan Franklin.

OpenCog: A software framework for integrative artificial general intelligence. Archived 2016-03-04 at the Wayback Machine David Hart and Ben Goertzel. *2008 Proceedings of the First Conference on Artificial General Intelligence*, eds. Pei Wang, Ben Goertzel, and Stan Franklin.

Inside OpenAI, Elon Musk's wild plan to set artificial intelligence free Archived 2016-04-27 at the Wayback Machine Cade Metz, Wired 27 April 2016.

"P7001 – Transparency of autonomous systems." P7001 – Transparency of autonomous systems. IEEE. Archived from the original on 10 January 2019. Retrieved 10 January 2019.

Thurm, Scott. July 13, 2018. "Microsoft calls for federal regulation of facial recognition." Wired. Archived from the original on 2019-05-09.

Bastin, Roland, and Georges Wantz. June 2017. "The general data protection regulation cross-industry innovation (PDF)." *Inside Magazine.* Deloitte. Archived (PDF) from the original on 2019-01-10.

"UN artificial intelligence summit aims to tackle poverty, humanity's 'grand challenges'." *UN News.* 2017-06-07. Archived from the original on 2019-07-26.

"Artificial intelligence – Organisation for economic co-operation and development." www.oecd.org. Archived from the original on 2019-07-22.

Anonymous. June 14, 2018. "The European AI alliance." Digital single market – European Commission. Archived from the original on 2019-08-01.

European Commission High-Level Expert Group on AI. June 26, 2019. "Policy and investment recommendations for trustworthy Artificial Intelligence." Shaping Europe's digital future – European Commission. Archived from the original on 2020-02-26.

"EU tech policy brief: July 2019 recap." Center for Democracy & Technology. Archived from the original on 2019-08-09.

Curtis, Caitlin; Nicole Gillespie, and Steven Lockey. May 24, 2022. "AI-deploying organizations are key to addressing "perfect storm" of AI risks." AI and Ethics: 1–9. doi:10.1007/s43681-022-00163-7. ISSN 2730-5961. PMC 9127285. PMID 35634256.

Gabriel, Iason. March 14, 2018. "The case for fairer algorithms – Iason Gabriel." Medium. Archived from the original on 2019-07-22.

"5 unexpected sources of bias in artificial intelligence." TechCrunch. 10 December 2016. Archived from the original on 2021-03-18.

Knight, Will. "Google's AI chief says forget Elon Musk's killer robots, and worry about bias in AI systems instead." *MIT Technology Review.* Archived from the original on 2019-07-04.

Villasenor, John. January 3, 2019. "Artificial intelligence and bias: Four key challenges." *Brookings.* Archived from the original on 2019-07-22.

Lohr, Steve. February 9, 2018. "Facial recognition is accurate, if you're a White guy." *The New York Times.* Archived from the original on 2019-01-09. Retrieved 29 May 2019.

Koenecke, Allison, Andrew Nam, Emily Lake, Joe Nudell, Minnie Quartey, Zion Mengesha, Connor Toups, John R. Rickford Dan Jurafsky, and Sharad Goel. April 7, 2020. "Racial disparities in automated speech recognition." *Proceedings of the National Academy of Sciences* 117.14: 7684–7689. Bibcode:2020PNAS..117.7684K. doi:10.1073/pnas.1915768117. PMC 7149386. PMID 32205437.

"Amazon scraps secret AI recruiting tool that showed bias against women." Reuters. 2018-10-10. Archived from the original on 2019-05-27. Retrieved 2019-05-29.

Friedman, Batya, and Helen Nissenbaum. July 1996. "Bias in computer systems." *ACM Transactions on Information Systems* 14.3: 330–347. doi:10.1145/230538.230561. S2CID 207195759.

"Eliminating bias in AI." techxplore.com. Archived from the original on 2019-07-25.

Olson, Parmy. "Google's DeepMind Has An Idea For Stopping Biased AI." Forbes. Retrieved 2019-07-26.

"Machine learning fairness | ML fairness." *Google Developers.* Archived from the original on 2019-08-10.

"AI and bias – IBM Research – US." www.research.ibm.com. Archived from the original on 2019-07-17.

Bender, Emily M., and Batya Friedman. December 2018. "Data statements for natural language processing: Toward mitigating system bias and enabling better science." *Transactions of the Association for Computational Linguistics* 6: 587–604. doi:10.1162/tacl_a_00041.

Gebru, Timnit, Jamie Morgenstern, Briana Vecchione, Jennifer Wortman Vaughan, Hanna Wallach, Hal Daumé III, and Kate Crawford. 2018. "Datasheets for datasets." arXiv:1803.09010 [cs.DB].

Pery, Andrew. October 6, 2021. "Trustworthy artificial intelligence and process mining: Challenges and opportunities." *DeepAI.*

Knight, Will. "Google's AI chief says forget Elon Musk's killer robots, and worry about bias in AI systems instead." *MIT Technology Review.* Archived from the original on 2019-07-04.

"Where in the world is AI? Responsible & unethical AI examples." Archived from the original on 2020-10-31.

Evans, Woody. 2015. "Posthuman rights: Dimensions of Transhuman worlds." *Teknokultura* 12.2. doi:10.5209/rev_TK.2015.v12.n2.49072.

Sheliazhenko, Yurii. 2017. "Artificial personal autonomy and concept of robot rights." *European Journal of Law and Political Sciences* 17–21. doi:10.20534/EJLPS-17-1-17-21. Archived from the original on 2018-06-14.

The American Heritage Dictionary of the English Language, Fourth Edition.

"Robots could demand legal rights." *BBC News.* December 21, 2006. Archived from the original on 2019-10-15.

Henderson, Mark. April 24, 2007. "Human rights for robots? We're getting carried away." *The Times Online.* The Times of London. Archived from the original on 2008-05-17.

McGee, Glenn. "A robot code of ethics." *The Scientist.* Archived from the original on 2020-09-06.

Cerqui, Daniela, and Kevin Warwick. 2008. "Re-designing humankind: The rise of cyborgs, a desirable goal?" *Philosophy and Design*, pp. 185–195. Dordrecht: Springer Netherlands. doi:10.1007/978-1-4020-6591-0_14, ISBN 978-1-4020-6590-3, Archived from the original on 2021-03-18.

Cave, Stephen, and Kanta Dihal. 6 August 2020. "The whiteness of AI." *Philosophy & Technology* 33.4: 685–703. doi:10.1007/s13347-020-00415-6.S2CID 225466550.

10 Ethical Issues Using AI in the Field of Pediatrics

Allyson Lim-Dy, Surovi Saikia, and Yashwant Pathak

10.1 INTRODUCTION

Artificial intelligence (AI) in healthcare is continuously transforming numerous facets of pediatric medicine. Several prominent applications of AI include precision medicine and electronic health databases. Precision medicine helps children through recommended treatments which accommodate their lifestyle. These treatments are generated by comparing the patient's data with other patients that illustrate similar conditions and situations. Meanwhile, health databases allow both clinicians and parents to efficiently access medical information and notes from previous visits. Many established programs in healthcare are deemed beneficial; however, many still question the practical issues for current and emerging AI applications. The potential of artificial intelligence to increase accurate predictions is primarily reinforced by large, real-world data from healthcare system databases [1]. As the relationship between real-world data and technology intensifies, increasing concerns towards misused information and patient privacy are inevitable. This will remain a complication since proper AI applications can only function through extensive amounts of data. Further underlying concerns persist, especially in pediatrics. Compared to adults, pediatrics encompasses unique factors, like developmental milestones and parental concerns, that can affect the implementation of efficient AI programs. To assess the potential of artificial intelligence in pediatrics, it is crucial to understand the ethical factors, concerns, and clinical elements that correspond with pediatric care.

10.2 ETHICAL FACTORS IN DESIGNING AI PROGRAMS

The ethical elements that contribute to the development of AI programs in medicine are data management, confidentiality, and validation, which heavily depend on the intended objective of the program. For instance, an algorithm created for physician scheduling would contain unrelated inputs for systems produced for creating clinical decisions for children [2]. Moreover, distinctive issues that encompass one program may be identical or different for another. This factor also explains why the data for AI programs in adults cannot be directly translated to pediatrics. A primary example would be that children possess different physiological characteristics than adults [2]. Children experience developmental milestones within a various range of ages, which can alter the accuracy and precision previously established in machine learning (ML) algorithms for adults. The complexities of applying these algorithms in pediatric medicine therefore must be monitored. If an algorithm is to be clinically implemented, it should encompass a legal and ethical framework that respects all involved parties. For pediatrics, those within these parties include designers, researchers, providers, patients, and parents. Each party must follow their own unique responsibility to ensure that their contribution is proven valuable and not detrimental. Accomplishing this would require each to understand the principal notions of data, communication, privacy, and validation.

DOI: 10.1201/9781003353751-10

10.2.1 Respecting Data Holistically

Artificial intelligence would cease to exist without data. The power that data contains is immense and must be respected. Clinical data is influential to society and should be utilized for favorable contributions and outcomes [3]. Nevertheless, debates around using clinical data for designing AI algorithms remain contradictory. Some suggest that employing data for creating new algorithms is unethical, while many argue that not utilizing data is also considered unethical since it disregards the capacity to improve healthcare systems [4]. Despite these arguments, the manner data is handled should not be negatively manipulated. Those who engage in the use of clinical data in healthcare should be morally responsible to provide beneficial care for current and future patients [3]. Harmfully employing data will create more distrust between technology, patients, and parents. Parents will feel less inclined to have their children's medical information utilized for algorithms if they perceive any potential future risks.

10.2.2 Communication and Confidentiality

Legal and ethical issues will subsequently arise without proper communication. Maintaining confidentiality is crucial when utilizing clinical data for the future safety of pediatric patients. Many concerns arise from when, where, and how data is collected. Parents should be made aware of how their children's data would contribute to AI programs and big data—as well as benefits, possible risks, or vulnerabilities [5]. Health professionals should be aware of these liabilities and are responsible for properly conveying these to parents or guardians. Parents must be told these factors so they can provide proper informed consent and decisions for their children. A significant component for regulating confidentiality is data anonymization. Data anonymization is where patients' personal data is encrypted or eliminated to ensure those patients stays anonymous [5]. Applying this process will help prevent future risks and privacy violations for pediatric patients.

10.2.3 Validation through Implementation

The customary method to validate an AI program is to adequately evaluate and then integrate it within the healthcare system. An algorithm should only be implemented if it is screened for any potential risks or consequences. Silent trials have been conducted to help perform proper evaluations. These trials utilize models to measure how they would function and influence provider outcomes [2]. A clinical evaluation can be conducted after another assessment of the silent trials' results and if it is supported by research ethics boards [2]. Once a program is clinically implemented, it should be monitored by all involved parties. Adjustment to novel programs can cause natural disturbances but should be anticipated and tackled to improve integration [2]. Both clinicians and parents should provide feedback to ensure it is valuable and properly functioning. This feedback would help guide designers and researchers focus their attention to any discrepancies that may arise.

10.3 PRECISION MEDICINE

Precision medicine meticulously tailors the treatment of a patient's condition through considering their genomics and lifestyle. Personalized medicine requires vast amounts of data, which correlates with deep learning since numerous, similar records can be compared simultaneously [6]. Furthermore, precision medicine can encompass other beneficial aspects like promoting health literacy and education through accessible databases and chatbots. The acquired data can also be analyzed by digital imaging systems that can analogize scans like CTs (computed tomographies) and MRIs (magnetic resonance imaging). Once similar data is evident, clinicians would be able to recommend diagnostic and therapeutic treatments that was administered to other pediatric patients [6].

Larger amounts of medical data exists for adults compared to children. It cannot be directly compared due to physiological factors or biomarkers that do not prominently appear in children. For example, pediatric oncology patients cannot always obtain the same drug given to adult oncology patients. The etiology of tumors originates differently in children, and biomarkers in numerous pediatric cancers are not reliable [7]. These instances illuminate the importance of precision medicine programs and AI in pediatrics. Incorporating deep learning AI algorithms in precision medicine would be favorable since they can be applied for unique or unfamiliar cases. Researchers and clinicians in academia should collaborate with medical and pharmaceutical companies to provide better opportunities and successes in precision medicine programs.

10.3.1 TELEHEALTH AND DIGITAL HEALTH DATABASES

Several of the most prominent artificial intelligence systems in medicine are Telehealth and digital health databases. Telehealth allows providers to assess and communicate with their patients remotely. It is convenient for both parties since it offers more flexibility. Nevertheless, clinicians must be careful when considering a Telehealth appointment since certain concerns cannot always be assessed or controlled through a teleconsultation [8]. Generally, within a Telehealth system there are digital health databases which include electronic medical records (EMRs) and electronic health records (EHRs). Information can also be exchanged via these databases throughout numerous healthcare facilities, allowing diverse health professionals and specialties to efficiently collaborate. A patient's medical information can be accessed by multiple facilities if they share the same EMR system. Additionally, the digital database eases parents' worries since they can arrive at their child's appointments without large packets of documents or image recordings. The visits are digitally documented, and parents can easily access their doctor's notes from previous visits, including both remote and in-person appointments.

Fundamentally, both providers and parents are responsible for directly promoting and educating the health of the child. Providers must inform the parents of the child's conditions and treatments for them to properly encourage their child's well-being. Health professionals should be empathetic and inclusive when conveying medical advice since it can enhance the parent's comprehension [9]. Through this relay of information, they can gain a sufficient level of health literacy. Health literacy is crucial since it indicates that the parent understands and can distinguish how to utilize health information for themselves and their children. Technology has the potential to improve health literacy since it can easily be utilized by the parent via phone or computer [10]. Diverse factors that accompany parenthood such as lack of sleep or fixed schedules can also be digitally accommodated through easily accessible web interfaces and apps.

10.3.2 CHATBOTS FOR PERSONALIZED HEALTHCARE

Chatbots are another pronounced constituent of precision and personalized medicine. Chatbots are managed by AI where patients can communicate with a messaging software that produces natural language and responses [8]. Such software can either be offered through a Telehealth portal, social media, or mobile apps. Thus, they are also easily accessible and can help provide valuable well-being opportunities for both parents and children.

Researchers and designers for AI medical apps and chatbots must carefully create their system to adapt and incorporate potential factors that surround children's developmental milestones at various ages. Factors are abundant and range from environmental, mental, physical, and sociological. These elements should be prioritized since personalized medicine requires genuine inputs to generate appropriate responses and treatments. Protecting user inputs and users is also crucial as information can become extremely detailed and confidential [11].

Social media is an influential network that can affect an individual's perception towards themselves and others. It is a dynamic platform for the youth since social media is popular among pre-adolescents and teenagers. Utilizing social media to promote health awareness on social media can truly be influential, but designers and health professionals must be aware of its openness and repercussions. Social media is a sharing community that enables individuals and groups of people to express and communicate but also increases their susceptibility to addiction and cyber bullying [8]. Addiction can also influence adolescents to feel more inclined to compare themselves to widespread trends that potentially create an unhealthy barrier between media and reality. Social media addiction can also affect an individual's daily habits, such as eating and sleeping. Sleep inversion commonly occurs alongside unrestricted social media, which can modify the body's metabolism and can increase overweightness or obesity in the adolescent population [8]. Researchers, designers, and health professionals should cooperate to create solutions to the unhealthy habits that arise from excessive social media and other associated conditions.

Despite potential difficulties, creating various applications and chatbots are favorable for promoting general health at a young age. Several systems have been developed to target children with special and congenital conditions. Two prevalent conditions in which systems have been implemented are diabetes mellitus and pediatric obesity. Many children with diabetes mellitus have inadequate comprehension and training to control their glucose levels, which enhances their susceptibility to future risks [12]. Mobile applications can help alert and monitor glycemic control which can interconnectedly be communicated with healthcare professionals [13]. Teddy bears with designated patches of finger pricking and insulin injection spots are a common learning tool for children with diabetes mellitus. Corresponding mobile applications have been developed where they can learn about each designated patch by using their phone to scan the areas. Learning about these areas allows both parents and children to understand how to properly regulate their glucose levels.

Diabetes and pediatric obesity are comorbid since children can potentially develop the former with improper food and weight management. Behavioral intervention technology has been created to address and assist pediatric obesity in the United States. This technology functions through combining technology and psychology to offer holistic health support for the youth [14]. The first behavioral intervention chatbot intended for pediatric obesity and weight management is Tess. Tess is available 24/7 and can be accessed through communication mediums like SMS and Facebook messenger [14]. This chatbot encompasses the behavioral and sociological factors that affect the youth at varying ages to sufficiently provide integrative support. In a study executed by Stephens et al., they evaluated the outcomes of implementing Tess, and results illustrated that approximately 81% of adolescents showed positive feedback [14]. Tess is a prime example demonstrating how current and arising chatbots and applications can provide beneficial support for preventative and qualitative health in adolescents.

10.4 CURRENT APPLICATIONS IN VARIOUS SUBSPECIALTIES

Artificial intelligence assistance in clinical settings is emerging rapidly as systems become more established. Though more applications exist for adult medicine, many current applications for pediatrics offer numerous opportunities for enhanced treatment and care. These applications are demonstrated in pediatric subspecialties such as pediatric radiology, cardiology, and gastroenterology.

10.4.1 PEDIATRIC RADIOLOGY

As AI becomes significantly more advanced, a common misconception is that these algorithms will replace the role of the clinician. This idea is most prominent in radiology since this specialty contains more applications compared to others. However, artificial intelligence is far from replacing a clinician due to the extent of current programs. The applications only assist rather than replace the radiologist's interpretative role and skills. For example, most AI applications in pediatric radiology assist radiologists by improving image quality and detection. Several tools include chest radiography, bone assessment, and fracture assessment.

10.4.1.1 Chest Radiography

Chest radiography is one of the most common screenings in children that encompasses interconnected aspects of pediatric medicine. It branches out to other subspecialties since these screenings are readily available and cost-efficient. Generally, the conditions that artificial intelligence can assist with are challenging due to the physiological features and anomalies naturally found in children. Pathologies also tend to deviate from established diagnoses in adults [15]. AI implemented within chest radiograph algorithms can potentially lessen these challenges. One example is that chest radiographs may help determine the positioning of medical devices [16]. Although currently underdeveloped, AI has been utilized to assess catheters and tubes for neonatal radiographs [16]. Several studies and efforts have also been executed to train artificial intelligence for diagnoses. Pneumonia is a prominent illness in which artificial intelligence has been analyzed and classified. Data for pneumonia is extensive yet still requires supervision from radiologists or professional reports to guarantee quality [16]. A dataset that has gained substantial success in detecting pneumonia is the GWCMC dataset. These radiographs were primarily based on adult patients, but many studies have utilized them for the pediatric population [16]. Nonetheless, the radiographs from the GWCMC dataset only help guide AI models in pediatrics to a certain extent. More data and models must be created and collected within the pediatric population to guarantee novel algorithms perform optimally.

10.4.1.2 Bone Assessment

Bone assessment AI applications are relatively new and have primarily existed as an AI-assist tool rather than an AI-replace tool. Currently, there are three bone assessment AI tools available, which includes the forerunner BoneXpert [17]. This tool is relatively quick to use, approximately 15 seconds per scan, and follows the Tanner-Whitehouse and Greulich and Pyle methods to estimate bone age [17]. BoneXpert was originally released as an AI-replace application, but certain circumstances prevent it from functioning as that modality. For example, BoneXpert can recognize radiographs with remarkable irregularities but not elusive differences. This disadvantage suggests that radiologists cannot fully rely on this system and, subsequently, they must review the radiographs to determine accuracy.

10.4.1.3 Fracture Assessment

Studies have been conducted on AI applications for fracture assessment. Analyzations primarily focused on detecting fractures through images in the elbow joint region. Appendicular fractures are the most common types of fractures found in children [18]. This region is unique since children have multiple unossified epiphyseal centers in their elbow joints [17]. In a literature review by Shelmerdine et al., the diagnostic accuracy for specific regions (elbow joint) was higher compared to general regions (whole arm) [18]. This founding implies that higher diagnostic accuracies for artificial intelligence algorithms are more likely to occur for concentrated areas. Input for improving bone and fracture assessment should be interpreted not only by radiologists, but also other interconnected subspecialties.

10.4.2 Pediatric Cardiology

Pediatric cardiology is another subspecialty that has gained substantial advances through AI. This is an important specialty to focus on as cardiac disease continually remains a leading complication globally. The main applications of AI in pediatric cardiology relate to cardiac imaging, wearable technology, and precision medicine.

10.4.2.1 Cardiac Imaging

Cardiac imaging has been improved by AI in adult medicine and has the potential to expand to pediatrics. For example, cardiac MRIs have certain stages that are automated and sped up by artificial intelligence. Automatic segmentation during evaluation of MRIs could influence how precise and accurate the arteries, chambers, and veins function [19]. Additionally, techniques like parallel imaging and compressed sensing have allowed cardiovascular MRI sequences to exhibit shorter waiting times [20]. Echocardiograms are also improving in this aspect and can save the cardiologists' time and energy to review each image. Convolutional neural networks (CNNs) in echocardiograms are also being trained to read echocardiograms and alert the cardiologist for any abnormal or unique cases [21].

10.4.2.2 Wearable Technology

Medical wearables are the most well-known AI application in cardiology. They are convenient, cost-efficient, and can monitor a child's vitals in outpatient settings. Wearable technologies are extremely beneficial for children with chronic illnesses that must be constantly monitored. These applications also help alert the parents or bystanders if the child's vitals are potentially alarming and dangerous. Children wearing these technologies must be properly taught its usage for any emergency that could arise.

10.4.2.3 Precision Medicine

Precision medicine in pediatric cardiology is a crucial area for artificial intelligence due to the array of imaging and data. It is difficult for cardiologists to determine an appropriate treatment plan for the child when clinical factors are abundant and scattered. Precision medicine gathers a patient's medical information and appropriately organizes them to create proper treatment plans. Other current applications within precision medicine include augmented reality (AU) and virtual reality (VR). These current applications allow hearts to be modeled and viewed. This is an aspect of precision medicine since the models are personally designed and various layers of the heart can be observed. Another benefit with these cardiac models is the teaching ability it offers to parents and children. The 3D images can easily be rotated and shown, allowing health professionals to provide a visual for explaining unique structural abnormalities [21].

10.4.3 PEDIATRIC GASTROENTEROLOGY

Pediatric gastroenterology is a subspecialty that requires collaboration with other specialties like radiology and pathology. Most procedures in gastroenterology are endoscopic and heavily rely on computer systems. Endoscopes for the pediatric population tend to be smaller due to the different anatomy and physiology in children. These endoscopes are connected to a system that projects the camera's imaging onto a screen. Moreover, artificial intelligence has significant potential for algorithms to aid the assessment of common conditions found in the digestive tract. These conditions include celiac disease and inflammatory bowel disease.

10.4.3.1 Celiac Disease

Celiac disease occurs in the intestine and is commonly diagnosed through endoscopic procedures. Biopsies are obtained during the procedure and then evaluated through pathology. Nonetheless, obtaining biopsies in this region can be complicated since it may require multiple samples, which can further irritate the intestinal tract. A possible AI application for celiac disease was suggested by Wimmer et al. [22]. They utilized CNNs to identify celiac disease through 1,661 luminal endoscopic images. The accuracy they obtained was 90.5% [22], which is favorable and indicates that their CNN could potentially be an alternative for diagnosis when biopsies are difficult to obtain.

10.4.3.2 Inflammatory Bowel Disease

An AI classification system has been developed to help clinicians assess patients with inflammatory bowel disease (IBD). This system uses a CNN and was created by Ozawa et al. through over 20,000 endoscopic images [23]. The algorithm was trained to discern normal mucosa and healing mucosal states while classifying each image with Mayo endoscopic scores. Their network efficiently identified broad ranges of inflammation stages (Mayo score of 2–3) and mucosal healing states (Mayo score of 0–1) [23]. The results from this developed network were overall successful and suggest a beneficial AI application for both adult and pediatric gastroenterology.

10.5 PARENTAL CONCERNS

Increased enthusiasm about the possibilities of precision medicine is transforming healthcare which has been made possible with recent advancement in AI technologies such as deep learning and machine learning [24, 25]. In the pediatrics field, AI-backed technologies are starting to enter the clinical trials for the management of conditions such as cancer, diabetes mellitus, and asthma [26, 27], but these AI-backed technologies also come with practical and ethical concerns particularly in pediatrics [28]. As precision medicine requires large genetic, lifestyle, and environmental datasets from various populations of children where parents might have concerns about the privacy of the child's data and the transparency level of how these data are utilized. Another problematic issue for parental concern is the impact of this information in the relationship between clinician and family relationship and the rate of increased dependence of children on computers, and medical reports without the use of real-time human involvements [29]. To reap the benefits of precision medicine, developers and researchers need access to huge amounts of diverse data, and these technologies are required to be used by children. The openness and acceptance showed by parents in this regard will surely help to expedite the process, and so it is critical to understand the ethical and practical concerns of parents [30]. The various parental concerns for AI-driven healthcare technologies are explored in the following subsections.

10.5.1 Cost

AI has been successful with stunning results in visual pattern recognition [28], and the major challenge within this is to convert these AI-derived results or recommendations into effective working principles. However, the most critical problem here is to change the behavior of millions of people and clinicians. Eighty percent of the healthcare costs can be attributed to the behaviors of physicians, ordering test kits, pharmaceuticals, other procedures [31]. In addition to clinical effectiveness, the evaluation of cost effectiveness of AI is important, and huge investments in this line with promises of efficiency and cost reductions are similar to robotic surgery. Cost reduction cannot be assured with AI techniques, which include data storage, model maintenance, data curation, data visualization and updating, etc. These tools have the potential to replace the current cost tables but with higher costs, and answers to lowering medical cost using AI still remain unanswered [32].

10.5.2 Convenience

An algorithm developed by Andre Esteva et al. [33] which can classify lesions from images taken by a mobile phone is very convenient and can also be used outside the clinic as a visual screening for cancer. An automatic diagnostic CNN was developed for analysis of MODS (Microscopic Observed Drug Susceptibility) digital images, which has the potential to assist laboratory personal for automatic diagnosis [34]. A custom Android application was developed which can synchronize external LED (light-emitting diode) illumination and image capture for AFI (autofluorescence imaging) and WLI (white light imaging). Oral cancer data is uploaded to a cloud server which are diagnosed by a remote specialist using a web app, transmitting triage instructions back to the patient and device very conveniently [35]. Another convenience factor offered by AI is the development of wearable BMDs (Biometric Monitoring Devices), which has enabled remote analysis and measurement of patient data in real time [36]. AI platforms have also been used extensively for operational delivery care cardiology patients [37].

10.5.3 HUMAN CARE ELEMENT

Parental concerns often arise for using these technologies, and the cumulative effects of using these reflect on the relationship between patients/families and clinicians. As healthcare is built upon relationships between individuals apart from the clinician and patient, other relationships exist such as relatives/families, between clinicians, inpatient facility etc., and AI technologies has the potential to modify these relationships. A two-way flow of information takes place between the clinicians and patient; however, a cache or an autonomous decision-making algorithm collected by the patient themselves brings a new set of including that of AI and the clinicians and AI and the patient [28].

10.5.4 PRIVACY

This involves the concerns about the usage of the child's information and how this information will be accessed and used by users. A survey showed that 63% of the adult population is uncomfortable with allowing personal data to improve healthcare and is unfavorable on using AI systems which may replace doctors and nurses in typical tasks [38]. Increased use of big data and AI techniques demands a re-examination of the Hippocratic oath and Belmont report, which articulate the foundational principles by which physicians interact with research subjects and patients. This is related to the potential issues related to confidentiality, privacy, informed consent, inequities, data ownership, and epistemology, with strong opinions from patients regarding these issues [39].

10.5.5 QUALITY AND ACCURACY

The fidelity and effectiveness in using these technologies is related to quality and accuracy. A DCNN (deep convolutional neural network) was developed as an efficient tumor classifier trained on a small dataset which was capable of classifying skin tumors better than board-certified dermatologists. A deep learning algorithm was developed with an accuracy of specialist level to detect diabetic retinopathy (DR); in Thailand, with 25,326 gradable retinal images of patients, the algorithm was found to be more sensitive as compared to human graders [40]. Amazon's Mechanical Turk (Mturk) is a new website which contains major aspects for research for psychology and other social sciences with high-quality data both inexpensive and rapid [41].

10.5.6 SOCIAL JUSTICE

This is related to the use of new technologies and may potentially affect the distribution of the benefits and burden by using the AI in healthcare [42]. The use of ML in care practices for complicated cases requires correct diagnosis in a certain case, and the best employed can be disputable. Introducing a specific practice or diagnosis at a premature stage may hint towards legitimacy unsupported by data [43]. Governmental institutes and healthcare bodies as a part of governance should have established standards in order to apply AI in healthcare. These standards should include the process through which the AI model is developed and implemented in the context of healthcare, and justice becomes the requirement for a basic classical biomedical ethics. The principal of social justice directs to the fairness in the access in healthcare, and the design of AI should warrant procedural and justice allocation of resources [44].

10.5.7 SHARED DECISION MAKING

This concern is related to the parental involvement and authority in determining the positive effects of using these technologies for the benefit of the child's care.

10.6 TRUSTING AI AS A CLINICIAN AND PATIENT

Trust in technical systems and interpersonal trust has many folds of differences, as the intention of a technical system is unknown pertinent to benevolence and honesty. The level of trust in AI leaves a significant impact on how these technologies are used for efficiency in healthcare, but the level of trust does not always have a positive correlation with patient or clinical outcomes. Advances in AI technology have moved from the automation of well-defined and repetitive tasks to guide decision-making under uncertainty currently carried out by medical professionals. Healthcare professionals trusting AI more, also termed as calibrated trust, is required for effective decision [45].

Trust factor in AI is influenced by several human factors such as education of the user, biases of the user, previous experiences, and impression for automation including different properties of the AI system such as transparency, complexity of the model, controllability, risk associated, etc. Among all these factors, the factor of reliability stands tall in healthcare due to the changes that occur in the reliability of AI systems when new data are introduced [46]. Trust for initial diagnoses in cases of life-threatening diseases was lower as compared to low-risk conditions. This trust gap can be eliminated if enforcing AI diagnosis can be eliminated to the *libertarian paternalism* giving the patient their freedom of choice [47].

10.7 TRAINING FOR HEALTH PROFESSIONALS

Health professionals nowadays have the chance to leverage AI tools for improving the care of patients. HPE (Health Professional Education) communities are now in need of incorporating training in AI across health professionals or otherwise it may risk health professionals becoming unprepared to support the promise of AI. It is important for the leaders of HPE to familiarize themselves with AI in order to guide the training of the current and future learners. Training in AI is currently at a position for inter-professional education approach, as most of the fields are at an early stage and the relationship of work across professions will help health professionals to minimize the demerits of AI. In the following subsection, the keystones for training AI are outlined.

10.7.1 Overload of Information

Due to the explosion of data from imaging, multi-omics, biometrics, EHRs (electronic health records), remote monitoring via sensors requires the application of algorithms sophisticated in nature to handle such big data. The traditional syllabus cannot prepare health professionals for such challenges of the future which require knowledge of effective knowledge management and utilization of ML and data analytics [48]. Also, healthcare interns and trainees need preparation for precision healthcare using digital tools such as Apple Watch.

10.7.2 Fundamentals of AI

All healthcare professionals must have a basic understanding of the fundamentals of AI in order to know the underlying mechanism and how it differs from the working of AI. The various ways in which AI can be applied to work of health professionals is essential in the early stage of the training. During the advancement of the training process, additional specific training may be required such as understanding the providers' role in AI application such as concerns relating to regulation and ethics.

10.7.3 New Capabilities

Recent literature suggests the use of AI training into HPE [49] and capabilities for broader understanding of AI in a professional aspect is to present as a responsible guardian for patient data to maintain the trust between the patient and the provider. Another capability is to understand the risk associated with privacy and healthcare, and so healthcare professionals should be able to advocate for the development and distribution of equitable and ethical systems. Some experts are of the opinion to incorporate skills such as computer programming and understanding good practices of software design for explicit training of health professionals [49, 50].

10.8 THE FUTURE OF AI IN PEDIATRICS

Artificial intelligence in pediatric medicine is promising and rapidly progressing. As algorithms gradually improve, more conditions could be potentially surveilled and prevented. AI algorithms are indispensable as they help identify, organize, and diagnose various illnesses, therefore assisting community and preventative medicine [51]. Nevertheless, further human inputs and holistic assessments are necessary for enhancing AI in healthcare. Artificial intelligence currently remains in a state where human support is still required. With logically created manmade-algorithms, AI can only assist to a certain extent. AI alleviates nuances associated with one's job yet cannot fully replace the role of health professionals nor emotions and sympathy related to patient care. Chatbots and other messaging platforms can resemble "human care" elements but will never compare to long-standing relationships between a provider and patient.

AI will continue to prevail if practical and ethical concerns remain properly monitored and addressed. Much concern revolves around protecting the safety and privacy of children, and global efforts are being executed to address these worries. A project initiated by UNICEF's Office of Global Insight and Policy focuses on understanding how AI can protect and strengthen children. It is crucial since artificial intelligence naturally surrounds their everyday lives but may unknowingly affect them. With concerns monitored, AI has the potential to create a positive effect for providing essential and preventative care in pediatrics. People must be prepared to accept the transformations AI will produce in coming years.

10.9 CONCLUSION

Ethical concerns for artificial intelligence in pediatrics mainly relate to how data is used, implemented, and protected. Many parental concerns also relate to these notions and must be addressed to ease parents' apprehension towards AI. Additionally, these common yet inevitable issues are critical to consider when designing and evaluating novel AI algorithms. Those involved with development are responsible for these factors, and they must also properly convey any associated complications and information to providers and parents. Doing so will allow providers and parents to gain a stronger understanding of AI's potential and perhaps increase trust. Moreover, numerous AI applications have been implemented and/or trialed, and many illustrate significant progress and success. These applications range over various subspecialties in pediatrics and continue to expand as more data becomes gradually incorporated. This is especially eminent in precision medicine since more datasets are being utilized to provide personalized treatment for children. Artificial intelligence also offers lower cost, convenience, higher accuracy, and quality of care. These systems are often flexible but may require additional training for providers, parents, and children. Promoting proper comprehension is crucial for ensuring the application runs accurately and smoothly. Overall, using AI in pediatric medicine is advantageous and will continue to positively influence providers and the pediatric population if concerns remain communicated and addressed.

CONFLICT OF INTEREST

The authors declare no conflict of interest, financial or otherwise.

ACKNOWLEDGEMENTS

We thank the Vice Chancellor, Bharathiar University, Coimbatore-641046, Tamil Nadu for providing the necessary facilities. We also thank UGC-New Delhi for Dr. D S Kothari Fellowship ((No.F-2/2006 (BSR)/BL/20-21/0396)).

REFERENCES

1. Lonsdale H, Jalali A, Ahumada L, Matava C. Machine learning and artificial intelligence in pediatric research: Current state, future prospects, and examples in perioperative and critical care. *The Journal of Pediatrics.* 2020; 221, S3–S10.
2. Nagaraj S, Harish V, McCoy LG, et al. From clinic to computer and back again: Practical considerations when designing and implementing machine learning solutions for pediatrics. *Current Treatment Options in Pediatrics.* 2020; 6(4), 336–349.
3. Larson DB, Magnus DC, Lungren MP, Shah NH, Langlotz CP. Ethics of using and sharing clinical imaging data for artificial intelligence: A proposed framework. *Radiology.* 2020; 295(3), 675–682.
4. Knoppers BM, Thorogood AM. Ethics and big data in health. *Current Opinion in Systems Biology.* 2017; 4, 53–57.
5. Anom BY. Ethics of big data and artificial intelligence in medicine. *Ethics, Medicine and Public Health.* 2020; 15, 100568.
6. Shu L-Q, Sun Y-K, Tan L-H, Shu Q, Chang AC. Application of artificial intelligence in pediatrics: past, present and future. *World Journal of Pediatrics.* 2019; 15(2), 105–108.
7. Salzer E, Hutter C. Therapy concepts in the context of precision medicine for pediatric malignancies—Children are not adults. *Memo – Magazine of European Medical Oncology.* 2021; 14(3), 273–277.
8. Chong NK, Chu Shan Elaine C, de Korne DF. Creating a learning televillage and automated digital child health ecosystem. *Pediatric Clinics of North America.* 2020; 67(4), 707–724.
9. Aarthun A, Øymar KA, Akerjordet K. How health professionals facilitate parents' involvement in decision-making at the hospital: A parental perspective. *Journal of Children's Health Care.* 2018; 22(1), 108–121.
10. Mörelius E, Robinson S, Arabiat D, Whitehead L. Digital interventions to improve health literacy among parents of children aged 0 to 12 years with a health condition: Systematic review. *Journal of Medical Internet Research.* 2021; 23(12), e31665–e31665.
11. Thompson D, Baranowski T. Chatbots as extenders of pediatric obesity intervention: An invited commentary on "Feasibility of pediatric obesity & pre-diabetes treatment support through Tess, the AI behavioural coaching Chatbot." *Translational Behavioral Medicine.* 2019; 9(3), 448–450.
12. Fernandez-Luque L, Al Herbish A, Al Shammari R, Argente J, Bin-Abbas B, Deeb A, Dixon D, Zary N, Koledova E, Savage MO et al. Digital health for supporting precision medicine in pediatric endocrine disorders: Opportunities for improved Patient Care. *Frontiers in Pediatrics.* 2021; 9, 715705–715705.
13. Al Hayek AA, Robert AA, Al Dawish M, Ahmed RA, Al Sabaan FS. The evolving role of short-term professional continuous glucose monitoring on glycemic control and hypoglycemia among Saudi patients with type 1 diabetes: A prospective study. *Diabetes Therapy: Research, Treatment and Education of Diabetes and Related Disorders.* 2015; 6(3), 329–337.
14. Stephens T, Joerin A, Rauws M, Werk LN. Feasibility of pediatric obesity and prediabetes treatment support through Tess, the AI behavioral coaching chatbot. *Translational Behavioral Medicine.* 2019; 9(3), 440–447.
15. Padash S, Mohebbian MR, Adams SJ, Henderson RDE, Babyn P. Pediatric chest radiograph interpretation: How far has artificial intelligence come? A systematic literature review. *Pediatric Radiology.* 2022; 52(8), 1568–1580.
16. Arthur R. The neonatal chest X-ray. *Paediatric Respiratory Reviews.* 2001; 2(4), 311–323.
17. Offiah AC. Current and emerging artificial intelligence applications for pediatric musculoskeletal radiology. *Pediatric Radiology.* 2022; 52(11), 2149–2158.
18. Shelmerdine SC, White RD, Liu H, Arthurs OJ, Sebire NJ. Artificial intelligence for radiological paediatric fracture assessment: A systematic review. *Insights into Imaging.* 2022; 13(1), 94–94.
19. Van den Eynde J, Kutty S, Danford DA, Manlhiot C. Artificial intelligence in pediatric cardiology: Taking baby steps in the big world of data. *Current Opinion in Cardiology.* 2022; 37(1), 130–136.
20. Taylor A. The role of artificial intelligence in paediatric cardiovascular magnetic resonance imaging. *Pediatric Radiology.* 2022; 52(11), 2131–2138.

21. Gaffar S, Gearhart AS, Chang AC. The next frontier in pediatric cardiology: Artificial intelligence. *Pediatric Clinics of North America*. 2020; 67(5), 995–1009.
22. Wimmer G, Vécsei A, Uhl A. CNN transfer learning for the automated diagnosis of celiac disease, 2016 6th International Conference on Image Processing Theory Tools and Applications (IPTA). 2016.
23. Ozawa T, Ishihara S, Fujishiro M, Saito H, Kumagai Y, Shichijo S, Aoyama K, Tada T. Novel computer-assisted diagnosis system for endoscopic disease activity in patients with ulcerative colitis. *Gastrointestinal Endoscopy*. 2019; 89(2), 416.e1–21.e1.
24. Rajkomar A, Dean J, Kohane I. Machine learning in medicine. *New England Journal of Medicine*. 2019; 380(14), 1347–1358.
25. Topol EJ. High-performance medicine: The convergence of human and artificial intelligence. *Nature Medicine*. 2019; 25(1), 44–56.
26. Chung WK, Erion K, Florez JC, et al. Precision medicine in diabetes: A consensus report from the American Diabetes Association (ADA) and the European Association for the Study of Diabetes (EASD). *Diabetes Care*. 2020; 43(7), 1617–1635.
27. Vo KT, Parsons DW, Seibel NL. Precision medicine in pediatric oncology. *Surgical Oncology Clinics of North America*. 2020; 29(1), 63–72.
28. Fenech M, Strukelj N, Buston O. *Ethical, Social, and Political Challenges of Artificial Intelligence in Health; Future Advocacy*. Welcome Trust: London, 2018.
29. Verghese A, Shah NH, Harrington RA. What this computer needs is a physician: Humanism and artificial intelligence. *JAMA*. 2018; 319(1), 19–20.
30. Sisk BA, Antes AL, Burrous S, DuBois JM. Parental attitudes toward artificial intelligence-driven precision medicine technologies in pediatric healthcare. *Children (Basel)*. 2020; 7(9), 145.
31. Emanuel EJ, Wachter RM. Artificial intelligence in health care: Will the value match the hype? *JAMA*. 2019; 321(23), 2281–2282.
32. Maddox TM, Rumsfeld JS, Payne PRO. Questions for artificial intelligence in health care. *JAMA*. 2019; 321(1), 31–32.
33. Esteva A, Kuprel B, Novoa RA, Ko J, Swetter SM, Blau HM, Thrun S. Dermatologist-level classification of skin cancer with deep neural networks. *Nature*. 2017; 542(7639), 115–118.
34. Lopez-Garnier S, Sheen P, Zimic M. Automatic diagnostics of tuberculosis using convolutional neural networks analysis of MODS digital images. *PLOS One*. 2019; 14(2): e0212094.
35. Utho RD, Song B, Sunny S, et al. Point-of-care, smartphone-based, dual-modality, dual-view, oral cancer screening device with neural network classification for low-resource communities. *PLOS One*. 2018; 13 e0212094.
36. Tran VT, Riveros C, Ravaud P. Patients' views of wearable devices and AI in healthcare: Findings from the ComPaRe e-cohort. *npj Digital Medicine*. 2019; 2, 53.
37. Tsay D, Patterson C. From machine learning to artificial intelligence applications in cardiac care. *Circulation*. 2018; 138(22), 2569–2575.
38. Vayena E, Blasimme A, Cohen IG. Machine learning in medicine: Addressing ethical challenges. *PLOS Medicine*. 2018; 15(11), e1002689.
39. Balthazar P, Harri P, Prater A, Safdar NM. Protecting your patients' interests in the era of big data, artificial intelligence, and predictive analytics. *Journal of the American College of Radiology*. 2018; 15, 580–586.
40. Raumviboonsuk P, Krause J, Chotcomwongse P, et al. Deep learning versus human graders for classifying diabetic retinopathy severity in a nationwide screening program. *npj Digital Medicine*. 2019; 2, 25.
41. Buhrmester M, Kwang T, Gosling SD. Amazon's mechanical Turk: A new source of inexpensive, yet high-quality, data? *Perspectives on Psychological Science*. 2011; 6(1), 3–5.
42. Topol EJ. High-performance medicine: The convergence of human and artificial intelligence. *Nature Medicine*. 2019; 25(1), 44–56.
43. Char DS, Shah NH, Magnus D. Implementing machine learning in health care—Addressing ethical challenges. *New England Journal of Medicine*. 2018; 378(11), 981–983.
44. Reddy S, Allan S, Coghlan S, Cooper P. A governance model for the application of AI in health care. *Journal of the American Medical Informatics Association JAMIA*. 2019;27(3):491–497.
45. Evans JSBT. Dual-processing accounts of reasoning, judgment, and social cognition. *Annual Review of Psychology*. 2008; 59(1), 255–278.
46. Parikh RB, Obermeyer Z, Navathe AS. Regulation of predictive analytics in medicine. *Science*. 2019; 363(6429), 810–812.
47. Juravle G, Boudouraki A, Terziyska M, Rezlescu C. Trust in artificial intelligence for medical diagnoses. *Progress in Brain Research*. 2020; 253, 263–282. [10.1016/bs.pbr.2020.06.006]. Epub. 2020.

48. Wartman SA, Combs CD. Medical education must move from the information age to the age of artificial intelligence. *Academic Medicine*. 2018; 93(8), 1107–1109.
49. McCoy LG, Nagaraj S, Morgado F, Harish V, Das S, Celi LA. What do medical students actually need to know about artificial intelligence? *NPJ Digital Medicine*. 2020; 3, 86.
50. dos Santos P, Giese D, Brodehl S, Chon SH, Staab W, Kleinert R, Maintz D, Baeßler B. Medical students' attitude towards artificial intelligence: A multicentre survey. *European Radiology*. 2019; 29(4), 1640–1646.
51. Kumar R. Future of pediatric practice — Artificial intelligence beckoning? *Indian Pediatrics*. 59.6. 2022, 443–444.

11 Understanding Data Analysis and Why Should We Do It?

Sapna Rathod and Bhupendra Prajapati

11.1 INTRODUCTION

As the information age progresses, a variety of sources besides people and servers are producing data, including sensors built into handsets and fitness trackers, surveillance cameras, MRI scanners, and set-top boxes. Mining in structured data is crucial for bioinformatics applications because the majority of biological data is not kept in databases with a singular, flat table (1, 2). In actuality, bioinformatics databases, or BDBs, are organised and linked items that are joined by relations that signify a complex internal structure. Protein (3), small molecule, metabolic, and regulatory network (4) libraries are a few examples of BDBs. Biological data representations are structured and heterogeneous; they include large sequences (for example, 106 gene sequences), 2D large structures (for example, 105 or 106 spots on DNA chips), 3D structures (for example, a model of DNA phosphate), graphs, networks, expression profiles, and phylogenetic trees.

The mining of biological data involves a number of topics, including kernel techniques for microarray time series data classification (5). This categorization of gene expression time series has a wide range of potential uses in medicine and pharmacogenomics, including the diagnosis of diseases, the prediction of drug responses, or the prognosis of disease outcomes, all of which support individualised medical care. To retrieve structure and biochemical data as well as forecast protein function, graph kernel representations of proteins have been developed.

Finally, since finding clusters in high-dimensional data is often difficult, it is important to create techniques for figuring out which high-dimensional data subspaces are home to density-based clusters. Furthermore, high-dimensional data may be assembled in various feature space subspaces in diverse ways. Finding all high-dimensional data subspaces with groups is the goal of subspace clustering. Biological data pre-processing activities like data cleaning and data merging; microarray classification and clustering methods; comparing RNA structures using energetics and string characteristics; discovering the sequence features of various genome regions; inference of the subcellular location of protein activity; sequencing of events leading to protein folding; haplotype mining to identify disease markers; structure-based categorization of chemical compounds; index structures and special purpose measures for phylogenetic applications; very quick indexing techniques for sequences and pathways; and query languages for protein finding based on protein shape.

11.1.1 LARGE DATA IN BIOINFORMATICS

To find novel disease biomarkers, biologists no longer depend on conventional labs; instead, they turn to the vast and continuously expanding genomic data made accessible by several research groups. This new age of big data in bioinformatics is being fueled by cheaper and more efficient bio data collection technologies like automated genome sequencers. In recent years, bioinformatics has experienced a sharp increase in data quantity. About 40 petabytes of data about genes, proteins, and small molecules were stored in the EBI, one of the biggest biology data repositories, in 2014, compared to 18 petabytes in 2013.

DOI: 10.1201/9781003353751-11

Numerous organisations, including EBI, the NCBI in the US, and the National Institute of Genetics in Japan, are storing, processing, and disseminating sizable collections of biological databases all over the globe. For more accurate analytics, high volume data availability is advantageous, particularly in the immensely complicated subject of bioinformatics.

The structure of bioinformatics data is extremely heterogeneous. For inference and validation, many bioinformatics analytics issues call for the use of numerous heterogeneous, independent datasets. Additionally, because there are numerous unregulated organisations that produce bioinformatics data, the identical kinds of data are represented by different sources in various ways. The world's bioinformatics data is dispersed geographically and is enormous in both size and number of cases. However, the remaining data cannot be transferred due to their size (which makes them inefficient) and expense; there are also privacy concerns and other ethical considerations (6). This occasionally necessitates conducting a portion of the analysis remotely and sharing the findings.

Cloud computing technologies have been employed with great success to address these big data issues in bioinformatics. The Beijing Genomics Institute offered Gaea, a cloud-based gene analytics tool, and Bina Technologies, a joint venture between Stanford University and UC Berkeley, offered the Bina box.

11.1.1.1 Types of Large Data

Gene expression data, DNA, RNA, and protein sequence data, PPI data, pathway data, and GO data are the main types of massively sized data that are heavily used in bioinformatics study. However, other types of data, such as the human disease network and the disease gene association network, are also used and crucial for several study paths, including disease diagnosis. Typically, the expression levels are recorded for study using microarray-based gene expression profiling. Gene-sample data is one of three kinds of microarray data. The findings of the study may be used, among other things, to recommend biomarkers for disease detection and inhibition. There are numerous open sources for microrarray datasets, including Stanford Microarray Database, Gene Expression Omnibus from NCBI, and ArrayExpress from EBI. Sequence analysis involves processing DNA, RNA, or peptide sequences using a variety of analytical techniques in order to comprehend their characteristics, purposes, structures, and development. Forensic identification, evolutionary biology, the study of genomes and proteins and their connections to diseases and phenotypes, the search for potential medications, and other disciplines all use DNA sequencing. Although RNA sequencing is primarily employed as a substitute for microarrays, it can also be utilised for other things like mutation identification, post-transcriptional mechanism identification, viral and exogenous RNA detection, and polyadenylation identification. RDP, miRBase, and the DNA Data Bank of Japan are a few significant sequence libraries.

Protein–protein interactions (PPIs) offer vital details about every biological activity. As a result, the creation and analysis of PPI networks can provide accurate information about how proteins operate. As a result, cancer and Alzheimer's disease are just two of the many illnesses that are caused by abnormal PPIs. The PPI libraries DIP, STRING, and BioGRID are significant ones.

Understanding the molecular causes of an illness can be done through pathway analysis. Pathway analysis also predicts drug targets, aids in targeted literature searches, and finds genes and proteins linked to the aetiology of a disease. Reactome (7), Pathway Commons (8), and KEGG (9) are the three most prominent sources of pathway data.

Controlled vocabularies are used in the GO database to support searching at various levels. The GO database is used by numerous instruments for bioinformatics analysis. The majority of these tools are third-party-based, but the GO project also supports some of them, including AmiGO, DAG-Edit, and OBO-Edit. The creation of timelines for model organisms, human illnesses, and plant growth environments has all made significant use of the GO database. It has also been used to verify the results of semi supervised and unsupervised analytics from data.

11.1.2 DATA ANALYTICS PROBLEM IN BIOINFORMATICS

There are various problems in data analysis in bioinformatics which are addressed in detail and depicted in Figure 11.1.

11.1.2.1 Microarray Data Analysis

The decreasing cost and widespread use of microarray experiments make this method popular. Additionally, gene-sample-time space has been used to monitor progress in expression levels as time passes or across different disease stages. In order to find disease biomarkers, it has become possible to determine the genes that are influenced by a disease as gene expression data are collected over time at various disease advancement stages.

11.1.2.2 Gene Regulatory Networks (GRNs)

The rebuilding of a single GRN is made possible by the blending of numerous large-scale GRNs from different sources. System biologists may benefit from the local reconstruction of GRNs and subsequent merging of those GRNs through cloud infrastructure to better analyse a diseased network. The result can be used in genomic therapy as well. These difficult problems require quick, dependable, and scalable architectural solutions in order to identify irregular networks and rate the specific proteins for druggability.

The connection between various gene–gene networks discovered through gene-expression analysis is estimated by gene co-expression network analysis. The modifications that the gene clusters exhibit over time or during various phases of illness are revealed by differential co-expression investigation. This aids in establishing links between interesting characteristics and gene complexes.

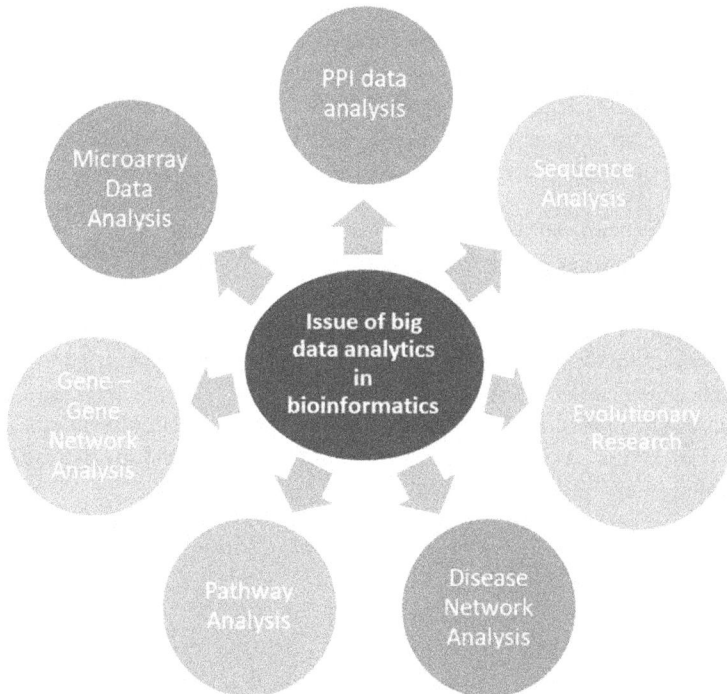

FIGURE 11.1 Different issues of big data analysis.

Massive data analytics systems are necessary for the examination of gene co-expression networks because this problem is difficult and extremely iterative.

11.1.2.3 PPI Data Analysis

PPI complexes and alterations in them make it difficult to provide extremely detailed information about various diseases. PPI Complex Analytics is a real big data issue because of the volume, velocity, and diversity of data. In order supply rapid and accurate PPI complex formation, authentication, and rating, it requires an effective and scalable design.

11.1.2.4 Sequence Analysis

The growing volume (in the order of petabytes) of DNA data deluge coming from countless sources has led to the discovery that the existing DNA sequencing tools are inadequate. Due to its more precise measurements of gene expression than microarray technology, RNA sequencing technology has arisen as a strong successor. Nevertheless, RNA sequence data also encompass supplementary information that must be extracted using sophisticated machine learning methods because it is frequently ignored. The knowledge on an individual's entire genome is provided by next-generation genome sequencing, which is orders of magnitude more comprehensive than techniques for genetic evaluation based on microarrays. Large-scale approaches are necessary to investigate the precise transition in genome sequences caused by specific illness and to compare with the results from that disease or from other diseases that are related to it.

11.1.2.5 Evolutionary Research

An important platform for the study and preservation of this vast amount of data is bioinformatics. The analysis of functional patterns of development and adaptation using microbial research and looking at ancient organisms has been a significant big data issue in bioinformatics.

11.1.2.6 Pathway Analysis

To forecast gene function, find biomarkers and traits, categorise patients and samples, and anticipate gene function, pathway analysis links genetic products with relevant phenotypes. Big data technologies are needed to conduct association analysis on enormous volumes of genetic, genomic, metabolomic, and proteomic data because these data sets have grown rapidly.

11.1.2.7 Disease Network Analysis

Understanding the relationships among diseases across networks requires the use of multi-objective links between diseases. Intelligent and effective analytics are needed because traditional network analytics methods cannot handle unstructured and heterogeneous data without compromising the quality of the information. Genes or mechanisms that are causal or predictive of disease-associated characteristics are characterised by complex networks of molecular phenotypes. Multiple, heterogeneous omics databases need to be analysed using the best integration techniques. To recognize and visualize complex data patterns for the purposes of analyzing and diagnosing illness genesis, large-scale machine computing tools are required.

11.2 TECHNIQUES FOR DATA ANALYSIS

11.2.1 Machine Learning

The use of machine learning methods in bioinformatics, network security, healthcare, banking and finance, and transportation has been found to be very practical and applicable. The study of algorithmic techniques that learn from data is known as machine learning. In machine learning, supervised and unsupervised learning methods are the two major categories of learning techniques (10). In supervised learning, a technique picks up information from a cluster of items with a class label, also known as a training set. Unknown objects, also known as test objects, are given labels based on the information that has been acquired. Unsupervised learning techniques, on the contrary, don't rely on the existence of prior information or training examples with class labels. The different categories of machine learning techniques are depicted in Figure 11.2.

For effective outcomes, all of these machine learning techniques need datasets to be pre-processed. One of the crucial pre-processing steps that improves results and cuts down on processing time is feature picking. Deep learning is one hybrid learning technique that has gained popularity recently and offers remarkably high precision. When using the available computational resources, conventional machine learning techniques are found to be insufficient for handling large amounts of data (11).

11.2.1.1 Feature Selection

The primary goal of feature selection is to choose a subset of the most pertinent and non-redundant characteristics that can improve a learning technique's performance. By eliminating unnecessary and redundant features, this method can enhance the presentation of prediction models while reducing the effects of the "curse of dimensionality," improving generalisation performance, accelerating the learning process, and upgrading the model interpretability. In order to extract the most crucial features from a large, ultrahigh dimensional collection, feature selection is crucial. The chosen feature set can be used to quickly and efficiently analyse a huge amount of data as to make a decision. By categorizing the traits as per their significance and selecting the most relevant features, the effectiveness of generalisation can be considerably increased. As a result of its characteristics as a SIP issue, feature selection is also regarded as being crucial for big data analytics (12). By iteratively choosing a subset of features to address the SIP problem, Tan et al. (13) suggested an effective feature selection algorithm that addresses a number of MKL subproblems.

When analysing protein sequences, a feature vector is crucial because it displays protein patterns with distinctive characteristics. However, a significant issue is that it has a large number of enormous characteristics, which not only make analysis more difficult but also make predictions less accurate. The feature selection issue in big data is addressed by Bagyamathi et al. (14) novel method to feature selection, which combines a more effective harmony search algorithm with rough set theory. For big data learning, Barbu et al. (15) suggest a novel feature selection strategy with annealing technique. In order to select incremental features for hybrid information systems, Zeng et al. (16) suggest the FRSA-IFS-HIS (AD) method, which applies fuzzy-rough set theory.

11.2.1.2 Supervised Learning

This model's goal is to anticipate the test instances' class labels grounded on the information learned from the accessible training instances. We can further differentiate between classification models, which concentrate on the projection of discrete (categorical) conclusions, and regression models, which foretell continuous outputs, within the supervised learning family. Decision trees,

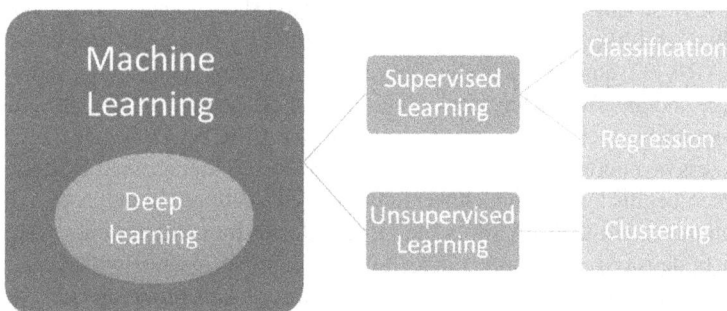

FIGURE 11.2 Types of machine learning techniques.

naive Bayes, SVMs, neural networks, and KNN are the most commonly employed techniques in various applications among the large number of models described in the literature (17). For parallel and distributed learning in big data analytics, we need more sophisticated supervised techniques like neural network classifiers, divide-and-conquer SVMs, and MM categorization models (18, 19). SVM is one of the most effective and popular guided learning techniques.

For the classification of big data, Nie et al. suggest a modified SVM called New Primal SVM (20). The technique employs two loss functions named L1-norm and L2-norm in ALM, a novel linear computational cost primal SVM solver. Haller et al. suggested using SVM analysis to identify Parkinson disease individuals on an individual basis (21). For the automatic diagnosis of diabetics, Giveki et al. (22) suggest a weighted SVM based on mutual information and modified cuckoo search. Bhatia et al. suggested another SVM-based decision support system for categorizing cardiovascular diseases that uses an integer-coded genetic algorithm to choose key features. SVM is used by Son et al. (23) to categorise individuals with heart failure.

GBDT can be distributed and parallelized using the methods proposed by Ye et al. (24). This approach used a MapReduced-based GBDT for horizontal data partitioning because it is very simple to convert GBDT to MapReduced model. Hadoop is unsuitable for this algorithm, according to the writers, because of the high communication overhead of HDFS (25). The fast, scalable, and distributable rxDTree decision tree proposed by Calaway et al. (26) can effectively predict decision trees on big data. This method is frequently employed in big data classification and regression issues.

If the dataset contains classifications for various user classes as class labels, a DT can be created to categorize unknown data. However, the training set will be much larger than normal, making the DT rule generation process difficult and time-consuming. Hall et al proposed (27) to modify DT learning to generate rules from a collection of DTs built simultaneously on manageable-sized training datasets seeks to solve this problem.

11.2.1.3 Unsupervised Learning

The class names of the objects are not used in unsupervised learning (28). Unsupervised clustering groups objects with the intention of maximising the space between objects within an identical cluster and minimising the distance between objects in various clusters (29). Calculating the distance between two items is a significant problem in clustering. For this, a number of closeness metrics, including Euclidean, Cosine, and city block distance, have been employed.

Two-dimensional data are analysed using clustering and bi-clustering, where each feature represents a characteristic of the objects. On such datasets, a different type of grouping known as tri-clustering (30) creates tri-clusters. Tri-clustering encourages simultaneous grouping of items, features, and time points.

11.2.1.3.1 *Existing Clustering Methods*

Partitional clustering, hierarchical clustering, density-based clustering, graph theoretic clustering, soft computing-based clustering, and matrix operation-based clustering are the major categories under which these clustering techniques can be categorised (31). Another, PAM, substitutes medoids for centroids. Two well-liked partitional clustering techniques that make use of sampling for big datasets are CLARA (32) and CLARANS (33). Various clustering methods are shown in Table 11.1.

Agglomerative and divisive techniques of hierarchical clustering can be distinguished (34). The well-liked agglomerative hierarchical clustering technique BIRCH (35) builds the CF tree first and extracts the clusters in a bottom-up manner. Another well-liked hierarchical clustering technique is CURE (36), which begins by creating clusters out of a few scattered items. A divisive hierarchical clustering technique called DIANA splits the largest cluster into smaller ones to identify splinter groups. The use of density-based clustering techniques identifies clusters with dense areas and low-density zones separating them (37). A very well-liked density-based grouping technique is DBSCAN (38). Another well-liked density-based grouping technique that makes use of the kernel density function is DENCLUE (39). Methods for grouping using graph theory are called "graph theoretic"

TABLE 11.1

Clustering Methods for Analysis

Method	Description
Hierarchical Method	Creates a cluster by partitioning in either a top-down and bottom-up manner
Partitioning Method	Relocate partitions by shifting from one cluster to another, which makes an initial partitioning
Density Model	Locating regions of higher density in a cluster
Grid-Based Clustering Method	Quantifying the object space into a finite number of cells
Model-Based Clustering Methods	Every cluster is hypothesized so that it can find the data which is best suited for the model
Constraint-Based Clustering Method	Application or user-oriented constraints are incorporated to perform the clustering

(40). CLIQUE is a grouping technique based on graph theory. The graph theoretic agglomerative hierarchical CA Chameleon (41) also makes use of the KNN graph.

Soft computation-based clustering techniques make use of fuzzy sets and neural networks, two examples of soft computation instruments. A clustering technique built on soft computing is fuzzy c-means (42). Another well-liked soft computing-based grouping technique, SOM (43), maps high-dimensional vectors to a two-dimensional grid space.

There are various bi-clustering and tri-clustering algorithms (44) proposed by research like OPSM (45), BIMAX (46), SAMBA (47), FLOC (48), ISA (49), TRICLUSTER (50) gTRICLUSTER (51), ICSM.

11.2.1.3.2 Clustering Methods for Big Data

While incremental clustering techniques manage high velocity data, parallel clustering techniques appear to be a solution for large volumes of data. Large size data can be handled by the DBSCAN, DENCLUE, CLARA, CLARANS, and CURE methods. K-mode and K-prototype techniques (52) are two variations of k-means that are used for large-scale categorical and mixed-type data, respectively. To reduce RAM usage and the quantity of scans performed on the dataset, Ordonez et al. (53) suggest a variant of the k-means algorithm. Bradley et al. offer a framework (54) to iteratively sample from a large dataset, improving the model with each iteration to ultimately create clusters. Using a kernel function, WaveCluster (55) applies the wavelet transform to transfer data from the spatial domain to the frequency domain.

On SIMD computers, Li et al. suggest parallel partitional and parallel single-linkage hierarchical CA. A concurrent MapReduce-based k-means clustering technique is proposed by Zhao et al. (56). Similar to this, PDBCSCAN (57) merges the findings after finding clusters from data spread across multiple machines. To reduce error, P-cluster (58) divides the items into groups. A parallel variant of BIRCH is called PBIRCH (59).

When computing new cluster centres, Chakraborty et al proposed incremental k-means clustering (60) uses only the current cluster centres and the recently arrived object. IGDCA (61) is a grouping technique based on incremental density. A multi-view clustering technique (62) based on the DBSCAN clustering method is proposed by Kailing et al.

According to a framework proposed by Zeng et al. (40), clustering is carried out independently in various feature spaces, and the findings are then iteratively projected and propagated in various graph-based link layers until they converge. A multi-view grouping technique is suggested by Chaudhuri et al. (63). In their multi-view spectral clustering proposal (64), Kumar et al. calculate eigen vectors in various feature spaces and utilise those eigen vectors to iteratively enhance the graph structures of other views.

11.2.1.4 Deep Learning

Deep learning uses supervised and/or unsupervised learning algorithms to characterize high-level concepts in data in an effort to learn from various levels of abstraction. It classifies data using hierarchical models. Pattern identification, computer vision, natural language processing, and speech recognition are just a few of the applications where deep learning techniques have been put to use. Deep model training has recently benefited from the development of efficient and scalable parallel methods (65). Multiple levels of data are represented by deep learning. Deep learning is also well suited for analysing diverse and unstructured data gathered from a variety of sources. A deep learning architecture called DBN (66) was created, allowing it to learn from both labelled and unlabelled data. When learning representations, Ngiam et al has proposed deep learning technique (38) integrates both audio and video data.

11.2.2 CLUSTERING ALGORITHMS

Through the process of clustering, data objects are gathered into a collection of disjointed classes where objects throughout a class have a high degree of resemblance and objects in different classes have a greater degree of dissimilarity. Clustering is a component of exploratory data analysis, where rules are ultimately discovered as a creative induction technique that highlights the need for theoretical and experimental model validations. The fact that it makes sense to combine both genes and samples is one of the traits of gene expression data (67). The differences between clustering assignments for gene expression data serve as the foundation for the dissimilarity between gene-based clustering and sample-based clustering.

11.2.2.1 *K*-means and Hierarchical Approaches

Both classifying genes and dividing up data can be done using these approaches. A partition-based grouping technique is K-means (68). Provided a pre-specified number K, K – implies divides the data collection into K distinct clusters if K is a predetermined number, minimising the sum of the squared distances of the components from their cluster centres. On the basis of gene expression data, K-means has been used to identify clusters (69) that comprise a sizable number of genes with related functions. Additionally, 18 motifs from upstream DNA-gene sequences within the same cluster were able to be extracted. These motifs represent potential candidates for new cis-regulatory elements. Compact clusters, which limit clusters to having a diameter of no more than d, are described in (70), while Adapt Cluster, a more effective method, is suggested in (71). In any event, either the number of clusters or a certain coherence threshold is necessary for the K-means algorithm and its derivatives. Like a mysterious black box, the clustering procedure. As a result, they are not adaptable to the data set's local structures and hardly allow for collaborative searching for coherent expression patterns.

11.2.2.2 SOM

SOMs are created using a single-layered neural network as their foundation (72). Each data object serves as a training sample for the reference vectors, which causes them to move towards the denser regions of the input vector space as they are trained to match the distributions of the input data set. All data points are mapped to the output neurons after the training process is complete in order to identify groups. In (73), hematopoietic differentiation is studied using the SOM method.

1,036 human genes' expression profiles are assigned to a 6 × 4 SOM. The SOM creates new hypotheses about hematopoietic differentiation by grouping genes into biologically significant clusters. This provides fascinating new insights into the differentiation process. The system SOMBRERO was recently created by the School of Information Technology at the National University of Ireland, Galway. SOMBRERO searches a collection of DNA sequences for over-represented patterns to identify regulatory binding sites (74).

11.2.2.3   MBA

MBA provides a statistical paradigm to explain the clustering of gene expression data. The data set's fundamental probability distributions should be mixed up, with each component representing a different cluster. The fact that MBA approaches offer an estimated chance that a given element belongs to a given cluster is a significant benefit. Gene expression data is a particularly good match for MBA's probabilistic feature because it is typical for a gene to engage in numerous cellular processes. Therefore, a single gene may strongly correlate with two distinct groups. MBA, however, presupposes that the data collection conforms to a particular distribution. This might not be the case since there is currently no defined standard framework for data on gene expression. Many MBA strategies assert that they have a multivariate Gaussian distribution.

11.2.2.4   Hidden Markov Model (HMM)

In this Markov model extension, a state has a possibility of radiating some output. The present state determines the likelihood of the following state because an HMM is a finite state machine with alternatives for each change. An HMM is a machine with a finite that has likelihoods for each transition, which means that the likelihood of the current state affects the likelihood of the next state. Each state has a potential to generate one of the detectable outputs, so the states are not instantly observable. They are frequently utilized when analyzing biological sequences to take into consideration relationships in time-series bio-data (75).

11.2.2.5   GBA (Graph Based Algorithm)

The classical graph theoretical issues that gave rise to the graph-based algorithms. Despite having a strong quantitative foundation, they might not be appropriate for gene expression data without modification. A large number of intermediate genes, for instance, may be strongly linked to groups of co-expressed genes in gene expression data. The CLICK technique creates a statistical framework for determining the level of coherence within a group of genes and the cut-off for the recursive splitting process.

11.2.2.6   HAs

Agglomerative, or bottom-up method, which views each data object at first as a separate cluster. The closest set of clusters are merged into a single cluster using agglomerative approaches at each stage. In an illustration of agglomerative hierarchical clustering is provided; it blends partitive K-means clustering and tree-structured vector quantization.

A divisional strategy that begins with a cluster holding all the data objects and works its way down. Divisive methods repeatedly split clusters until either a predetermined stop criterion is satisfied or each cluster includes just one data object (76).

The system is less susceptible to data normalisation and pre-processing. Additionally, the outcomes are strikingly comparable to those of self-organizing maps.

11.2.2.7   Clique Graph

The merger of distinct complete graphs results in an undirected graph. The goal of clustering is to distinguish as accurately as possible between the original and corrupted versions of a clique network. The CAST Algorithm is an illustration of a graph theoretic strategy that makes use of a divisive clustering strategy and depends on the notion of a clique graph. E-CAST, an improved variant of CAST, is discussed in (77). With CAST, a dynamic threshold is used, which is the primary distinction.

11.2.2.8   PBCA

These have been suggested as a way of evaluating the level of coherence displayed by a subgroup of genes on a subset of attributes. This method takes into consideration the fact that in molecular biology, only a small subset of the attributes is necessary for any cellular process to occur. The concept of bi-clustering, which was first introduced in (78), involves locating a division of the vectors and a subset of the dimensions such that the vector projections along those directions of each cluster are comparatively close to one another. On yeast and lymphoma data sets, bi-clustering of gene expression has been implemented using evolutionary methods (79). Simulated annealing (80) is the foundation of a stochastic algorithm (81) that is presented and verified on various data sets, demonstrating that Simulated Annealing frequently finds important bi-clusters.

11.2.2.9   ECA

Recently, this method for analysing biological data has been suggested in order to reduce the computational complexity of greedy algorithms and to enhance the space solution scan.

11.2.2.10   *GenClust* Algorithm

This has two key advantages: (a) an inventive search space coding that is simple, concise, and straightforward to update; and (b) the ability to work seamlessly with internal validation methods that are data-driven (82).

11.2.2.11   IGKA

To enhance the computation of the K-means algorithm, IGKA is an expansion of a genetic algorithm that has previously been suggested. The primary goal of IGKA (83) is to gradually cluster centroids and determine the objective value of Total Within-Cluster Variation when the likelihood of mutation is low.

11.2.3   PCA and SVD

11.2.3.1   PCA

A thorough method to gene expression analysis requires an upfront description of the data's structure. PCA and SVD are effective methods for getting this kind of characterization. One of the earliest methods of dimension reduction is PCA (84). Principal components (PCs), which are linear combinations of the original measurements, are sought after because they are capable of accurately representing the impacts of the original measurements. PCs may have much lower dimensions than the initial measurements because they are orthogonal to one another. PCA has been widely used in numerous statistical fields due to its ease of calculation and favourable statistical characteristics. Most recently, it has been used to lessen the dimensionality of high-throughput measurements in bioinformatics studies, especially gene expression studies (85).

PCs have been called "metagenes," "super genes," "latent genes," and other terms in gene expression research. The following are just a few examples of where PCA may be used in gene expression research: (i) exploratory analysis and data display (86), (ii) analysis based on clustering, and (iii) analysis based on regression. All genes are measured, and PCA is performed in research like (87). Additionally, PCA-based analysis has the option of including the hierarchical organisation of genes. Examples of available PCA packages include: (i) R: the prcomp function, (ii) SAS: procedures PRINCOMP and FACTOR, (iii) SPSS: factor function (data reduction), (iv) MATLAB®: princomp, (v) NIA array analysis tool (http://lgsun.grc.nia.nih.gov/ANOVA/) and others.

11.2.3.2   SVD

The SVD of any n × m matrix A (gene expression matrix) has the form,

$$A = USV^T, (32.1)$$

where U is an n × m orthonormal matrix, whose columns are called the left singular vectors of A (gene coefficient vectors), and V is an m × m orthonormal matrix, whose columns are called the right singular vectors of A (expression level vectors). S is a diagonal matrix $S = \text{diag}(s1,s2,...,sm)$. The diagonal elements of matrix S are, as a convention, listed in a descending order, $s1 \geq s2 \geq \cdots \geq sm \geq 0$, and are called the singular values of A.

Prior to clustering, SVD is occasionally used to extract the cluster structure from the data and minimise its dimensionality. The first few most important characteristic modes, which comprise the majority of the variations in the data, are typically used because they are uncorrelated and ordered.

Gene coefficient vectors (columns of matrix U) corresponding to the collection of the most important characteristic modes are examined by the gene selection algorithm. Each coefficient is compared to the threshold value, which is equivalent to a cutoff of 3 for statistical significance and equal to W n1/2, where n is the number of genes and W is a weight factor with a suggested value higher than 3. The corresponding gene is chosen for the clustering set if the element's magnitude exceeds the threshold number. In reality, we choose genes whose values of the index for the most important categories are sufficiently high. In other words, we select genes with suitably large coefficients for the most significant characteristic modes. As a consequence, we are able to identify a group of genes that display patterns that are "comparable" to the dominant modes.

11.2.4 MATHEMATICAL OPTIMIZATION

This is also known as operations research, combines a variety of mathematical modelling strategies to handle issues with the most effective planning or allocation of limited resources and, more generally, to effectively support decision-making.

11.2.5 GENE–GENE NETWORK ANALYSIS

Gene expression databases are very large and continue to grow. The FastGCN (88) tool takes advantage of parallelism in GPU architectures to efficiently identify co-expression networks. Arefin et al. (89) and McArt et al. (90) both suggest similar GPU-accelerated co-expression networks research techniques. To identify disease gene correlations, the UCLA Gene Expression Tool (UGET) (2) conducts extensive co-expression analysis. In disease networks, gene correlations are noticeably greater, and UGET calculates the correlations between all potential gene pairs. When evaluated on Celsius (91), the largest co-normalized microarray dataset of the Affymetrix-based gene expression data warehouse, UGET was found to be effective. WGCNA (92) is a well-known R package that can be utilised in an R-Hadoop distributed computing environment to conduct weighted gene co-expression network analysis.

BioPig (93), which can be immediately ported to many Hadoop infrastructures and scales automatically with the amount of information, is a notable Hadoop-based tool for sequence analysis. Another one of these tools is SeqPig (94). The Crossbow (95) utility performs large-scale whole genome sequence analytics on cloud platforms or on a native Hadoop cluster by combining Bowtie (96), an incredibly fast and memory-efficient short read aligner, and SoapSNP (97), an accurate genotyper. Stormbow (98), CloVR (99), and Rainbow (100) are additional cloud-based tools that have been created for large-scale sequence analysis. Other tools, like Vmatch (101) and SeqMonk23, are available for large-scale genome analysis without the use of big data technologies. Several tools have been created to support pathway analysis, including PP (102) to analyse expression data concerning metabolic pathways, GO-Elite (103) to express specific genes or metabolites, PathVisio (104) for investigation and drawing, directPA (105) to implement investigation in a high-dimensional space for recognizing pathways, Pathway-PDT (106) to conduct study employing raw genotypes in general nuclear families, and Pathview. To achieve high scalability, these tools don't make use of distributed computing platforms or build as cloud-based applications.

11.2.6 MINEPATH

MinePath, an online application (www.minepath.org), significantly enhances and expands the discriminant analysis method for identifying differentially expressed GRN submodules (107). The MinePath methodology involves three steps: (1) breaking down the target GRNs into their component subpaths; (2) interpreting and transforming each subpath into its binary active state (represented in binary (0,1) format); and (3) aligning and matching the binary representative of each subpath against the discretized binary representatives of the input gene-expression samples. Lastly, the differential power of each subpath is calculated using a metric that takes into account the percentage of matched samples (108).

The Weka API (www.weka.org) is seamlessly used to realise the following machine learning techniques: C4.5/J48, Support Vector Machines, and NaiveBay. The system's basic input consists of choosing from pre-stored records or uploading new (user-specified) gene-expression data, selecting preferred GRNs (all homo sapiens KEGG pathways are presented), and optionally parameterizing the subpath ranking metric and validating the algorithm. Additionally, MinePath has features that let you remove genes, edge connections, and other network components, as well as rearrange the network's structure, to reduce network complexity.

11.2.7   METATRANSCRIPTOMIC DATA ANALYSIS

Drug discovery and human health will both benefit greatly from metatranscriptomic analysis, which offers complementary insight into the gene expression profiles and even regulatory processes (109). In the meantime, a number of effective Web sites and pipelines for metatranscriptomic analysis have been created in recent years (20). Data collections for the analysis of metatranscriptomes can be used with MG-RAST. MetaPhlAn, the ChocoPhlAn pangenome library, and DIAMOND are used to speed up functional profiling and translated searches in HUMAnN2, which is intended for analysis in both metagenomics and metatranscriptomics (110). Various tools utilised for analysis of metatranscriptomic and metagenomic data are depicted in Table 11.2.

Transfer RNA (tRNA) reads are removed and aligned using SortMeRNA and BLASTN, and mRNA reads are assigned using MegaBLAST. The assignment bit scores are then classified by Leimena-2013 to identify the phylogenetic origin (111). For the study of gut microbiome data, SAMSA is a comprehensive workflow with four stages: preprocessing, annotation, aggregation, and analysis (112). To expedite taxonomic and gene expression analyses of active microbial communities, MetaTrans makes use of multi-threading processors.

TABLE 11.2
Various Bioinformatic Tools Utilized for Analysis of Metatranscriptome and Metagenome

Tools	Description	Step
FastQ Screen	Match a library with libraries expectation DB	
FastQC	Reads quality, seq length distribution and GC%	
Cutadapt	Find and remove adapters, primers, poly-A tails and others	Quality report
MultiQC	Summarise results	
Trimmomatic	Trimming tool for Illumina	
Diginorm	Downsampling reads	
Bbtools	Trims and filters by k-mers and entropy and downsampling reads	Trimming
Khmer	k-mer error trimming	
kneadData	Host sequences removal	
miARma	Quality, trimming and host sequences removal	Host removal
Meta SPAdes	Multiple k-mers better assemblies with different abundances	
MEGAHIT	Iterative k-mer fast and co-assembly robust metagenomic tool	Assembly
CheckM	Evaluate quality of assemblies and contamination	

11.2.8 METAGENOMIC DATA ANALYSIS

11.2.8.1 16S rRNA Analysis

This is one of the most well-liked and reasonably priced methods for profiling the species makeup of microbiota. Due to the lack of tools for taxonomy assignment and library demultiplexing as high-throughput pyrosequencing has developed (113), QIIME aims to conduct downstream analyses. To deal with the issue of analysis and database deposition using raw sequencing data, QIIME utilises the PyCogent toolkit (114). In order to create OTUs from scratch using next-generation reads, UPARSE was created (115). It improves richness estimates on mock communities and gets high accuracy in biological sequence recovery. UPARSE clusters the remaining reads after quality-filtering, trimming to a set length, and, if desired, removing singleton reads.

MOTHUR aims to be a complete software suite that enables users to analyse community sequence data using just one piece of software. The methods used in earlier tools like DOTUR, SONS, TreeClimber, LIBSHUFF, -LIBSHUFF, and UniFrac are implemented in MOTHUR. DADA2 is a model-based method for PCR error correction without OTU construction. While producing few erroneous positives, it seeks to detect subtle differences in 454-sequenced amplicon data (116). To enhance the DADA algorithm, DADA2 employs a novel quality-aware model of Illumina amplicon errors. In order to overcome the limitation of fine-scale resolution descriptions of microbial communities, MED is used to differentiate high-throughput sequencing data sets using de facto similarity criteria and pairwise sequence alignments for similarity assessment. MED iteratively separates amplicon sequence data into homogenous OTUs that serve as the input for alpha- and beta-diversity studies (117).

11.2.8.2 Species Level Metagenomic Data Analysis

There are six instruments for metagenomic analysis, including MG-RAST (20), MetaPhlAn2 (118), Kraken (119), CLARK (120), FOCUS (121), SUPERFOCUS (122), and FOCUS. Bowtie2 (123) and UCLUST (124) are specifically used by MetaPhlAn2 as its main algorithms, Kraken uses exact k-mer alignment; when reporting organism profiles, FOCUS employs NNLS, while CLARK utilises reduced sets of k-mers (i.e., DNA words of length k); SUPER-FOCUS reports the subsystems found in metagenomic data sets and profiles their abundances using a reduced reference database, and MG-RAST isolates multiple features to assist users in assessing sequence quality and resolving some of the more frequent issues. Kraken is capable of genus-level sensitivity and accuracy comparable to those attained by Megablast, the fastest BLAST programme (125). MG-RAST provides post annotation analysis and visualisation directly through the Web interface thanks to tools like matR (126), which utilise the MG-RAST API to rapidly download data from the processing pipeline. Each sequence categorised by Kraken produces a single line of output in Kraken (http://ccb.jhu.edu/software/kraken/MANUAL.html), which includes the letter code of classification, sequence ID, taxonomy ID, length of the sequence in bp, and LCA mapping findings of each k-mer. The hit count in the target, length of the object, gamma ratio, target with the highest hit count, confidence score, etc. are all included in CLARK's findings. FOCUS can create the STAMP format, which has a graphical user interface and enables straightforward statistical result analysis and the creation of publishable quality plots for determining the biological significance of features in a metagenomic profile. It is available at http://kiwi.cs.dal.ca/Software/STAMP.

11.2.8.3 Strain Level Metagenomic Data Analysis

Numerous human diseases, including cancer, have been linked to a specific microorganism or collection of microorganisms (127). Additionally, various strains of the identical species may have various effects on human health, such as the extremely virulent *Escherichia coli* strain O157:H7 (128). The majority of strains within a genus are not pathogenic, though. Determining human-microbial interactions thus relies on the identity and characteristics of microbial strains in the surroundings and among human hosts (129).

To analyse breeds at the metagenome level, there are five tools: StrainPhlAn (130), PanPhlan (131), Constrains (132), Sigma (133), and LSA (134). They seek to characterise microbes' functional potential and recognise microbes, two tasks that are crucial for pathogen discovery, epidemiology, population genomics, and biosurveillance. To map the provided reads to the MetaPhlAn2 marker database, StrainPhlAn utilises MetaPhlAn2. For its metagenomic read mapping, quality filtering, and per-base coverage calculations, PanPhlan utilises Bowtie2 and SAMtools (135). The Sigma algorithm employs a read mapping approach and a probabilistic model to sample reads with defects from genomes with unfamiliar abundances. LSA employs a deconvolution algorithm to recognise groups of hashed k-mers that characterise potential variables.

11.3   CHALLENGES AND ISSUES IN BIG DATA ANALYSIS

Big data are incremental and dispersed in addition to having volume, velocity, and diversity. These characteristics of large data make it very challenging for conventional data analytics to operate quickly and correctly. Machine learning techniques may be utilised for managing big data analytics because they were created in the field of computer science with rationales such as speed and efficiency.

11.3.1   CHALLENGES

Big data cannot be effectively analysed and visualised using the methods used for conventional databases. The traditional techniques for data analytics are challenged by the volume, velocity, variety, distributedness, and incremental nature of such data. According to a 2014 study from Napatech, a manufacturer of high-speed network accelerators, network traffic will grow by 23% annually through 2018. The expanding rise in the utilisation of mobile devices and the instruments they are connected to has largely been responsible for the expansion of huge data in current years. The speed of data generation and transmission is growing along with the volume of data. The typical mobile network connection speed in 2014 was 1,683 kbps, and the Cisco report (136) predicts that it will increase to about 4.0 Mbps by 2019. High data velocity makes real-time analytics on big data more challenging. Even though distributed and parallel computing techniques can scale batch mode analytics to high data velocity, the performance of the analytics is badly hampered by the slow I/O operations.

Additionally, the nature of these constantly generated data is very heterogeneous. Traditional databases are built up according to a number of predefined schemas. After the processes of extraction, transformation, and loading, data is stored and updated in data warehouses. An organized database, like a data warehouse, is entirely unsuitable for dynamic storage and real-time retrieval because big data systems are constantly acquiring new data at high speeds and with considerable variety from heterogeneous sources.

Due to these difficulties, conventional data analytics methods like statistical analysis and machine learning are ineffective when used with big data in their raw form. As a result, it is necessary to examine the issue of machine learning enabled analytics from a big data viewpoint.

Data privacy is a main problem with big data analytics, specifically in the bioinformatics and healthcare sectors. Data sources may utilise data anonymity or only share a portion of the total data in order to protect sensitive information. Analytics performed on incomplete or private data may be more difficult and ineffective.

11.3.2   ISSUES

Big data analytics call for handling enormous amounts of continuously expanding structured, semi-structured, poly-structured, and unstructured data. Time-bound computation is an extra requirement of real-time analytics. In unstructured material, patterns and relations can be discovered using AI techniques. However, scalability and performance problems with traditional data analytics on big data are addressed below.

1) There is still a demand for a comprehensive big data analytics infrastructure that is highly reliable and capable of managing vast amounts of diverse data both in batches and constantly in real time.

2) The best way to deal with the enormous amount of big data is through distributed computing. However, the majority of methods to artificial intelligence, data mining, and statistical analysis were not initially intended for distributed computing. Though distributed algorithms have been suggested in the literature, most of it is academic research, and none of the MapReduce frameworks have a robust implementation for them (137).

3) There is no standard data format in a large data store. Big data analytics must process heterogeneous data that has been gathered by various kinds of sensors. Therefore, to extract a coherent meaning from disparate data, intelligent programmes are needed. The intricacy of analytics rises as a result.

4) Additional issues like data duplication and inconsistency are introduced by unstructured, semi-structured, and poly-structured data. Due to the heterogeneity and amount of the data, pre-processing is expensive. Traditional data analytics methods have been found to be expensive in terms of time and area complexity when handling noisy and inconsistent data.

5) Big data analytics must mine databases at various abstraction levels. As a result, analytics methods are now considerably more complex, but allowing biologists to analyse the data at different levels of abstraction aids in their understanding of the significance of semantic biological data.

6) A critical examination problem implicitly asks for (1) co-articulation and administrative organisations for large and varied human and other small cluster datasets to be developed more quickly and (2) common and unique highlights to be similar and broken down more quickly.

7) Creating a cost-effective, flexible design that enables extensive information analysis to query a variety of natural information sources on any disease group is a challenging task.

8) Another research problem is the creation of an integrated system for the quicker examination of big and varied quality articulation information collections over the GST region. This will allow for the identification of common and unique patterns that aid in the diagnosis of infections. In order to validate the mining findings, the structure should also allow the use of articulation, semantic, geographic, and succession similarities both locally and remotely.

9) Evaluation of large-scale diseased-compared GRN in the TF-target prediction of both networks and prioritised patient care are based on a topological study of both networks.

10) A multifaceted view of a group that is frequently distinct in terms. A significant information investigation problem is the dynamic GRN portrayal.

11) The majority of computer-intensive and insatiable derivation methods. They occasionally fall short for larger companies. Without jeopardising the accuracy of the induction findings, the appropriated figuring model utilising MapReduce and Hadoop could be investigated as a possibility.

12) The re-establishing of the bound-together GRN is made possible by the integration of a huge number of GRNs from various sources. Framework scientists are now able to collect terabytes of data thanks to the dramatically improved sequencing technology. It is frequently difficult to move such large papers.

11.3.2.1 Privacy Issues

Individual privacy protection is a long-standing issue in biomedical study. Data anonymization and informed consent are common safeguards for preserving participant privacy in research and clinical settings. Personal data is typically kept in facilities with limited access in addition to them. Big data analysis has been made possible by the switch from locked filing boxes to digital databases, but there are also new dangers.

A genome can disclose private information about family members while also being uniquely identifiable. Given that we each share 50% of our genome with our parents, siblings, and children (and 25% with our grandparents, grandchildren, aunts/uncles, etc.), any privacy violation involving genome-based information about an individual's (past or future) health status may have an impact on other family members. According to Chow-White et al. (138), it creates a new group of privacy concerns for familial networks.

Different tactics have been used in this situation to safeguard the privacy of those who take part in genomic research projects. For example, the ELSI group established several anonymization practises to protect privacy in the 1000 Genomes Project. These practises included oversampling (i.e., seeking more people than the final count needed to participate, so that not even participants could be sure of their inclusion in the study), and not collecting any personal data besides sex. In order to facilitate data sharing through a web tool and safeguard participant privacy, other initiatives encourage the publication of aggregated data, such as allele frequency or allele-presence information.

The Global Alliance for Genomics and Health's (GA4GH, http://genomicsandhealth.org/) Beacon Project, which only gives allele-presence data, is a prime example of this strategy. Despite our best efforts, in the age of big data, internet access, and data mining algorithms, we cannot promise complete privacy to participants in genomic initiatives. Chow-White et al. noted that de-identifying and aggregating data is "the ability to conceal a person's unique genome profile in a DNA haystack. Nevertheless, computational data mining techniques are excellent at locating needles in haystacks and connecting them to other needles in related haystack".

DTC genetic services are a possible source of privacy risks. For the majority of them, the demand for genetic ancestry services is rising, particularly in the USA. These services reconstruct genetic ancestry primarily by identifying genetic markers that are uniparentally inherited, i.e., those that can only be passed down through the maternal line (mitochondrial DNA) or the paternal line (Y chromosome). Along with search tools to look for possible relatives, databases of information on genetic ancestry are also accessible online. Consider that Ysearch (www.ysearch.org), one of the largest and most well-known genetic genealogy databases, has more than 190,000 records and more than 100,000 unique surnames to give you an idea of the scale of these databases. In 2013, Melissa Gymrek and associates came to the realisation that the data in these databases could be used to track down anonymous participants in open-access sequencing initiatives (139). By matching the Y-chromosome markers from three known community genomes with the information from some genetic datasets, they first identified the surname of one of the genome. The authors then concentrated on the privacy of existing de-identified public datasets because personal genomes could be identified. By employing a similar strategy, the authors were able to completely identify five anonymous Utahns with multigenerational European ancestry who had been involved in some of the most significant genetic projects, including the 1000 Genomes Project and the HapMap. In total, the private of close to 50 people from the CEU pedigrees was violated.

Reidentification efforts may be possible even with combined data. Shringarpure and Bustamante, for instance, recently showed that it was possible to locate a specific genome, or "beacon," in a dataset that held only allele-presence data (140). A person's membership in a beacon may disclose health-related information about them or members of their family because most beacons summarise the genomic data of cohorts with a particular illness of interest. Despite the fact that some tactics can be used to successfully reduce privacy risk (50), there is no such thing as 0% risk, not even in aggregated datasets.

Most nations have strict privacy laws in place, but only a small number of them have complete policies in place to control genomic data. The Genetic Information Nondiscrimination Act (GINA) in the USA expressly forbids genetic information-based discrimination in jobs and health insurance, and some states have expanded protection to include other types of discrimination. For instance, California promoted CalGINA to shield people from genetic discrimination in a variety of settings, including housing and schooling. Regarding genetic information, the position in the EU is not uniform and can vary from one country to another. The EU Data Protection Directive, however, governs the safety of all types of data, comprising data pertaining to health.

11.3.2.2 Ethical Issues

Today, a number of nations keep DNA databases of criminals. Sadly, there have been instances where the DNA of individuals who were detained but not found guilty was mistakenly added to the database. In this situation, it is possible to see DNA fingerprinting as a tool that invades people's privacy and makes their private information readily accessible to others. Genetically engineered new species might cause ecological issues. The desire to produce more crop goods has led to the genetic modification of plants, which has sparked intense debates about political, ethical, and social issues. The effects of genetically engineered animals on the environment are unpredictable. Additionally, a lot of genetically modified foods use donors that are microorganisms whose potential to cause allergies is either unstudied or unclear. Additionally, novel gene combinations and genes derived from non-food sources may cause allergic reactions in some people or exacerbate already-existing ones. The misuse of this technology in the creation of biological weapons or warfare is the greatest drawback in the context of applications of genetic engineering in human life. Transgenic animals and plants hold great promise, but they also pose serious concerns about how far we should push the boundaries of genetic engineering. Political concerns are raised by this, and they have been discussed in judicial proceedings, legislative hearings, and regulatory actions. The first gene-edited infants, named Lulu and Nana, who are innately immune to the human immunodeficiency virus were recently produced by Chinese scientist Jian-kui He (HIV). He altered the babies' germline genes using the CRISPR-Cas9 method. Because of ethical and scientific concerns, China's guidelines and laws forbid germline genome editing on human embryos for clinical use. In addition to breaking other ethical and regulatory standards, Jian-kui He's human experiments broke these Chinese laws 122. According to Chinese scientists, CRISPR-Cas should not be used to create genetically altered offspring because of significant off-target risks and related ethical concerns. Therefore, this gene modification may not provide the babies with many significant advantages while presenting unknown and uncontrollable risks to them and their future generations (68).

11.4 CONCLUSION

The world's bioinformatics data is vast, heterogeneous, incremental, and spatially dispersed. As a result, big data analytics methods are needed to address the issues in biology. A complete big data analytics solution that is quick, fault-tolerant, large-scale, incremental, distributed, and optimised for iterative and complicated computations is not offered by the current big data architectures.

In essence, bioinformatics is a puzzling field of study. There would not be a singular computational approach that was best for all datasets and organisations. Globally, bioinformatics is extensive, diverse, incremental, and geologically adapted. Therefore, it's crucial to use large-scale knowledge study techniques to address the issues with bioinformatics. In the field of bioinformatics, issues, knowledge sources, and informational formats are complex in character.

Detailed knowledge examination and appropriate registration, or distributed computing, are becoming more common in bioinformatics applications, and a community of figures is used to organise and break down information. The greatest innovation that can transform bioinformatics uses is machine learning techniques. Deep learning is also applied to biology to solve even more astounding issues. It is common practise to use in-depth bioinformatics knowledge to assist research efforts to understand the human genome and find cures for various diseases.

For some bioinformatics issues, there is still not enough knowledge available on the newest instruments. Similar to the PPI network analysis, study into disease networks, and other significant bioinformatics issues, huge information devices in Hadoop or the cloud also fall short. Immense studies on knowledge in bioinformatics ought to be acceptable given the rapid advancements in information production, useful applications like machine learning, and study opportunities as a result of the abundance of data in the area.

REFERENCES

1. Matsuda T, Motoda H, Yoshida T, Washio T, editors. *Mining Patterns from Structured Data by Beam-Wise Graph-Based induction*: Springer; 2002.
2. Day A, Dong J, Funari VA, Harry B, Strom SP, Cohn DH, Nelson SF. Disease gene characterization through large-scale co-expression analysis. *PLOS One*. 2009;4(12):e8491.
3. Schäffer AA, Aravind L, Madden TL, Shavirin S, Spouge JL, Wolf YI, et al. Improving the accuracy of PSI-BLAST protein database searches with composition-based statistics and other refinements. *Nucleic Acids Research*. 2001;29(14):2994–3005.
4. Karp PD, Riley M, Saier M, Paulsen IT, Paley SM, Pellegrini-Toole A. The EcoCyc and MetaCyc databases. *Nucleic Acids Research*. 2000;28(1):56–9.
5. Vert JP. Support vector machine prediction of signal peptide cleavage site using a new class of kernels for strings. *Pacific Symposium on Biocomputing*. 2002:649–60.
6. Marx V. Biology: The big challenges of big data. *Nature*. 2013;498(7453):255–60.
7. Croft D, O'Kelly G, Wu G, Haw R, Gillespie M, Matthews L, et al. Reactome: A database of reactions, pathways and biological processes. *Nucleic Acids Research*. 2011;39(Database issue):D691–7.
8. Cerami EG, Gross BE, Demir E, Rodchenkov I, Babur O, Anwar N, et al. Pathway commons, a web resource for biological pathway data. *Nucleic Acids Research*. 2011;39(Database issue):D685–90.
9. Kanehisa M, Goto S. KEGG: kyoto encyclopedia of genes and genomes. *Nucleic Acids Research*. 2000;28(1):27–30.
10. Bhattacharyya DK, Kalita JK. *Network Anomaly Detection: A Machine Learning Perspective*: CRC Press; 2013.
11. Floridi L. Big data and their epistemological challenge. *Philosophy and Technology*. 2012;25(4):435–7.
12. López M, Still G. Semi-infinite programming. *European Journal of Operational Research*. 2007;180(2):491–518.
13. Tan M, Tsang IW, Wang L. Towards ultrahigh dimensional feature selection for big data. *Journal of Machine Learning Research*. 2014 15:1371–1429.
14. Bagyamathi M, Inbarani HH. A novel hybridized rough set and improved harmony search based feature selection for protein sequence classification. In: Hassanien AE, Azar AT, Snasael V, Kacprzyk J, Abawajy JH, editors. *Big Data in Complex Systems: Challenges and Opportunities*: Springer International Publishing; 2015:173–204.
15. Barbu A, She Y, Ding L, Gramajo G. Feature selection with annealing for computer vision and big data learning. *IEEE Transactions on Pattern Analysis and Machine Intelligence*. 2017;39(2):272–86.
16. Zeng A, Li T, Liu D, Zhang J, Chen H. A fuzzy rough set approach for incremental feature selection on hybrid information systems. *Fuzzy Sets and Systems*. 2015;258:39–60.
17. Duda RO, Hart PE. *Pattern Classification*: John Wiley & Sons; 2006.
18. Hsieh C-J, Si S, Dhillon I, editors. *A Divide-and-Conquer Solver for Kernel Support Vector Machines*: PMLR; 2014.
19. Djuric N. *Big Data Algorithms for Visualization and Supervised Learning*: Temple University; 2013.
20. Meyer F, Paarmann D, D'Souza M, Olson R, Glass EM, Kubal M, et al. The metagenomics RAST server–a public resource for the automatic phylogenetic and functional analysis of metagenomes. *BMC Bioinformatics*. 2008;9(1):1–8.
21. Haller S, Badoud S, Nguyen D, Garibotto V, Lovblad KO, Burkhard PR. Individual detection of patients with Parkinson disease using support vector machine analysis of diffusion tensor imaging data: Initial results. *American Journal of Neuroradiology*. 2012;33(11):2123–8.
22. Giveki D, Salimi H, Bahmanyar G, Khademian Y. Automatic detection of diabetes diagnosis using feature weighted support vector machines based on mutual information and modified Cuckoo search. arXiv Preprint ArXiv:12012173. 2012.
23. Son YJ, Kim HG, Kim EH, Choi S, Lee SK. Application of support vector machine for prediction of medication adherence in heart failure patients. *Healthcare Informatics Research*. 2010;16(4):253–9.
24. Ye J, Chow J-H, Chen J, Zheng Z, editors. *Stochastic Gradient Boosted Distributed Decision Trees*: Association for Computing Machinery, New York City, United States; 2009.
25. Borthakur D. The hadoop distributed file system: Architecture and design. *Hadoop Project. Website*. 2007;11(2007):21.
26. Calaway R, Edlefsen L, Gong L, Fast S. Big data decision trees with r. Revolution. 2016.
27. Hall LO, Chawla N, Bowyer KW, editors. *Decision Tree Learning on Very Large Data Sets*: IEEE; 1998.
28. Kluger Y, Basri R, Chang JT, Gerstein M. Spectral biclustering of microarray data: coclustering genes and conditions. *Genome Research*. 2003;13(4):703–16.

29. Tan P-N, Steinbach M, Kumar V. Data mining cluster analysis: Basic concepts and algorithms. *Introduction to Data Mining.* 2013;487:533.
30. Ahmed HA, Mahanta P, Bhattacharyya DK, Kalita JK, Ghosh A, editors. *Intersected Coexpressed Subcube Miner: An Effective Triclustering Algorithm*: IEEE; 2011.
31. Jain AK, Murty MN, Flynn PJ. Data clustering: A review. *ACM Computing Surveys (CSUR).* 1999;31(3):264–323.
32. Kaufman L, Rousseeuw PJ. *Finding Groups in Data: An Introduction to Cluster Analysis*: John Wiley & Sons; 2009.
33. Ng RT, Han JCLARANS : A method for clustering objects for spatial data mining. *IEEE Transactions on Knowledge and Data Engineering.* 2002;14(5):1003–16.
34. Berkhin P; 2006. A survey of clustering data mining techniques. *Grouping Multidimensional Data: Recent Advances in Clustering*: Springer-Verlag, Berlin Heidelberg; 2006:25–71.
35. Zhang T, Ramakrishnan R. Livny M. Birch: An efficient data clustering method for very large databases. *ACM Sigmod Record.* 1996;25(2):103–14.
36. Guha S, Rastogi R, CURE, Shim K. An efficient clustering algorithm for large databases. *ACM Sigmod Record.* 1998;27(2):73–84.
37. Kriegel HP, Kröger P, Sander J, Zimek A. Density-based clustering. *Wiley Interdisciplinary Reviews: Data Mining and Knowledge Discovery.* 2011;1(3):231–40.
38. Ngiam J, Khosla A, Kim M, Nam J, Lee H, Ng AY, editors. *Multimodal Deep learning*: 2011.
39. Hinneburg A, Keim DA. *An Efficient Approach to Clustering in Large Multimedia Databases with Noise*: Bibliothek der Universität Konstanz; 1998.
40. Hubert LJ. Some applications of graph theory to clustering. *Psychometrika*: Springer Nature; 1974;39(3):283–309.
41. Karypis G, Han E-H, Kumar V. Chameleon: Hierarchical clustering using dynamic modeling. *Computer.* 1999;32(8):68–75.
42. Höppner F, Klawonn F, Kruse R, Runkler T. *Fuzzy Cluster Analysis: Methods for Classification, Data Analysis and Image Recognition*: John Wiley & Sons; 1999.
43. Kohonen T. The self-organizing map. *Proceedings of the IEEE.* 1990;78(9):1464–80.
44. Cheng Y, Church GM, editors. *Biclustering of Expression Data*: American Association for Artificial Intelligence (AAAI) Press, USA; 2000.
45. Ben-Dor A, Chor B, Karp R, Yakhini Z, editors. *Discovering Local Structure in Gene Expression Data: The Order-Preserving Submatrix problem*: Association for Computing Machinery, New York City, United States; 2002.
46. Prelić A, Bleuler S, Zimmermann P, Wille A, Bühlmann P, Gruissem W, et al. A systematic comparison and evaluation of biclustering methods for gene expression data. *Bioinformatics.* 2006;22(9):1122–9.
47. Tanay A, Sharan R, Kupiec M, Shamir R. Revealing modularity and organization in the yeast molecular network by integrated analysis of highly heterogeneous genomewide data. *Proceedings of the National Academy of Sciences of the United States of America.* 2004;101(9):2981–6.
48. Yang J, Wang H, Wang W, Yu P, editors. *Enhanced Biclustering on Expression Data*: IEEE; 2003.
49. Bergmann S, Ihmels J, Barkai N. Iterative signature algorithm for the analysis of large-scale gene expression data. *Physical Review. Part E.* 2003;67(3):031902.
50. Raisaro JL, Tramer F, Ji Z, Bu D, Zhao Y, Carey K, et al. Addressing Beacon re-identification attacks: Quantification and mitigation of privacy risks. *Journal of the American Medical Informatics Association.* 2017;24(4):799–805.
51. Jiang H, Zhou S, Guan J, Zheng Y, editors. *gTRICLUSTER: A More General and Effective 3D Clustering Algorithm for Gene-Sample-Time Microarray data*: Springer; 2006.
52. Huang Z. Extensions to the k-means algorithm for clustering large data sets with categorical values. *Data Mining and Knowledge Discovery.* 1998;2(3):283–304.
53. Ordonez C, Omiecinski E. Efficient disk-based K-means clustering for relational databases. *IEEE Transactions on Knowledge and Data Engineering.* 2004;16(8):909–21.
54. Bradley PS, Fayyad U, Reina C. Scaling clustering algorithms to large databases. *Knowledge Discovery Data Mining*: AAAI Press; 1998.
55. Sheikholeslami G, Chatterjee S, Zhang A. Zhang A. WaveCluster: A wavelet-based clustering approach for spatial data in very large databases. *The VLDB Journal.* 2000;8(3–4):289–304.
56. Zhao W, Ma H, He Q, editors. *Parallel k-Means Clustering Based on Mapreduce*: Springer; 2009.
57. Xu X, Jäger J, Kriegel H-P. A fast parallel clustering algorithm for large spatial databases. *High Performance Data Mining: Scaling Algorithms, Applications and Systems.* 2002:263–90.

58. Judd D, McKinley PK, Jain AK. Large-scale parallel data clustering. *IEEE Transactions on Pattern Analysis and Machine Intelligence*. 1998;20(8):871–6.
59. Garg A, Mangla A, Gupta N, Bhatnagar V, editors. *Pbirch: A Scalable Parallel Clustering Algorithm for Incremental Data*: IEEE; 2006.
60. Chakraborty S, Nagwani NK, editors. *Analysis and Study of Incremental k-Means Clustering Algorithm*: Springer; 2011.
61. Chen N, Chen A-z, Zhou L-x. An incremental grid density-based clustering algorithm. *Journal of Software*. 2002;13(1):1–7.
62. Kailing K, Kriegel H-P, Pryakhin A, Schubert M, editors. *Clustering Multi-Represented Objects with Noise*: Springer; 2004.
63. Zeng H-J, Chen Z, Ma W-Y, editors. *A Unified Framework for Clustering Heterogeneous Web Objects*: IEEE; 2002.
64. Chaudhuri K, Kakade SM, Livescu K, Sridharan K, editors. *Multi-View Clustering via Canonical Correlation Analysis*: Association for Computing Machinery, New York City, United States; 2009.
65. Kumar A, Daumé H, editors. *A Co-training Approach for Multi-View Spectral Clustering*: Association for Computing Machinery, New York City, United States; 2011.
66. Hinton GE, Salakhutdinov RR. Reducing the dimensionality of data with neural networks. *Science (New York, NY)*. 2006;313(5786):504–7.
67. Ben-Dor A, Shamir R, Yakhini Z. Clustering gene expression patterns. *Journal of Computational Biology: A Journal of Computational Molecular Cell Biology*. 1999;6(3–4):281–97.
68. Jayanitha S, Kartheeswaran T. *Ethical, Legal and Social Issues in Bioinformatics Applications*: Association for Computing Machinery, New York City, United States; 2019.
69. Tavazoie S, Hughes JD, Campbell MJ, Cho RJ, Church GM. Systematic determination of genetic network architecture. *Nature Genetics*. 1999;22(3):281–5.
70. Heyer LJ, Kruglyak S, Yooseph S. Exploring expression data: Identification and analysis of coexpressed genes. *Genome Research*. 1999;9(11):1106–15.
71. De Smet F, Mathys J, Marchal K, Thijs G, De Moor B, Moreau Y. Adaptive quality-based clustering of gene expression profiles. *Bioinformatics*. 2002;18(5):735–46.
72. Kohonen T *Self-Organization and Associative Memory*: Springer Science & Business Media; 2012.
73. Tamayo P, Slonim D, Mesirov J, Zhu Q, Kitareewan S, Dmitrovsky E, et al. Interpreting patterns of gene expression with self-organizing maps: Methods and application to hematopoietic differentiation. *Proceedings of the National Academy of Sciences of the United States of America*. 1999;96(6):2907–12.
74. Mahony S, Golden A, Smith TJ, Benos PV. Improved detection of DNA motifs using a self-organized clustering of familial binding profiles. *Bioinformatics*. 2005;21(suppl_1):i283-i91.
75. Di Gesù V, editor. *Data Analysis and Bioinformatics*: Springer; 2007.
76. Sultan M, Wigle DA, Cumbaa CA, Maziarz M, Glasgow J, Tsao M-S, Jurisica I. Binary tree-structured vector quantization approach to clustering and visualizing microarray data. *Bioinformatics*. 2002;18(suppl_1):S111–S9.
77. Bellaachia A, Portnoy D, Chen Y, Elkahloun AG, editors. *E-CAST: A Data Mining Algorithm for Gene Expression Data*: Association for Computing Machinery, New York City, United States; 2002.
78. Mirkin B *Mathematical Classification and Clustering*: Springer Science & Business Media; 1996.
79. Chakraborty A, Maka H, editors. *Biclustering of Gene Expression Data Using Genetic Algorithm*: IEEE; 2005.
80. Kirkpatrick S, Gelatt Jr CD, Vecchi MP. Optimization by simulated annealing. *Science*. 1983;220(4598):671–80.
81. Bryan K, Cunningham P, Bolshakova N, editors. *Biclustering of Expression Data Using Simulated Annealing*: IEEE; 2005.
82. Di Gesú V, Giancarlo R, Lo Bosco G, Raimondi A, Scaturro D. GenClust: A genetic algorithm for clustering gene expression data. *BMC Bioinformatics*. 2005;6:1–11.
83. Lu Y, Lu S, Fotouhi F, Deng Y, Brown SJ. Incremental genetic K-means algorithm and its application in gene expression data analysis. *BMC Bioinformatics*. 2004;5(1):1–10.
84. Jolliffe IT. *Principal Component Analysis for Special Types of Data*: Springer; 2002.
85. Knudsen S *Cancer Diagnostics with DNA Microarrays*: John Wiley & Sons; 2006.
86. Hibbs MA, Dirksen NC, Li K, Troyanskaya OG. Visualization methods for statistical analysis of microarray clusters. *BMC Bioinformatics*. 2005;6(1):1–10.
87. Yeung KY, Haynor DR, Ruzzo WL. Validating clustering for gene expression data. *Bioinformatics*. 2001;17(4):309–18.

88. Liang M, Zhang F, Jin G, Zhu J. FastGCN: A GPU accelerated tool for fast gene co-expression networks. *PLOS One.* 2015;10(1):e0116776.

89. Arefin AS, Berretta R, Moscato P, editors. *A GPU-Based Method for Computing Eigenvector Centrality of Gene-Expression Networks*: Association for Computing Machinery, New York City, United States; 2013.

90. McArt DG, Bankhead P, Dunne PD, Salto-Tellez M, Hamilton P, Zhang S-D. cudaMap: A GPU accelerated program for gene expression connectivity mapping. *BMC Bioinformatics.* 2013;14(1):1–6.

91. Day A, Carlson MRJ, Dong J, O'Connor BD, Nelson SF. Celsius: A community resource for Affymetrix microarray data. *Genome Biology.* 2007;8(6):1–13.

92. Langfelder P. Horvath S. WGCNA: An R package for weighted correlation network analysis. *BMC Bioinformatics.* 2008;9(1):1–13.

93. Nordberg H, Bhatia K, Wang K, Wang Z. BioPig: A Hadoop-based analytic toolkit for large-scale sequence data. *Bioinformatics.* 2013;29(23):3014–9.

94. Schumacher A, Pireddu L, Niemenmaa M, Kallio A, Korpelainen E, Zanetti G, Heljanko K. SeqPig: Simple and scalable scripting for large sequencing data sets in Hadoop. *Bioinformatics.* 2014;30(1):119–20.

95. Langmead B, Schatz MC, Lin J, Pop M, Salzberg SL. Searching for SNPs with cloud computing. *Genome Biology.* 2009;10(11):1–10.

96. Langmead B, Trapnell C, Pop M, Salzberg SL. Ultrafast and memory-efficient alignment of short DNA sequences to the human genome. *Genome Biology.* 2009;10(3):1–10.

97. Li R, Li Y, Fang X, Yang H, Wang J, Kristiansen K, Wang J. SNP detection for massively parallel whole-genome resequencing. *Genome Research.* 2009;19(6):1124–32.

98. Zhao S, Prenger K. Smith L. Stormbow: A cloud-based tool for reads mapping and expression quantification in large-scale RNA-Seq studies. *International Scholarly Research Notices.* 2013. https://doi.org/10.1155/2013/481545

99. Angiuoli SV, Matalka M, Gussman A, Galens K, Vangala M, Riley DR, et al. CloVR: A virtual machine for automated and portable sequence analysis from the desktop using cloud computing. *BMC Bioinformatics.* 2011;12:1–15.

100. Zhao S, Prenger K, Smith L, Messina T, Fan H, Jaeger E, Stephens S. Rainbow: A tool for large-scale whole-genome sequencing data analysis using cloud computing. *BMC Genomics.* 2013;14(1):1–11.

101. Kurtz S. The Vmatch Large Scale Sequence Analysis Software. *Ref Type: Computer Program.* 2003;412:297.

102. Grosu P, Townsend JP, Hartl DL, Cavalieri D. Pathway Processor: A tool for integrating whole-genome expression results into metabolic networks. *Genome Research.* 2002;12(7):1121–6.

103. Zambon AC, Gaj S, Ho I, Hanspers K, Vranizan K, Evelo CT, et al. GO-Elite: A flexible solution for pathway and ontology over-representation. *Bioinformatics.* 2012;28(16):2209–10.

104. van Iersel MP, Kelder T, Pico AR, Hanspers K, Coort S, Conklin BR, Evelo C. Presenting and exploring biological pathways with PathVisio. *BMC Bioinformatics.* 2008;9(1):1–9.

105. Yang P, Patrick E, Tan S-X, Fazakerley DJ, Burchfield J, Gribben C, et al. Direction pathway analysis of large-scale proteomics data reveals novel features of the insulin action pathway. *Bioinformatics.* 2014;30(6):808–14.

106. Park YS, Schmidt M, Martin ER, Pericak-Vance MA, Chung R-H. Pathway-PDT: A flexible pathway analysis tool for nuclear families. *BMC Bioinformatics.* 2013;14:1–5.

107. Koumakis L, Moustakis V, Zervakis M, Kafetzopoulos D, Potamias G, editors. *Coupling Regulatory Networks and Microarays: Revealing Molecular Regulations of Breast Cancer Treatment Responses*: Springer; 2012.

108. Koumakis L, Potamias G, Tsiknakis M, Zervakis M, Moustakis V. Integrating microarray data and GRNs. *Microarray Data Analysis: Methods and Applications*; 2016:137–53.

109. Bashiardes S, Zilberman-Schapira G, Elinav E. Use of metatranscriptomics in microbiome research. *Bioinformatics and Biology Insights.* 2016;10:BBI-S34610.

110. Buchfink B, Xie C, Huson DH. Fast and sensitive protein alignment using DIAMOND. *Nature Methods.* 2015;12(1):59–60.

111. Leimena MM, Ramiro-Garcia J, Davids M, van den Bogert B, Smidt H, Smid EJ, et al. A comprehensive metatranscriptome analysis pipeline and its validation using human small intestine microbiota datasets. *BMC Genomics.* 2013;14(1):1–14.

112. Westreich ST, Korf I, Mills DA, Lemay DG. SAMSA: A comprehensive metatranscriptome analysis pipeline. *BMC Bioinformatics.* 2016;17(1):1–12.

113. Cole JR, Wang Q, Cardenas E, Fish J, Chai B, Farris RJ, et al. The ribosomal Database Project: Improved alignments and new tools for rRNA analysis. *Nucleic Acids Research*. 2009;37(suppl_1):D141–D5.

114. Caporaso JG, Kuczynski J, Stombaugh J, Bittinger K, Bushman FD, Costello EK, et al. QIIME allows analysis of high-throughput community sequencing data. *Nature Methods*. 2010;7(5):335–6.

115. Edgar RC. UPARSE: Highly accurate OTU sequences from microbial amplicon reads. *Nature Methods*. 2013;10(10):996–8.

116. Callahan BJ, McMurdie PJ, Rosen MJ, Han AW, Johnson AJA, Holmes SP. DADA2: High-resolution sample inference from Illumina amplicon data. *Nature Methods*. 2016;13(7):581–3.

117. Eren AM, Morrison HG, Lescault PJ, Reveillaud J, Vineis JH, Sogin ML. Minimum entropy decomposition: Unsupervised oligotyping for sensitive partitioning of high-throughput marker gene sequences. *The ISME Journal*. 2015;9(4):968–79.

118. Truong DT, Franzosa EA, Tickle TL, Scholz M, Weingart G, Pasolli E, et al. MetaPhlAn2 for enhanced metagenomic taxonomic profiling. *Nature Methods*. 2015;12(10):902–3.

119. Wood DE, Salzberg SL. Kraken: Ultrafast metagenomic sequence classification using exact alignments. *Genome Biology*. 2014;15(3):1–12.

120. Ounit R, Wanamaker S, Close TJ, Lonardi S. CLARK: Fast and accurate classification of metagenomic and genomic sequences using discriminative k-mers. *BMC Genomics*. 2015;16(1):1–13.

121. Silva GGZ, Cuevas DA, Dutilh BE, Edwards RA. FOCUS: An alignment-free model to identify organisms in metagenomes using non-negative least squares. *PeerJ*. 2014;2:e425.

122. Silva GGZ, Green KT, Dutilh BE, Edwards RA. SUPER-FOCUS: A tool for agile functional analysis of shotgun metagenomic data. *Bioinformatics*. 2016;32(3):354–61.

123. Edgar RC. Search and clustering orders of magnitude faster than BLAST. *Bioinformatics*. 2010;26(19):2460–1.

124. Langmead B, Salzberg SL. Fast gapped-read alignment with Bowtie 2. *Nature Methods*. 2012;9(4):357–9.

125. Morgulis A, Coulouris G, Raytselis Y, Madden TL, Agarwala R, Schäffer AA. Database indexing for production MegaBLAST searches. *Bioinformatics*. 2008;24(16):1757–64.

126. Keegan KP, Glass EM, Meyer F. MG-RAST, a metagenomics service for analysis of microbial community structure and function. *Microbial Environmental Genomics (MEG)*; 2016:207–33.

127. Turnbaugh PJ, Hamady M, Yatsunenko T, Cantarel BL, Duncan A, Ley RE, et al. A core gut microbiome in obese and lean twins. *Nature*. 2009;457(7228):480–4.

128. Karch H, Tarr PI, Bielaszewska M. Enterohaemorrhagic Escherichia coli in human medicine. *International Journal of Medical Microbiology*. 2005;295(6–7):405–18.

129. Tu Q, He Z, Zhou J. Strain species identification in metagenomes using genome-specific markers. *Nucleic Acids Research*. 2014;42(8):e67-e.

130. Truong DT, Tett A, Pasolli E, Huttenhower C, Segata N. Microbial strain-level population structure and genetic diversity from metagenomes. *Genome Research*. 2017;27(4):626–38.

131. Scholz M, Ward DV, Pasolli E, Tolio T, Zolfo M, Asnicar F, et al. Strain-level microbial epidemiology and population genomics from shotgun metagenomics. *Nature Methods*. 2016;13(5):435–8.

132. Luo C, Knight R, Siljander H, Knip M, Xavier RJ, Gevers D. ConStrains identifies microbial strains in metagenomic datasets. *Nature Biotechnology*. 2015;33(10):1045–52.

133. Ahn T-H, Chai J, Pan C. Sigma: Strain-level inference of genomes from metagenomic analysis for biosurveillance. *Bioinformatics*. 2015;31(2):170–7.

134. Cleary B, Brito IL, Huang K, Gevers D, Shea T, Young S, Alm EJ. Detection of low-abundance bacterial strains in metagenomic datasets by eigengenome partitioning. *Nature Biotechnology*. 2015;33(10):1053–60.

135. Li H, Handsaker B, Wysoker A, Fennell T, Ruan J, Homer N, et al. The sequence alignment/map format and SAMtools. *Bioinformatics*. 2009;25(16):2078–9.

136. Dateki T, Seki H, Minowa M. From LTE-advanced to 5g: Mobile access system in progress. *Fujitsu Scientific and Technical Journal*. 2016;52(2):97–102.

137. Choudhury A, Nair PB, Keane AJ, editors. *A Data Parallel Approach for Large-Scale Gaussian Process Modeling*: SIAM; 2002.

138. Chow-White PA, MacAulay M, Charters A, Chow P. From the bench to the bedside in the big data age: Ethics and practices of consent and privacy for clinical genomics and personalized medicine. *Ethics and Information Technology*. 2015;17(3):189–200.

139. Gymrek M, McGuire AL, Golan D, Halperin E, Erlich Y. Identifying personal genomes by surname inference. *Science*. 2013;339(6117):321–4.

140. Shringarpure SS, Bustamante CD. Privacy risks from genomic data-sharing beacons. *The American Journal of Human Genetics*. 2015;97(5):631–46.

12 Disease Diagnostics, Monitoring, and Management by Deep Learning
A Physician's Guide in Ethical Issues in NMR Medical Imaging Clinical Trials

Rakesh Sharma

12.1 INTRODUCTION

The term "deep learning" was coined in the year 2009 and used in medical image analysis in year 2015 [1]. After machine learning artificial neural networks began outperforming, deep neural networks are now the state-of-the-art machine learning models used in image segmentation, tissue classification by natural language processing in disease diagnosis, monitoring, and management in academia and healthcare industry. The deep learning (DL) algorithm with convolution neural network architecture automatically infers the contrast of MRI scans based on the image intensity of multiple slices. Convolutional neural network (CNN) models are used for genetic brain masking method for Alzheimer's disease, deeply seated brain tumor texture extraction from deep spaced objects by magnetic resonance images (MRIs) of high-grade tumors such as gliomas. Now, transfer learning (TL) approaches for meningioma, glioma, and pituitary tumors using six pre-trained TL classifiers InceptionV3, Xception, Resnet50, EfficienNetB0, VGG16, and MobileNet automatically identify and classify brain tumors. With rapid advancement of deep learning approach, it has become mandatory to use deep learning information carefully with ethical approval for medical imaging analysis and clinical practice.

12.1.1 NMR-Biochemical Correlation

The concept of NMR-biochemical correlation was proposed by authors in the nineties as noninvasive diagnostic monitoring tool [1]. Initially, correlation of longitudinal T1, transverse T2* with ex vivo NMR and in vivo Magnetic Resonance Spectroscopy (MRS) data along with serum analytes and tissue histology were integrated as established NMR relaxation-biochemical biomarkers [1]. Over years of continued efforts, Nuclear Magnetic Resonance (NMR) now showed great potential towards "deep learning" features by integrating multimodal T1-, T2- molecular images, and 2D/3D spectra applying regression CNN framework, trained dataset decision making along with measuring metabolite concentrations at different locations in human tissues for "physiological and functional£ metabolic screening with parallel development in ultrahigh resolution NMR imagers [2]. Simultaneously, robust automated magnetic resonance spectroscopic imaging (MRSI) segmentation and registration processing have been in routine for extracting focal lesion features (size,

DOI: 10.1201/9781003353751-12

chemical composition, disease-specific metabolism) extraction as trained non-invasive MRI and spectroscopy supervised dataset (biosignals) along with tissue-specific psychological, organ function lab tests to define the structure and physiochemical or behavioural nature of diseased soft tissues such as heart, muscle, kidney, breast, prostate, liver, and brain for confirming disease behavior, differential diagnosis, and monitoring purposes.

12.1.2 ETHICAL ISSUES IN NMR MEDICAL IMAGING

The author shares his opinion that T1, T2 relaxation constants are tissue- and disease-specific, showing up as hypo-, normal, and hyper-intensities in MRI images. MRS peaks and MRSI produce NMR spectral diagnostic peaks and metabolite maps to monitor the metabolite distribution in tissues based on "spectromics" peak ratio, features of peak area FHWM, peak relaxivities, peak shape, peak cross-connectivity, peak dynamics [2]. However, these factors don't absolutely represent enough to confirm the disease to make a descion in diagnosis and therapy. It creates ethical concern in the interest of public as well as uncertainty. Over years, clinicians now routinely use the integrated in vivo MR imaging and spectroscopy to define quantitative changes of in vivo organ anatomy, tissue histology, digital histochemistry, perfusion, and metabolism for theranostic physiochemical specificity as clinical efficacy while ignoring ethics and morals [3–5]. Further, due to time-consuming data analysis, some research centers adopted deep learning (integrated extraction and calculated biosignal changes as specific disease status footprint) data processing methods using 3D metabolic peaks/maps to reveal metabolite concentration profiles or metabolomics as disease-specific fingerprints of focal lesions in human heart, muscle, kidney, breast, prostate, liver, and brain diseases with no ethical consideration [3–7]. Thus, "disease burden" as spatially and temporarily averaged metabolite information is measured by 3D- and 4D-tissue metabolite concentrations distributed in human brain tumors, muscle, bone, and glands without any ethical standards [8–11]. Despite all, low NMR visible metabolite concentrations by in vivo NMR spectra and poor metabolic maps of tissues still show limited clinical value in federal guidelines. It needs ethical clearance. Perhaps, "deep learning" features after ethical clearance may enhance the theranostic accuracy in disease diagnosis and treatment towards precision medicine, as illustrated in Figure 12.1.

12.1.3 MEDICAL MR IMAGING AND DEEP LEARNING

The MRSI signal extracts the metabolic information from spatial MRI maps of large areas by an optimization process to select diseased tissue volume with no external artifacts. The chemical shift encoding and readout methods generate metabolite signals using suppression of water and lipids signals by fat saturation bands. Fourier-transformation and Fast Pace transform signal processing visualize chemical shift data from non-zero gyromagnetic ratio nuclei of less abundant metabolites, temperature, pH, tissue oxygenation [12]. Now high-speed localized 2D NMR-TOCSY, LCOSY, and high rotation MAS spectra (HRMAS) acquisition with added second spectroscopic dimension can generate greater spectral peak discrimination and separate overlapping resonances at ultrahigh NMR spectrometers for deep learning, as shown in Figure 12.2 [13].

Using MRI processing, robust MRSI co-registration, atlas matching, and tissue segmentation methods diagnostic sensitivity improved using "deep learning" by spatial transformation of diseased tissue normalized MRSI data coordinates relative to normal or control tissue metabolite concentrations in the matched region. "Atlas matching" identify brain regions for analysis of matching MRSI voxels with database of normal metabolite values [14].

The "segmentation" delineates the disease-specific tissue pixel areas with change in metabolites using boundary-based, thresholding, feature plots, subtle points, volume rendering, filtering, and interpolation methods [15]. The "registration" permits connected disease-specific pixels across tissue slices (coordinates) to extract the disease volume based on edge detection, morphmetry match from validated segmented out trained data sets of disease burden by the observer [16]. In the

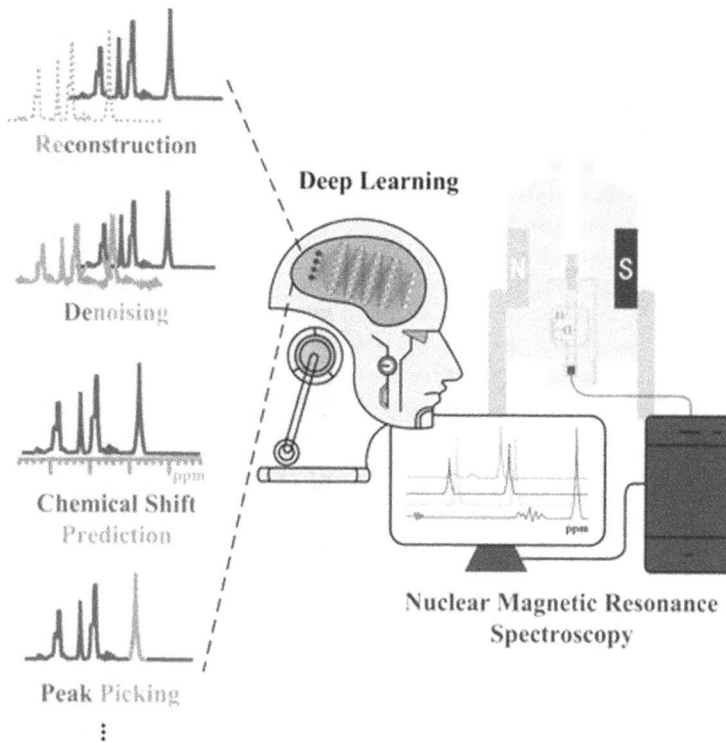

FIGURE 12.1   Deep learning and precision medicine.

brain, "deep learning" now is done by a trained operator on selected brain tissue voxels or brain regions [17].

The present chapter with a focus on ethical issues has an objective to interpret MRSI and hybrid imaging features in physiochemical imaging and spectroscopy as tool of "deep learning" in clinical trials on theranosis in the light of meta-analyses and previous clinical trials if suitable for precision medicine after ethical clearence.

12.2 CLINICAL DISORDERS AND MRSI APPLICATIONS TOWARDS DEEP LEARNING

How deep learning by MR imaging is useful within ethical norms

MRSI is now used in clinics to help diagnose, treat, and follow cancer and lesion progress in patients. "Deep learning" of MRSI data further visualizes metabolite peaks as "spectromics" fingerprints (relaxation constants + metabolite ratio or concentrations) from different diseased brain regions as true biosignals, as shown in Figures 12.1 and 12.2. Integrated with other imaging CT, IR, PET, US, MALDI modalities, MRS imaging offers anatomical, physical, physiological, and molecular behavior of disease progress and recovery (Table 12.1). However, ethical issues remain uncertain.

12.2.1 SEGMENTATION, COREGISTRATION, VALIDATION, AND FEATURE ANALYSIS BY ARTIFICIAL INTELLIGENCE

Artificial intelligence, Kernel Null Foley-Sammon Transform (KNFST), Support Vector Machines (SVM) and third (PC3) and fourth (PC4) degree polynomial, fuzzy c-means (FCM) clustering alhgorithm, KNN, CNN, regression, pattern for the segmentation trained data set, image registration,

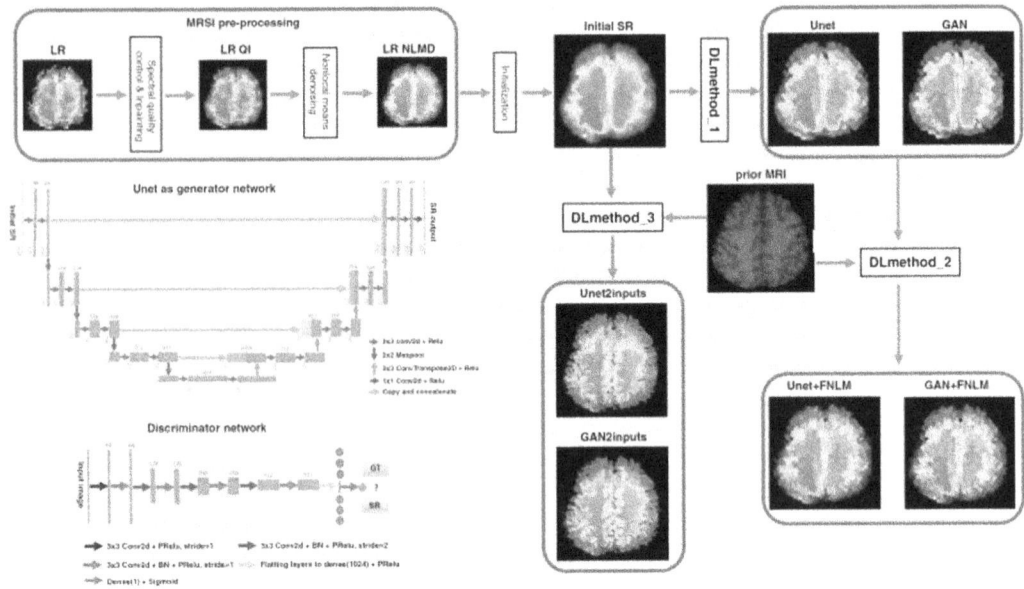

FIGURE 12.2 A flowchart diagram of deep learning for MRSI shows three blocks: 1. Filtered low resolution metabolite maps using denoising, painting and sptrum quality criteria; 2. Interpolation of denoised maps to make SR maps, and 3. Implementing deep neural network (Unet or GAN) to produce final SR image. Use of SR images in last block to run three deep learning methods as: 1. Deep neural network applied to initial MRSI data; 2. Feature nonlocal means (FNLM) with prior MRI input applied to results of deep learning method 1; 3. Deep neural network applied to initial MRSI and pror MRI input both. The sketch of generator network (Unet) and discriminator GAN network is shown in bottom left [18]. Courtesy: Dr Xianqui Li, MGH.

pixel coordinates, delineation and localized NMR peak characteritics extract the tissue pathology or disease burden or tumor stage using color-coding and biophysical signal measurements [18]. However, all these methods need optimization or trimming out the input data by one or other method. Thus, maximum outcome of tissue disease burden can be measured by "combined artificial intelligence outcome" from multimodal signal processing methods reaching close to real disease specific etiology or pathology obtained from biopsies.

12.2.2 "SPECTROMICS" FINGERPRINTS IN PRECISION MEDICINE BY MULTIMODAL MR SPECTROSCOPIC IMAGING WITH PHYSIOLOGICAL SCREENING

Deep Learning by Integrated MRI with Multiparametric Modalities answers metabolic spectral distribution. "Spectromics" fingerprint is proposed by the author to define tissue metabolite distribution as fingerprint of disease burden by deep learning. *Spectromics* is an integration of multiple patient data of clinical, molecular imaging, and genomics modalities towards "Precision Medicine Initiative" is precise interpretation to decide patient-care management plan specific to individual patient as shown in Figures 12.2–12.4.

Now, image analysis by computer-aided diagnosis (CAD) using quantitative breast lesion metabolites and NMR relaxation constants and MRI image analysis can reflect the associations of clinical, pathologic, and genomic data (genomic measurements) in tumor phenotypes. For example, brain tumors, multiple sclerosis lesions, Alzheimer's disease, epilepsy episodes, and effective cancer "theranosis" rely on the combined information from multiple patient tests including molecular, clinical imaging and genomic data towards patient-specific coutcome for precision medicine. Adapting the "Precision Medicine Initiative" includes current radiological interpretation from the

TABLE 12.1
Clinical Trials on NMR-Biochemical Correlation (MRI+MRS and Other Biomarkers) Using "Deep Learning"

Clinical Trial	Outcome of MR spectroscopy Clinical Trials and Reviews	Reference
Sleep apnea	Echo-planar J-resolved spectroscopic imaging using compressed sensing: validation in obstructive sleep apnea.	[20]
Multiple Sclerosis	Tissue-specific metabolic contents in secondary progressive multiple sclerosis: a MRSI study.	[21]
Multiple Sclerosis	Clinical trials and clinical practice in multiple sclerosis	[22]
	High field MR imaging and 1H-MR spectroscopy in multiple sclerosis: diagnostic MR imaging criteria.	[23]
		[24]
	The role of MRS and fMRI in multiple sclerosis	[25]
	Guidelines for MR monitoring the treatment of multiple sclerosis. US National MS Society	
Brain Tumor	Reproducibility of localized 2D correlated MR spectroscopy	[26]
Prostate Cancer	Prostate endorectal MRSI: standard evaluation system	[27]
SCA2/MSA-C	Differentiation of SCA2 from MSA-C using proton MRSI	[28]
Obesity	Effect of Low fat diets on Intramyocellular and Hepatocellular Lipids: A Randomized Clinical Trial	[29]
Brain Tumor	MRI biomarkers in neuro-oncology	[30]
Prion Disease	Combined diffusion imaging and MR spectroscopy in human prion diseases.	[31]
Glioblastoma	Impact of MRSI on response assessment: glioblastoma clinical trials.	[32]
Neurodegeneration	In vivo imaging markers of neurodegeneration of the substantia nigra	[33]
Glioblastoma	The Theranostic Ability of Tamozolomide Follow-Up MRI+MR Spectroscopy and Apparent Diffusion Coefficient Mapping	[34]
Amylotropic Lateral Sclerosis	Diffusion tensor MRI and MR spectroscopy in upper motor neuron in amyotrophic lateral sclerosis.	[35]
Non-Hodgkin lymphoma	MRS in multi-centre trials and early results in non-Hodgkin's lymphoma.	[36]
Breast cancer	Multiparametric MRI+MRS+(^{23}Na)MRI monitoring in advanced breast cancer.	[37]
	Feasibility of 7 Tesla breast MRI for intrinsic sensitivity, diffusion-weighted imaging, and (1)H-MRS of breast cancer patients receiving neoadjuvant therapy.	[38]
Low Back Pain	In vivo intervertebral disc MRS and T1ρ imaging: discography and Oswestry Disability Index-36 Health Survey.	[39]
Ischemia Syndrome	Assessment of women with suspected myocardial ischemia: (WISE) Study.	[40]
Neurofibrometosis	Randomised controlled trial for autism neurofibromatosis type 1 (SANTA).	[41]
Aorta Coarctation	Perfusion imaging of surgical repair of Coarctation of Aorta study.	[42]
Brain pH	Detecting activity-evoked pH changes in human brain.	[43]
GABA	Thalamic Gamma Aminobutyric Acid Levels in depressive Disorder	[44]
Glioma	Volumetric and metabolic evolution of diffuse intrinsic Pontine gliomas	[45]
	Multiparametric MRI and FDG-PET predict outcome in diffuse brainstem glioma? prospective phase-II study	[46]
	Dynamic susceptibility-weighted and spectroscopic MRI monitoring	[47]
Coronary Plaque	Multi-modality intra-coronary plaque characterization: a pilot study	[48]
HIV and Dementia	A in vivo proton-MRS of HIV- dementia and its relationship to age	[49]
Renal Disease	MRS to predict radiofrequency-ablated renal tissue.Phase 1 trial	[50]

(Continued)

TABLE 12.1 (CONTINUED)
Clinical Trials on NMR-Biochemical Correlation(MRI+MRS and Other Biomarkers) Using "Deep Learning"

Clinical Trial	Outcome of MR spectroscopy Clinical Trials and Reviews	Reference
Brain fMRI	Functional brain imaging: an evidence-based analysis.	*[51]*
Soft tissue Tumor	31Phosphorus-MRS of histologic tumor response after perfusion for soft tissue tumors.	*[52]*
Cerebral Astrocytoma	Multivoxel spectroscopy at short TE: choline/N-acetyl-aspartate ratio and the grading of cerebral astrocytomas.	*[53]*
		[54]
	201Thallium SPECT and 1H-MRS compared with MRI in chemotherapy monitoring of high-grade malignant astrocytomas.	
Colorectal Liver Metastasis	The pre-operative MRS of steatohepatitis in patients with colorectal liver metastasis.	*[55]*
Alzheimer Disease	Proton MRS of Alzheimer's disease and Binswanger's disease	*[56]*
Prostate Cancer	Prostate Cancer MRI Screening:Göteborg Prostate Cancer Screening 2 Trial	*[57]*
		[58]
	Dynamic MRI and CAD vs. choline MRS in prostate cancer?	
Bone Metastasis	3-T MR-guided high-intensity focused ultrasound (3 T-MR-HIFU) for bone metastases of solid tumors.	*[59]*
	Selective serotonin reuptake inhibitor in rostral anterior cingulate choline metabolite decrease	*[60]*
Depressive Disorder	Creatine for female adolescents with SSRI-resistant major depressive disorder: a 31-phosphorus MRS study.	*[61]*

"average patient" to interpret the precise patient-care management decision specific to an individual patient.

Chronology of ethical issues in medical image processing:

- First time deep learning was defined as artificial neural network [1].
- Feature discovery, feature learning, and performing task using CNN (convolution layer, pooling data, dropout control, batch normalization). For example, use of filter to make feature map of MS lesion→Pooling maps→Extracted pooled lesion feature→Vectorized lesion feature maps→Output of MS lesion composition and shape [2].
- Multimodal image processing to minimize misclassification using toolkit libraries.

12.3 EXAMPLES OF MRI AND MRS-BASED DEEP LEARNING

Three examples are illustrated towards understanding the value of "deep learning" as potential "precision medicine initiative" to evaluate disease evolution, correlation with disability, evaluating the occult disease to establish a prognosis, correlation with fatigue, and in vivo theranostic effect of therapies.

12.3.1 CLASSIFICATION OF BREAST TUMORS TO DETERMINE FUTURE RISK AND NORMAL BREAST TISSUE

The breast image analysis by computer-aided diagnosis (CAD) predicts the breast lesion character-istics on clinical images. The CAD is useful in the risk assessment, detection, screening, diagnosis, theranostic response, recurrence, and other clinical tasks using "virtual digital biopsies," as shown in Figures 12.4 and 12.5 [18].

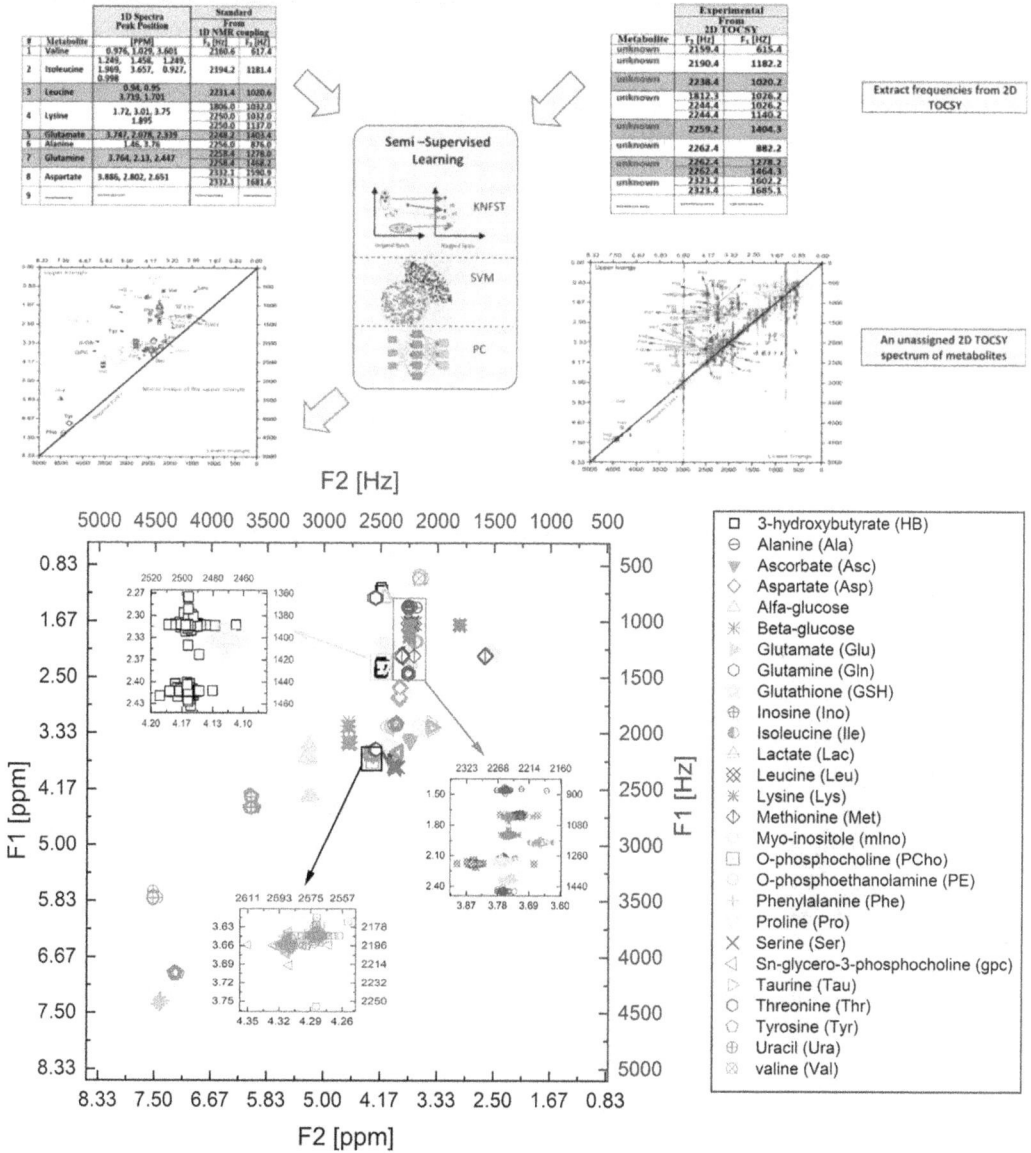

FIGURE 12.3 Deep learning (DL) extracts digital features using artificial intelligence or computation algorithms to analyze the pathology in voxels on images to process the MRS data with demonstrated results in computer metabolomic dataset, medicasl images, natural language processing, and so forth. Applications of deep learning in 2D NMR specxtroscopy are shown as feature space of 27 metabolite trained dataset derived from TOCSY spectrum of breast cancer tissue. The insets are selected enlarged peaks overlapped in (F2,F1) dimensions. [19]. Courtesy: Dr Migdadi.

The "deep learning" of MRSI interpretation explores digital metabolic lesion segmentation–based features of lesion volume, sphericity, texture, uptake and maximum enhancement variance (DCE), margin sharpness on ImageNet and texture (T2wt), and ADC(DWI) with minimum computer data processing, as shown in Figures 12.6 and 12.7.

The "deep learning" feature extraction on CADx tasks minimizes the use of extensive computing, as shown in Figure 12.8.

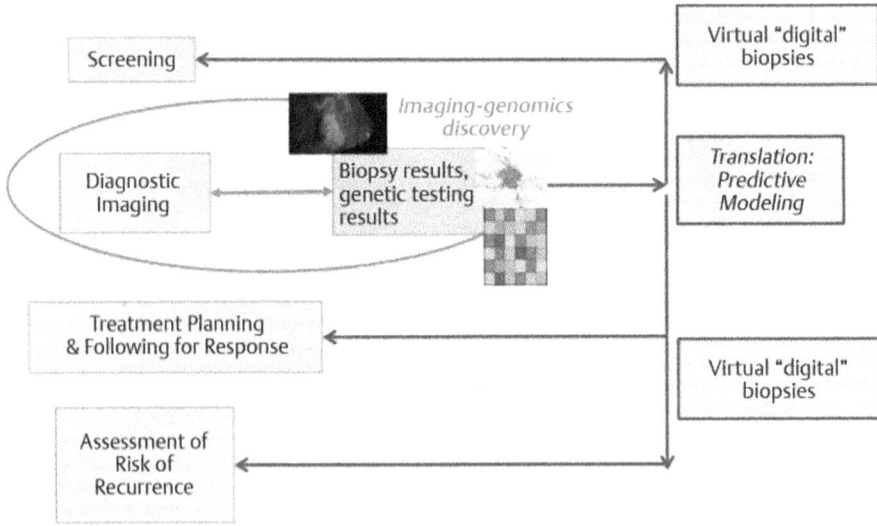

FIGURE 12.4 An illustration of multiple stage 'spectromics'discovery, process and translation for validation of quantitative imaging features based on discovery of new relationships to other "omics" and translation into clinical predictive models as reference "virtual digital biopsies," similar to actual biopsies to screen and assess therapeutic response [18]. Courtesy: Dr ML Giger.

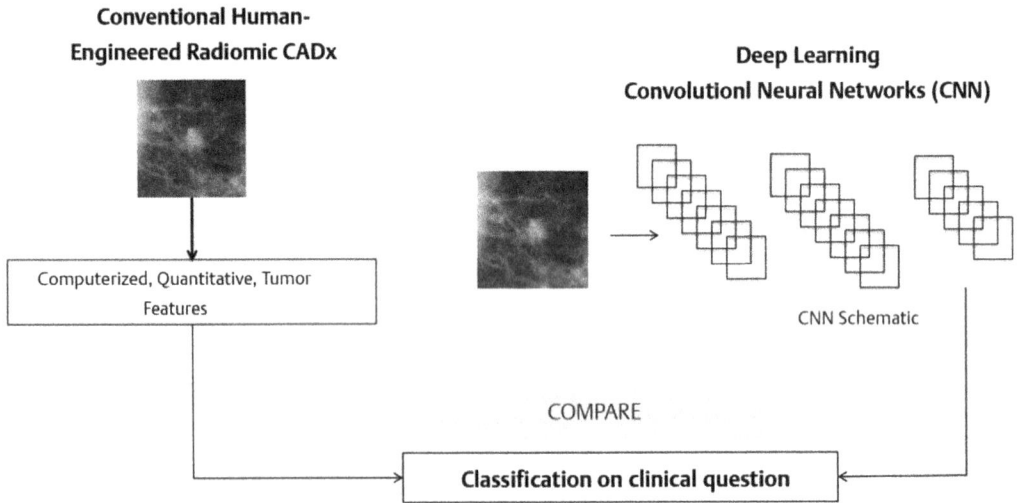

FIGURE 12.5 A scheme of CAD (on left) and deep learning (on right) shows the approach of quantitative spectromics using segmentation features of DCE-MRS of tumor. (18). Courtesy: Dr ML Giger.

The CNNs-based breast image features classify the malignant and benign breast tumors using mammography, ultrasound, and DCE-MRI visual likelihoods of malignant lesions with "deep learning" performance shown in Figure 12.9.

12.3.2 COMPUTERIZED IMAGE-BASED BREAST CANCER RECCURENCE RISK

Breast cancer risk or density measurement by DCE-MRI and mammogtaphy screens out high-risk women. Cancer-risk "spectromic" features are breast metabolite density, parenchyma texture

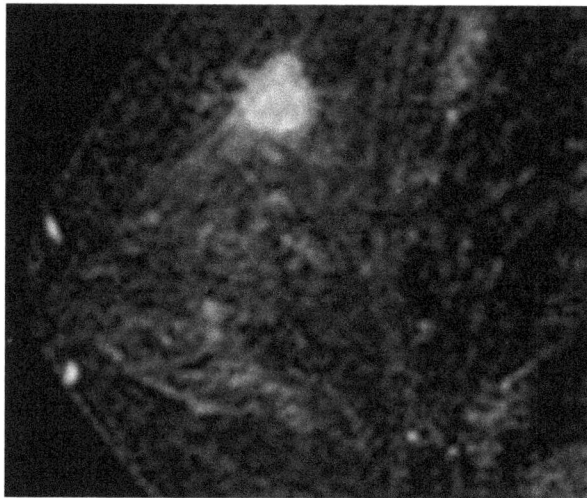

FIGURE 12.6 A case of luminal A, ER+ve,PR+ve,HER2-ve tumor stage II with negative lymph nodes shows the segmented tumor outline by 4D automatic computer segmentation algorithm. Computer analysis of the MRSI spectromics measured the tumor size of 13.6 mm with spectromics irregularity shape 0.49, and the spectromics enhancement texture energy 0.00185 [18]. Courtesy: Dr ML Giger.

pattern on digital mammograms, parenchymal enhancement (BPE) signal on DCE-MRI. MRI visible breast volume-growing algorithm classifies the fibroglandular and MRS visible fat depots. Breast fibroglandular enhanced kinetic curves in breast regions categorize the parenchymal enhancement using fuzzy c-means (FCM) clustering. The mammogram images show texture and metabolite

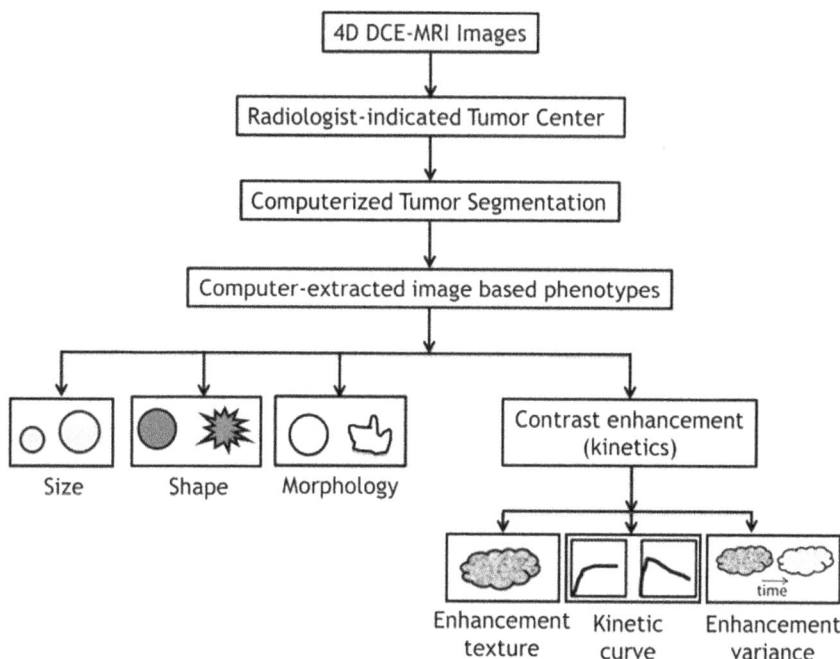

FIGURE 12.7 Schematic diagram illustrates the steps of computer-extraction of MRSI-based tumor phenotypes. Multiple mathematical descriptors calculate these phenotype features for specific clinical tasks [18]. Courtesy: Dr ML Giger.

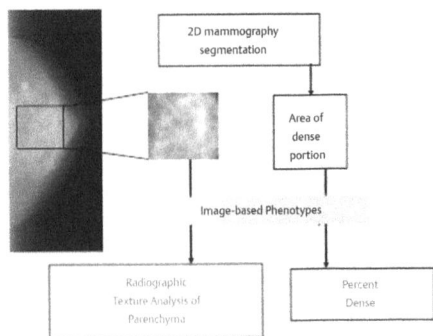

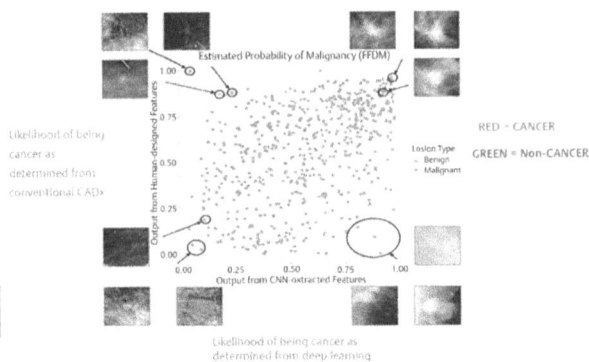

FIGURE 12.8 (on left) Image based extracted phenotypes on digital mammograms of breast parenchyma show the risk of breast cancer reoccurance. [18]. (on right) Distinct cancer and non-cancer features show the likelihood of malignancy by deep learning. Courtesy: Dr Xianqi Li, MGH.

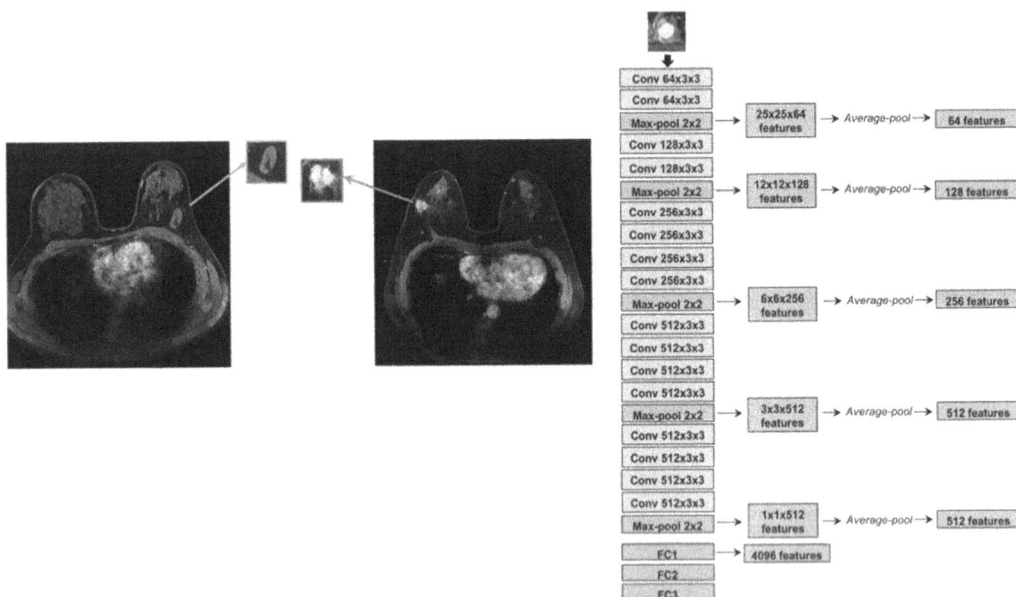

FIGURE 12.9 The ROIs from malignant and other benign breast tumor as input show convolution on VGGNet Neural Network. The VGG19 model is shown in 5 blocks. Each block has two or more convolution layers and max-pooling layer. These five blocks have three connected layers. From five max-pooling layers, features are extracted as average-pooled across the channel in third dimension and normalized with L2 norm. Extracted normalized features are shown to generate CNN feature vector. [18].Courtesy: Dr ML Giger.

density correlation. Dense breast tissues show a high Citrate and Cho peak enhancement than the fatty breast tissues.

12.3.3 BREAST TUMOR MRI PHENOTYPING RELATIVE TO MOLECULAR SUBTYPING FOR DIAGNOSIS AND PROGNOSIS

The "spectromics" needs extensive data analysis to transform the breast 4D DCE-MR images and spectromic features into phenotypic descriptors at breast MRI workstation. One spectromic feature

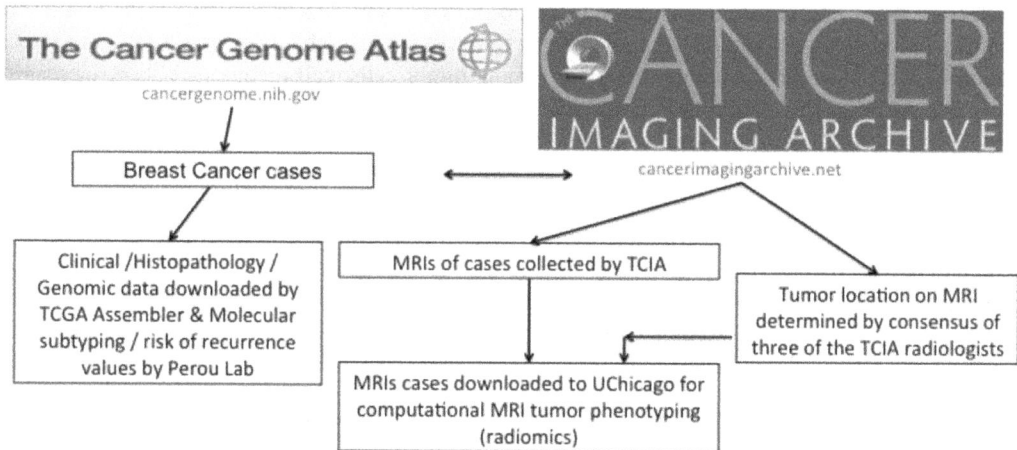

FIGURE 12.10 A TCGA/TCIA genomic atlas and imaging scheme illustrates the datasets and analysis of invasive breast carcinomas. Relationships of computer extracted radiomic MRI lesion features with molecular, clinical, genomic gene expression markers show likelihood of new carcinoma re-occurrence. Reproduced with permission [18]. Courtesy: Dr ML Giger.

is chosen for one clinical task only. Finally, breast biopsy confirms the tumor pathologic stage after in vivo MRI imaging of a breast cancer, to plan treatment. Recently, in place of TNM staging system, MRSI phenotyping resource advocated "spectromic biomarkers" by NCI investigators "The Cancer Genome Atlas/The Cancer Imaging Archive (NCI-TCGA/TCIA) dataset" predicts the pathological stage and quick patient management/treatment by neoadjuvant chemotherapy and/ or radiation therapy, surgery [18]. The TCGA/TCIA cancer research group established the relationships between computer-extracted quantitative MRSI spectromics with prognostic markers, clinical, molecular, gene expression "genomics" profiles [18]. The biopsy-proven invasive breast cancer tissue DCE-MRI image archieve and genomic data showed increased citrate and choline NMR peak analysis, as shown in Figure 12.10, to predict the power of spectromic MRS features vs pathological stage and cancer subtypes.

"Deep learning" characterizes the breast tumors based on MRI radiomics: (1) radiologist-measured tumor size; (2) computer-extracted quantitative "spectromic" features or models; (3) computer-extracted MRI tumor phenotype for lymph node damage and pathological stage. Today, estimation of tumor size is the best predictor of pathologic stage. Moreover, "spectromic" feature predicts clearly the metabolic role in tumor stage I (benign) and II (premalignant) vs stage III (malignant). Breast tumor receptor status (human epidermal growth factor receptor 2[HER2], estrogen receptor [ER], progesterone receptor [ER]) further classify tumors into subtypes. The profiles of breast cancer gene expression categorize tumors into molecular subtypes of basal-like, normal-like luminal A, luminal B, HER2-enriched tumors. Different breast cancer subtypes have different MRI radiomic features, prognostic indicators called PR+ vs PR−, ER+ vs ER−, HER2+ vs HER2−, and triple negative tumors or cancers. All these respond differently to different therapies and used in deep learning.

In short, "deep learning" information from computer-extracted MRI features of phenotypes distinguish breast cancer subtypes by quantitative signatures to predict prognosis for precision medicine.

12.3.4 CLASSIFICATION OF BRAIN GLIOBLASTOMA (GBM) TUMORS

Routinely, Brain Tumor Segmentation (BRATS), Ischemic Stroke Lesion Segmentation (ISLES), Mild Traumatic Brain Injury Outcome Prediction (mTOP), Multiple Sclerosis Segmentation

(MSSEG), Neonatal Brain Segmentation (NeoBrainS12), and MR Brain Swgmentation (MR BrainS) and MR Image Automated Processing (MRIAP) are available in segmentation and registration of MRI brain tumor images. In glioblastoma tumor cells, cell membranes show increased choline-containing phospholipids (Cho) concentration in proliferating cells while N-acetylaspartate (NAA) compound in neurons diminish in glioblastoma tissue samples after neuronal local destruction in neuronal milieu. The neurochemical choline to NAA Ch0/NAA ratio is elevated in GBM tumor. The T1w-CE and FLAIR pulse sequences generate distinct MRI scans. The echo-planar MRI pulse sequence with GRAPPA parallel scans on 3T scanner with 32 channel or 20 channel head and neck coil covering 314 MRI pulse sequence with GRAPPA parallel scans on Siemens 3T scanner with a 32-channel or 20-channel head and neck coil covered voxel size of 314 µL voxel size µL over FOV of 280 mm x 280 mm x 180 mm, and matrix size 50 x 50 x18 generated resolution of 4.5 mm x 4.5 mm x 5,6 mm. The T1 weighted images with 1 mm isotropic voxels are converted into co-registered spatial-spectral "spectromic" data using MRIAP or MIDAS software. All imaging data in the DICOM format are processed by MIDAS on custom-built Brain-Imaging-Collaboration-Suite plateform (BrICS) to coregister all images. The abnormal Cho/NAA ratio in a NAWM voxel vs contralateral NAWM on BrICS lateform identifies the contour region showing elevated ratio of Cho/NAA $\geq 2\times$ normal (see Figure 12.11A).

The glioblastomas (GBMs) brain tumors remain aggressive and hidden even after 60 Gy radiation therapy and temozolomide. Magnetic resonance spectroscopic imaging (MRSI) seems to highlight high-risk metabolite pattern of glioblastoma recurrence; otherwise, it remains nonvisible on contrast-enhanced (CE) MRI. Recently, a randomized phase II clinical trial reported efficacy, safety, feasibility of MRSI-guided escalated radiation therapy (60 Gy $\rightarrow$ 75 Gy) to manage newly diagnosed GBMs. The outcome of MRSI guided escalated radiation therapy, however, warrants additional testing [62].

The author presents the value of MRSI in brain cancer patients with HCG (glioblastoma multiforme [GBM] and anaplastic astrocytoma), LGG (oligodendrogliomas and astrocytomas) and brain metastatic tumors. MRS visible "spectromic" high ratio of neurochemicals confirm specific tumor core and peritumoral edema by elevated Choline/NAA, Choline/Cr (choline containing compounds/creatine–phosphocreatine complex) with low NAA/Cr ratio. NMR visible lipids/lactate ratio in the peritumoral and tumoral regions combined with high Cho/Cr, Cho/NAA ratio and low NAA/Cr ratio discriminate different HGG, LGG, gliomas and metastases [63].

12.3.5 CLASSIFICATION OF MULTIPLE SCLEROSIS LESIONS

Using "deep learning" methods, an accelerated MRIAP and APSIP post-processing method provides a reproducible and efficient assessment of white matter MS lesion volumes, WM-GM-CSF composition and metabolites using T1, T2* parametric and T1, T2* probability maps as "spectromic" fingerprints [64, 65]. The author believes that metabolite changes of neurochemicals in MR spectrum as low N-acetylaspartate peak (marker of neuronal and axonal integrity), high choline peak (marker of cell membrane metabolism), and high myoinositol peak (marker of gliosis) can be biochemical–NMR fingerprints. Diminished N-acetylaspartate peak represents dysfunction of neuronal/axonal loss. The elevated choline peak represents high cell-membrane turnover due to poor remyelination or demyelination; inflammation and gliosis elevated choline peak represents heightened cell-membrane turnover during demyelination, remyelination, inflammation, or gliosis. Thus, the MRI+MRS measure the lesion evolution and correlate the disability with lesion, assessment of occult disease. MRS establishes a prognosis, correlates with fatigue, monitors the effect of drug therapies [66]. At out lab, a training model in semi-supervised KNFST algorithm showed an optimized projection matrix, confidence bands values and class-wise MS lesion training data projections into the null space. Semi-supervised KNFST algorithm iteratively selected samples in learning phase from data. The classifier predicted new labels to accepted lesion data samples as shown in Figure 12.12 [64, 67].

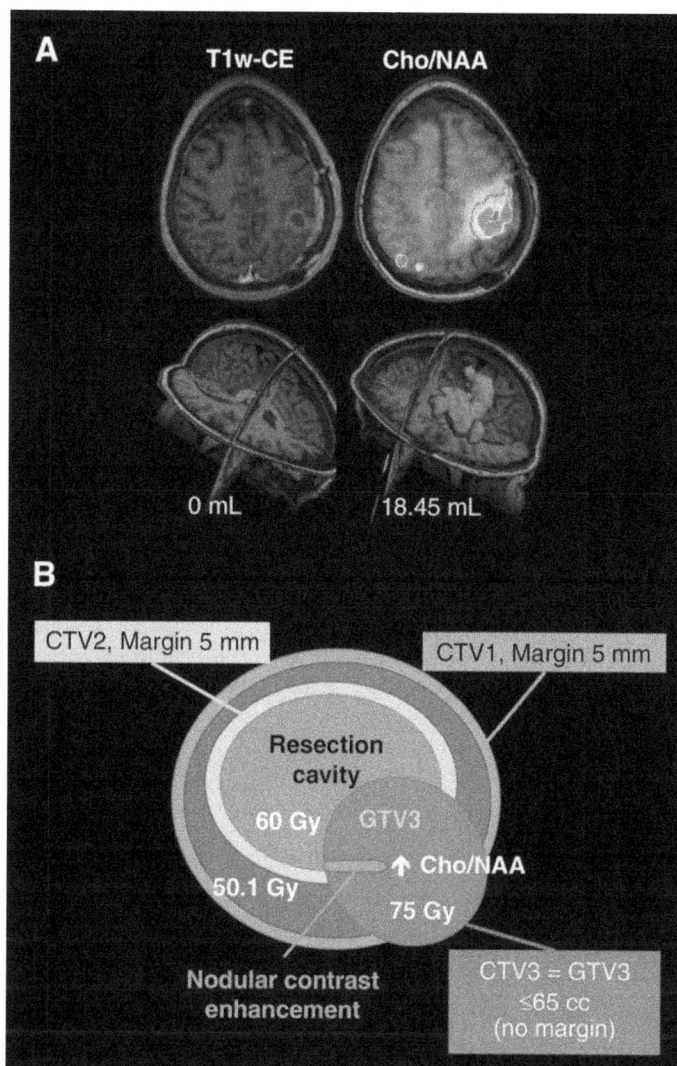

FIGURE 12.11 (A) on left, a glioblastoma postoperative subject shows nodular resection cavity high contrast on T1w-contrast enhanced MRI on BrICS plateform showing normal spectromic (Cho/NAA ≥2× normal) data in GTV3 tumor size 19 mL (right). Deep learning shows contours in GTV3 after radiation 75 Gy. (B) on right, radiation therapy target volumes at different doses show 30 dose fractions (concurrent dose-painted intensity-modulated radiation). Modified from reference [18].

12.4 NEW DEVELOPMENTS IN MRSI MULTIMODAL APPROACH, ETHICAL ISSUES AND LIMITATIONS

The potentials of 4D localized in vivo NMR image spectroscopy, slice selective NMR spectroscopy, biopsy HRMAS spectroscopy, multimodal, multinuclear and water-fat suppression NMR techniques have emerged for better tissue contrast with metabolomics on unethically high magnetic fields up to 11.4 tesla whole human body imagers and ultrahigh resolution at 21 tesla and 36 tesla microimagers. These need ethical clearance first before any trial. The localized tissue metabolism in diseased selected body parts or biopsies were studied using modified surface coils, multiple contrast media, modified radiofrequency pulses with high resolutions, and dynamic gradient fields for localized 2D spectral characterization or metabolic screening for spectrally resolved metabolomics imaging applications with ethical scepticism [20]. Due to noninvasive and nonradioactive NMR techniques being proven safe with no biological hazards at current level of SAR energy exposures, in next few years, NMR methods will be ethically safe, cost-effective alternate modality better than other diagnostic or therapeutic techniques using least ionized radiations.

MRI-digital histochemical correlation is now established to exhibit malignancy associated changes due to NMR T1 variations where T1 relaxivities depend on visible water contents, proton densities, environments in tissue and its pathology [1].

MRI and biochemical correlation initially proposed in nineties [1], now has been established in routine clinical assessment trials for tissue content chemical analysis, in vivo MR spectroscopy, ex vivo MR spectroscopy, and relaxivities using artificial metabolite solutions and MR database. MRI

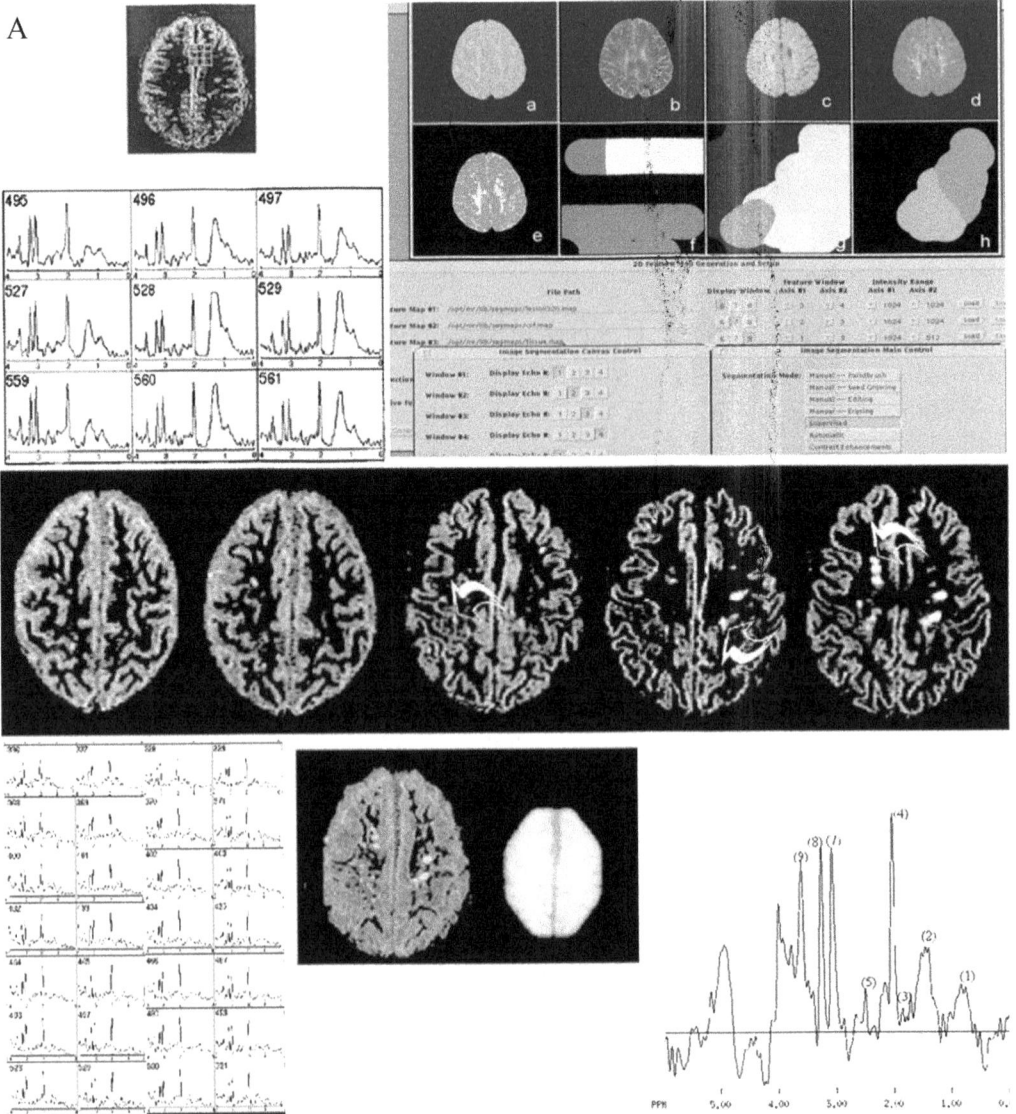

FIGURE 12.12A   Segmentation of multiple sclerosis lesions are showing trainined data set with yellow color coding and MRS peaks.

B

patient	age	sex	scan#	date	voxel	X	Y	NAA	Cr	Cho	Lac	Ace	Lipid	GM	WM	CSF	Lesion
16	35	F	1	101897	460	12	15	381	221	212	145	41	90	0.0025	0.0017	0	0
16	35	F	1	101897	461	13	15	380	249	260	181	34	93	0.0025	0.0017	0	0
16	35	F	1	101897	462	14	15	368	255	251	86	40	75	0.0034	0.0008	0	0.0026

Training Model Δ, Confidence band σ
training set X,
unlabeled data $Z = \{z_1, ..., \tilde{z}_n\}$
re-train flag t

End ← No — $\exists z_i \in Z, i = \{1..n\}$

Yes

Select sample $z_i \in Z$, compute K_{z_i} and $\overline{K}_{z_i}$

Use $\omega \in \Delta$ to determine the projection of test sample z_i.
compute $z' = K(z_i)^T \overline{\omega}^1, ... K(z_i)^T \overline{\omega}^{c-1}$

Use $D \in \Delta$ to compute the distance between the projected sample and the projections of training data
$Class(z_i) = \min_{1 \leq i \leq c} dist(z', D)$

Calculate the confidence band σ_{z_i} for z_i

add z_i and $Class(z_i)$ to X_{tmp}, remove z_i from Z ← Yes — $\sigma_{min}(Class(z_i)) \leq \sigma_{z_i} \leq \sigma_{max}(Class(z_i))$ — No

Size (X_{tmp})~t — No

Yes

Update/retrain Δ using X_{tmp} using the training algorithm

$i = n$ — No

Yes

Updated Training Model Δ
Updated Training dataset X_{new}

Training dataset X with labels

Compute the kernel of training data
$K_{Training} = \varphi(x_i)^T \varphi(x_i)$

Compute the centralized kernel
$\overline{K}_{Training} = (I - 1_N)K_{Training}(I - 1_N)$

Compute the eigenbasis
$B = \widehat{V}\varphi(X)$

Decompose $\overline{K}_{Training} = VEV^T$
Obtain the scaled eigenvector
$V = VE^{1/2}$

Compute the solution β using
$HH^T \beta = 0$

Compute the null projection direction
$\overline{\omega}^1 = ((I - 1_N)\widehat{V})\beta^j$
$\forall j = 1, ..., c - 1$

Compute confidence bands Δ for each sample X

Compute the class-wise projections of training data into the null space
$D = K(X)^T \overline{\omega}^1, ... K(X)^T \overline{\omega}^{c-1}$

Training Model Δ
includes ω, σ and D

- X_{tmp} are the accepted samples that will be added to the training dataset. X_{tmp} contains the confident predicted samples and their labels.
- Re-train flag t is the number of instance collected in X_{tmp} before retraining the classifiers.
- Class is the class label assigned to a sample.

- I is the $N \times N$ identity matrix
- 1_N is a $N \times N$ matrix with all elements equal to $\frac{1}{N}$
- $H = B^T X_w$
- $\varphi(x^i) = \varphi(x^i) - 1/N \sum_{j=1}^{N} \varphi(x^j)$

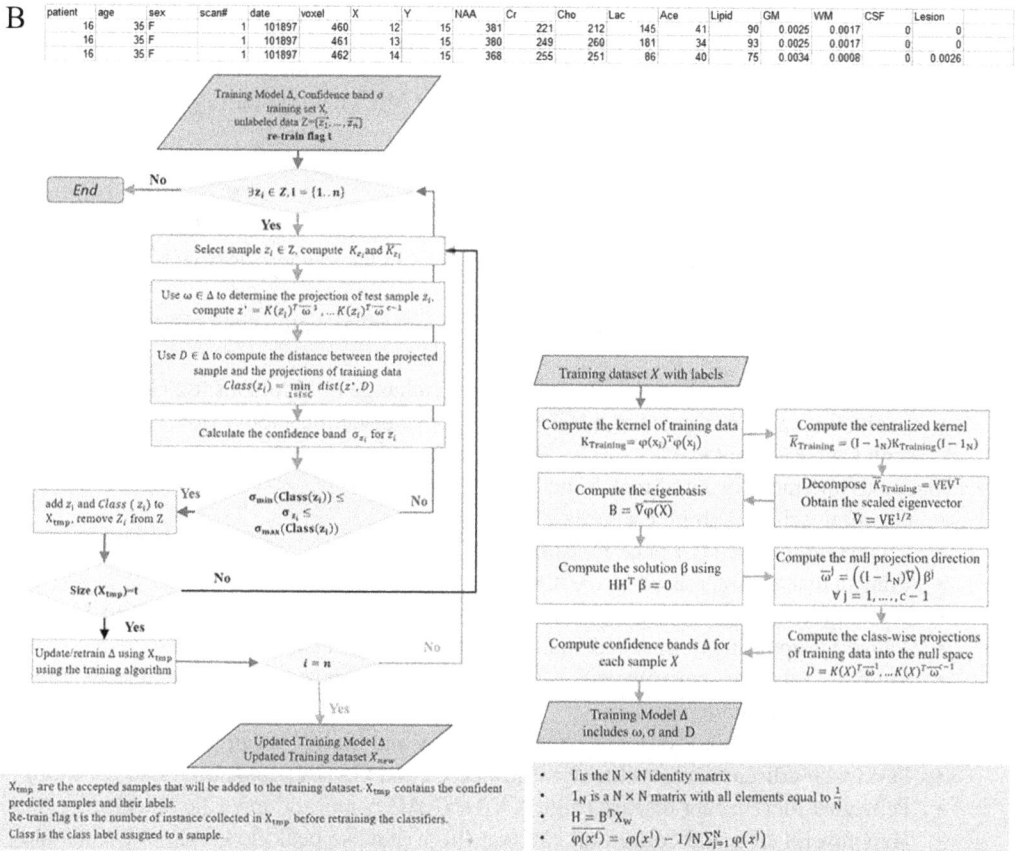

FIGURE 12.12B The semi-supervised KNFST algorithm procedure is shown to select data samples, accept the unlabeled data match with confidence band conditions till all processed data is predicted. Modified from references [64, 66, 67].

and 2D-spectral characterization of small size metabolites in diseased tissues are reported with COSY-MRI, NOESY-MRI, and HPLC-MRI with little success but greater potential in noninvasive molecular details in undefined molecular etiology of poorly understood diseases. Final outcome depends on significant information and cost-effect constraints [20–61].

Use of MRS in clinical use has been slow due to ethical issues and still remains unclear because of high technical demands of spectroscopic methods and difficult interpretations in the brain. However, MRI with MRS using ^{1}H and ^{31}P spectroscopic imaging provide some information of demyelination, neuron loss, glial tissue formation, changes in glygenolysis and lactate accumulation, energy metabolism and pH imbalance in brain diseases, and inborn errors of metabolism. In the skeletal muscles, ^{31}P MRS is used in dynamic metabolic disorders. ^{1}H MRI /MRS evaluates muscle degeneration with fat accumulation. ^{31}P MRS is inconclusive for liver metabolic function but ^{1}H MRS/MRI indicates metabolites, mobile lipids, bone marrow characteristics. In the heart, MRSI provides in vivo metabolites with global cardiomyopathy. The role of MRI/MRS in cancer differential diagnosis and treatment is still not clear because of lack of preclinical investigations on physiological and metabolic events on NMR spectra and very limited clinical trials. All these limitations pose ethical issues.

From a technical point of view, coil sensitivity with higher signal-to-noise ratio is crucial in enhanced metabolite quantification.

12.5   PRESENT ACCEPTANCE OF MRSI USE AND ETHICAL ISSUES IN CLINICAL TRIALS AND HEALTHCARE

The author considers the possibility of ethically promising MRS method for following disease indications collected over three decades by neurophysicians:

- Focal brain tissues:
  - Maturating brains in gestation (elevated NAA, Glx peaks, low Cho, Tau peaks)
  - Recurrent brain tumor distinction from radiation induced necrosis (low NAA, Cr with high Cho peaks)
  - Assessment of prognosis in hypoxic ischemic encephalopathy (low NAA, low NAA/Cho)
  - Grading the low-grade and high-grade glioma (change in MyoInositol, GABA peaks)
  - Evaluating indeterminate brain lesion to postpone resection/biopsy (low NAA, low Cr, high Cho, Lac peaks)
- Diagnosis and monitoring of metabolic diseases:
  - Cerebral ischemia (high Lac peak)
  - Creatine deficiency (Cr peak missing)
  - Canvan disease (increased NAA/tCr ratio)
  - Nonketotic hyperglycinemia (DWI, DTI and tractography, FA and diffusivity)
  - Maple Syrup Urine disease (edema, high Glx, tau, ala peaks)
  - Multiple sclerosis periventricular and parahippocampus hypointense lesions (low NAA/tCr and Cho/tCr ratios)
  - Matachromatic leukodystrophy (low NAA peak and high NAA in urine)
  - Parkinson's disease substantia nigra (low NAA and GABA peaks, high Lac/tCr ratio)
  - Pelizauus-Merzbacher disease (high NAA+NAAG peaks and low Cho peaks)
  - Hypomyelination and congenital cataract (high glucose peak, low valine, lysine, and tyrosine peaks)
  - Krabbe disease (Globoid Cell Leukodystrophy) (low NAA peak)
  - X-linked adrenoleukodystrophy (X-ALD, CALD)(low NAA/Cr & Glx/Cr, high Ino/Cr & Cho/Cr)
  - Mitochondrial disorders (Kearn-Sayre syndrome, Leigh's syndrome, MELAS) (low Cr peak)
  - Alexander disease (ALX, AXD, demyelinogenic leukodystrophy) (elevated Glx/tCr ratios)
  - Megalencephalic leukoencephalopathy with subcortical cysts (low NAA, low Cr peaks)
  - Wasted white matter disease (Leukoencephalopathy CACH/VWM syndrome) (low NAA peak)
  - Neuroboreliosis (low NAA peak, high mI/Cr, Lipid/Cr and Cho/Cr ratios and normal NAA/Cr and Lac/Cr ratios)

While MRI findings are inconclusive due to ethical concerns to decide the change in treatment, MRS becomes mendatory in disease monitoring evaluation by MRS-based spectromics in the following diseases (but needs ethical clearance and disease specific evidences of its metabolite specificity and accuracy):

- Cancers of breast, prostate, colon, esophagous, liver, brain, bone tissues
- Coma and cerebrovascular diseases/injuries and disorders
- Cognitive disorders, movement disorders and dementia (frontotemporal dementia, vascular dementia, Alzheimer's disease with Lewy bodies, motor neuron disease, Huntington's disease, motor neuron disease, Parkinson disease/Parkinsonian syndromes)

- Psychiatric disorders (autism disorder, attention-deficit/hyperactivity disorder, bipolar disorder, schizophrenia, emotional dysregulation, depression, obsessive-compulsive disorder)
- Multiple sclerosis
- Dermatomyositis
- Hepatic steatosis in liver donor survivors
- Central nervous system with autoimmune rheumatic diseases
- Esophagous squamous cell carcinoma
- Mesial temporal sclerosis
- Primary CNS lymphoma lesions
- Epilepsy juvenile myoclonic epilepsy, mesial lobe epilepsy, temporal lobe epilepsy
- Hepatic encephalopathy
- Migraine pathophysiology
- Trauma of head
- Low back pain
- Hepatic carcinoma
- Liver cirrhosis
- Lyme neuroborreliosis
- Mucopolysaccharidosis
- Multiple sclerosis
- Radiatio encephalopathy
- Polymyositis
- Sport injuries
- Substance abuse disorders
- Brain trauma injury

Physician's repertory and ethical concerns: Now, MRS (MRS) is established as a non-invasive analytical technique to study metabolite changes in depression, Alzheimer's disease, stroke, seizures, brain tumors, and other diseases affecting the brain. MRS examines the metabolism of other human organs. The role of MRS in medical diagnosis and therapeutic planning is not established due to ethical concerns for clinical trials or clinical practice. Now advanced techniques of high-resolution MRI and MRSI favor routine clinical practice with following evidences in literature. Still, MRI with MRSI use remains inconclusive as per new guidelines.

In Massachusetts, Tuft's Agency on Healthcare Research and Quality (AHRQ) indicated concern of diagnostic thinking and therapeutic decision-making in the light of paucity of high-quality evidence. Still, MRS spectra interpretation is not standardized due to ethical concerns. MRS technical feasibility, study plan flaws of inadequate sample size, retrospective design, and other limitations bias the results and decision-making. A review on MRS of brain tumor by BlueCross BlueShield Association Technology Evaluation Center (2003) indicated that weak evidence is not sufficient to draw conclusion on MRSI use on health outcome [67].

Center of Medicare and Medicaid Service (CMS, 2004) evaluated the insufficient evidence of MRS-based diagnosis of brain due to ethical concerns for brain lesion detection. However, CMS announced to continue its decision of current national non-coverage [67].

Patients with neuroborreliosis disease showed non-specific due to ethical concerns, MRS changes in assessing central nervous system tissue damage. 1H MRS used PRESS sequence with placing an 8 cm^3 voxel box on normal-appearing white matter region of the frontal lobe [68].

The 1H MRS was described for epilepsy surgery as a research tool with correlation of ipsilateral MRS abnormality as good outcome. However, prospective studies on both localized and non-localized ictal scalp electroencephalopathy in MRI-negative patients are needed for validation due to ethical concerns. Furthermore, 1H MRS is only an adjunct to MRI characterization of brain tumors to convince policy makers [69].

Male fragile X syndrome (FRAX) patients are at risk of significant cognitive, behavioral deficits and executive prefrontal systems. The cholinergic system damage secondary to FRAX mental retardation show protein deficiency and contribute to cognitive behavior impairments. The 1H MRS showed low choline/Cr ratio in the right dorso—lateral prefrontal cortex in male FRAX vs controls with negative correlation to intelligence and age in left cortex. The donepezil improved cognitive-behavioral function [70]357]. The author believes that biochemical-MRI neuroimaging approach has the potential after ethical clearance to design treatment for FRAX and other genetic disorders based on neurometabolite intervention for treatment efficacy.

Proton endorectal MRI with 1H MR spectroscopic imaging of prostate cancer lesions showed high Gleason score towards increased (Cho+Cr)/Citrate ratio and large tumor volume. For detecting prostate cancer, proton MRS visible citrate peak and citrate concentration in semen or prostatic secretions showed better results over prostate specific antigen testing [71]. The 3D chemical-shift imaging (CSI) spin-echo sequence with MRS of biopsy from proven prostate carcinoma were used for deep learning along with tumor volume and tumor voxels per slice counts. The MR spectroscopy differentiated marginally T2 and T3 tumor stages. MRI with MR spectroscopy slightly improved the tumor staging with no advantage or ethical concerns in diagnosis and tumor staging over MRI alone [72].

MRS is a technique for diagnosis and monitoring cancer of prostate, colon, breast, cervix, pancreas, and esophagus organs. The 3D MRS has emerged as a new and sensitive tool in metabolic evaluation as loss of citrate and elevated ratio of choline.citrate of prostate cancer [73]. American College of Radiology Imaging Network (ACRIN) considers MRS imaging as routine diagnostic technique with ethical concerns. MRS has high diagnostic accuracy. However, randomized controlled large scale trial studies are needed [73].

Mucopolysaccharidosis (MPS) patients show poor correlation of enzyme levels, urine mucopolysaccharides (GAG), and neuroimaging findings. The semi-automated and automated segmentation techniques analyzed the T2-fluid-attenuated inversion recovery (FLAIR) brain images for several MRI variables of normalized cerebral volume (NCV), normalized cerebrospinal fluid volume (NCSFV), normalized ventricular volume (NVV), and normalized lesion load (NLL). The point-resolved MRS spectroscopy annotation positioned at white and gray matter showed positive correlation to age, enzyme levels, urinary GAG, neuroimaging. However, metabolite ratios by MRS, MRI visible NCV, NCSFV, NVV, and enzyme activity or GAG levels were poorly coorelated to disease duration or age of patients. Patients with MPS II showed aggressive white matter lesion increases that remain non-visible by MRI and MRS findings neither correlated to enzymatic nor to glycosaminoglycan levels [74]361]. All these are ethical concerns.

The MRI protocol of a sagittal T1-weighted 3D fast low-angle shot (3D FLASH) sequence, transverse T2- and spin-density-weighted turbo spin-echo sequences with 2D 1H-MR chemical shift imaging (2D CSI) sequence at short and long TE MRSI images showed low Cho/Cr, NAA/Cr ratios in MSA-C and SCA-2 with no difference from normal controls. These data show ethical concerns of its use. The cerebellar lactate peak in SCA-2 patients was distinct due to cerebellar degeneration while no lactate peak was evident in MSA-C or control subjects [75].

The 3T MRI with (1)H MRSI measured the increased peri-tumoral metabolites choline, lactate, NAA/tCr ratio and decreased NAA, choline/NAA ratio at peri-tumoral tissues in glioblastoma multiforme(GBM) to assess the post-resection response in contralateral brain at 3 to 5 weeks of post-Gliadel therapy and surgery plan prior to radiation therapy [76]. The 1H MRSI localizes different regions showing heterogeneous response after Gliadel treatment given. The author suggests that MRS metabolic indicators can monitor presurgery and postsurgery resection after Gliadel Inplantation to assess the tumor regression and gauze the chemotherapy efficacy.

Clinical MRS in multiple sclerosis (MS) showed metabolic changes at MS lesion sites, and multicentered clinical trials of MS incorporated MRS into their MRI imaging diffusion tensor DT-MRI protocols to quantify the effect of therapeutic intervention on multiple sclerosis tissue damage [77, 78].

Biomarkers of disc degeration discogenic back pain are not available. It poses ethical concerns to use pain biomarkers in treatment. In frozen disectomy disc samples, quantitative ex vivo 1H high resolution magic angle spinning (HR-MAS) MR spectroscopy measured low proteoglycan/collagen and proteoglycan/lactate ratios; and high lactate/collagen ratio as biochemical markers associated with discogenic back pain vs scoliosis conditions. Several MRS visible metabolites proteoglycan, collagen, and lactate may serve as metabolic markers of disc degeneration back pain [79].

MRS-visible biomarkers distinguished the painful discs (PD) and quantifed the severity of disc degeneration to predict the surgical outcomes in chronic low back pain (CLBP) patients. MRS showed disc proteins (proteoglycan and collagen) and acidic metabolites (lactate, alanine, propionate) as pain MRS biomarkers in CLBP vs PD Pfirrmann grade or ratios of acidity metabolites to proteins. MRS correlates with PD may improve surgical outcomes in CLBP patients. In vivo MRS may define mechanisms of pain metabolites to design therapies [80, 81]. The author believes that MRS-derived pain analysis may be first-line method after ethical clearence.

ACR Guidelines on bone tumor musculoskeletal imaging reflect that MRS may differentiate benign lesions from malignant lesions. Still, ethical clearance and more efforts are needed [82].

With ultrahigh MRS availability and access, MRS is a clinical tool in oncologic management of patients [83]. MRS is still an experimental investigation tool in suspected breast cancer because of limited peer-reviewed clinical literature and ethical concerns on cancer MRS use in theranosis.

The diagnostic value of proton 1H-MRS in cancer is based on the detection of elevated choline levels as markers of an active tumor. The 1H MRS with breast MRI improves the specificity to distinguish benign from malignant lesions, and monitors to predict the neoadjuvant chemotherapy response to treat patients. Integrated MRI with 1H MRS evaluation cuts down the need of multiple benign biopsies, and MRS may predict response within twenty-four hours after first dose of neoadjuvant chemotherapy given [84].

The non-invasive in vivo proton (1)H-MRS differentiated the benign and malignant breast lesions as different increased levels of choline (Cho) compounds and increased Cho metabolism in breast pre-cancer cells showing infiltrating ductal carcinoma, infiltrating medullary, mucinous, lobular adenoid cystic carcinoma as active lesions by 1H-MRS. However, (1)H-MRS showed no change or normal metabolites in benign breast lesions including cysts, galactoceles, ductal carcinoma, papilloma, fibroadenoma, fibrocystic changes, tubular adenoma, and phyllodes tumors [85]. It is an ethical concern. The author suggests that stronger ultrahigh field MR imagers with advanced coils will increase the 1H-MRS sensitivity. Ultrahigh field (1)H-MRS will detect the smallest malignant lesions to characterize the malignant lesions into non-invasive or invasive towards disease progression monitoring. However, there is ethical concern.

The insufficient clinical evidence on MRS is an ethical concern in the evaluation of leukoencephalopathy and childhood white matter diseases because both disorders show similar MRI signal intensity changes despite different pathologies. The 1H-MRSI distinguished three conditions of white matter rarefaction, hypomyelination, and demyelination. Neurochemicals from six white matter rarefaction and intra-intervoxel (relative to gray matter) neurochemical ratios showed a variance as significant pathophysiological differences of high Cho/NAA, Cho/Cr ratio and low NAA/Cr ratio as accurate linear discriminant parameters to classify hypomyelinating conditions [86]. Combined MRI/MRS explored risk profiles in prostate cancer and benefits. MRI combined with MRS can be promising cost-effective screening test for low-risk patients [87].

In a prospective, multi-center trial on patients undergoing radical prostatectomy, endorectal 1.5T MRI and MRS reported an accurate sextant localization of peripheral zone (PZ) in prostate cancer tissue matched with biopsy-confirmed prostate adrenocarcinoma to schedule removal of prostate or radical prostectomy [88, 89]. T1-weighted, T2-weighted and 1.5T MRS with pelvic-phased array coil in combination with an endorectal coil improved diagnostic accuracy of MRI-MRS over MRI alone.

5.13 The 19F MRS used SR4554 (fluorinated 2 nitroimidazole) as a hypoxia intra-tumoral marker in phase I trial. The 1H/19F surface coils and localized 19F MRS acquisition spectra showed different signals in post-SR4554 infusion (MRS no.1) after 16 hrs (MRS no. 2) and 20 hrs (MRS no. 3) in both unlocalized and localized MRS indicating different grades of hypoxia [90].

12.6 ETHICAL CONCERNS, PRESENT CONSENSUS ON CLINICAL MRS ACCEPTANCE, GUIDELINES, META-ANALYSIS AND CLINICAL TRIAL REPORTS ON MRSI IN THE YEARS 2010–2022

MRI generates an image while MRS generates a graph or peak "spectrum" arrays of metabolite types to quantity them in the brain or other organs. Despite ethical concerns, the present consensus on MRS metabolite interpretation is in favor of theranosis and treatment planning in neurological and other human diseases. Documented meta-analysis and review studies in recent years (2010–2022) recommend the need for investigative and reproducible multicentric cost-effective clinical trials on large patient number with diagnostic accuracy and validation before clinical practice. Still, MRS evaluation in primary brain tumors or metastases remains investigational/experimental due to lack of peer-reviewed clinical literature on the effectiveness of MRS. The Food and Drug Administration (FDA) granted 510(k) clearance to prescribe MRS for cancer patients. However, MRI and MRS use different postprocessing softwares to acquire data and integrate and manipulate the spectroscopic image signal.

The clinical evidence is inconclusive and full of ethical concerns on the MRS of prostate cancer. The MR spectroscopy offers prostate gland choline and citrate metabolic information as prognostic information useful for treatment planning. Combined proton MRSI with T2-weighted MRI to T2-weighted MRI improves the tumor localization, volume estimation, staging, tissue characterization, and identification of recurrent disease after therapy. American College of Radiology Imaging Network showed the combined 1H-MRSI and T2-weighted MR images do not improve tumor detection in patients with low-grade, low-volume disease selected to undergo radical prostatectomy. Thus positive 1H-MRSI reflect only higher tumor grade and/or volume. In a retrospective trial, MRI and MRS predicted the normal prostate or suggestive of progressive prostate cancer malignancy risk of the Gleason score and subsequent biopsy. MRS imaging did not predict cancer progression due to ethical concerns [88].

The National Comprehensive Cancer Network (NCCN, 2016) gave guidelines on CNS cancer imaging. The MRS examination may be useful to differentiate anaplastic gliomas from radiation induced necrosis or "pseudoprogression." The MRS marginally differentiated the glioma recurrence from radiation necrosis. The MRS data statistics in glioma showed elevated Cho/Cr, Cho/NAA ratio in favor of high sensitivity, specificity, areas under receiver operative curves for heterogeneity test, threshold effect test as moderate diagnostic value to distinguish radiation induced necrosis from glioma recurrence [91]. Further, MR perfusion and MR spectroscopy differentiated the primary brain tumors from radiation induced necrosis and metastasis of brain in patients using cerebral blood volume (rCBV), ratio of Cho/tCr, Cho/NAA, chi square based heterogeneity test using Cochran's Q statistics of twelve patients having tumor reccurence showed high Cho/NAA and Cho/Cr ratio [92]. The MRS method shows accuracy of metabolites and rCBV in primary brain tumors or metastases, necrotic recurrent tumors in patients. The author suggests the MRS with MRI technique will improve diagnostic accuracy using multimodal multicentre trials in the future without any ethical concern.

In vivo single voxel 1H MR spectroscopy showed brain metabolite differences as low NAA in putamen due to loss of neuronal integrity in early HD, pre-HD vs controls; high glial myo-inositol cell marker in pre-HD and early-HD vs controls with Unifield Huntington Disease Rating Scale (UHDRS) motor score, tongue pressure task and disease burden score as neuropsychological biomarkers of HD onset and progression to establish an association with motor performance in TRACK-HD study [93]. The lower putaminal tNAA and high mI in early HD compared to controls established putamen MRS as HD onset and progression biomarker in a cross-section of subjects.

Non-invasive in vivo 31-phosphorous (31P) MRS showed metabolic state and biochemical score of the myocardial molecular imaging in ischemic heart disease, and valvular disease to assess the effectiveness of metabolic modulating agents. Limited research, low temporal and spatial resolution, poor reproducibility, and longer data collection time are major ethical concerns [91]. The author believes that MRS will be a multi-modal non-invasive cardiac assessment in future.

12.6.1 Malignancy and Tumors in Humans

The role of MRS is wide in the classification of human brain tumors, distinction of tumors vs non-neoplastic lesions, prediction of survival, treatment planning, monitoring of post-therapy effects in tumor diagnosis and treatment response [94]. A retrospective MRSI study with T2-weighted MRI images showed recurrent cancer of prostate after androgen therapy and definite external beam radiation therapy. A(Z) analysis of T2-weighted MRI + MRSI improved the diagnostic accuracy and detection of locally recurrent prostate cancer [95].

For prostate cancer evaluation, MRS-guided radiation administration of higher dose delivery is precise to treat metabolically active tumor areas. MRS compatible implants are available to deliver high radiation doses in metabolically active regions of tumors.

Breast proton 3T MRS differentiated the benign tumors from malignant lesions by 1D single-voxel spatially resolved MRS high choline peak high sensitivity and variable specificity [96].

MR spectroscopy with breast MRI enhances the cancer diagnostic accuracy to avoid repeat benign biopsies. MR spectroscopy clearly visualizes the choline peaks in the evaluation of suspicious non-cancer mass otherwise non-visible on breast MRI. MR spectroscopy still remains an investigational tool due to ethical concerns, but it monitors the outcome and response of therapy.

The diagnostic accuracy of MRS and contrast enhanced dynamic DCE-MRI and diffusion-weighted MRI localize malignancy and prostate abnormalities in negative prostate biopsies. Multiparametric T2-weighted MRI + MRS, transrectal ultrasound-guided biopsy T2-weighted MRI, with DCE/DW-MRI images and histopathological assessment of prostate biopsy tissue showed high sensitivity and specificity of MRS than T2-weighted MRI to detect moderate or high-risk cancer [97]. Few multi-parametric MRS, DCE-MRI/DW-MRI imaging approach with MR-guided biopsy and extended 14 core ultrasound-guided biopsy showed better diagnostic accuracy against a reference biopsy obtained by histopathology biopsy, template biopsy or prostatectomy specimens. MRS is useful in differential diagnosis of dermatomyositic and polymyositis in differential diagnosis.

12.6.2 Neurophysiological Diseases

Neuro-metabolic changes in glucose metabolism, ionic shifts, release of neurotransmitters, altered cerebral blood flow, and impaired axonal functions are associated with concussion, neuronal depolarization. 1H-MRS, or MRS measures brain metabolites and physiologic changes after sport-related concussion. MRS visible neurochemicals monitor the altered neurophysiology and recovery of post-concussive symptoms or injury returning to normative levels in adult athletes [98]. The author suggests that large cross-sectional, prospective and longitudinal studies will establish the high sensitivity and prognostic value of MRS in concussion. American Medical Society for Sports Medicine advocated the use of MRS in concussion in sport as a research tool not as management tool due to ethical concerns [99].

The "Current Lyme disease" and "Nervous system Lyme disease" guidelines mentioned the MRS as diagnostic tool not as management tool due to ethical concerns [100].

The MRS studies showed high specificity and sensitivity in children at short TE-chemical shift imaging of brain tumors [101].

American College of Radiology's Appropriateness Criteria of dementia and movement disorders favors the use of fMRI and MRS in neurodegenerative disorders but not for routine clinical practice due to ethical concerns [102]. Moreover, PET, SPECT, fMRI, DTI, and MRS may visualize head truma injury in children by standard imaging but insufficient in routine clinical use due to ethical concerns. In young children with attention-deficit/hyperactivity disorder (ADHD), autism spectrum disorders, emotional dysregulation, and schizophrenia, 1H-MRS showed increased glutamine/glutamate-related metabolites in the anterior cingulate cortex (ACC) and other regions. In major deression, bipolar disorders and obsessive-compulsive disorder children, 1H-MRS showed low glutamine/glutamate. Limited evidence showed a normal GLx levels after treatment of bipolar disorder and ADHD diseases to indicate the mechanism of glutamate dysregulation in these disorders [103].

12.6.3 AMYLOID PATHOLOGY IN ALZHEIMER'S DISEASE

1H MRS showed low N-acetyl aspartate (NAA), low NAA/Cr ratio in posterior cingulate and bilateral left/right hippocampus regions, and high myoInsoitol/Cr ratio in posterior cingulate and gray matter in Alzheimer's disease patients [104]. The author suggests that NAA/Cr ratio, NAA, mI are indicators of brain dysfunction while mI/NAA and Cho/Cr ratio are diagnostic indicators in Alzheimer's disease subjects.

The amylid pathology and proton MRS metabolites in non-dementia individuals show the cognitive decline as changes +2.9%/year of myo-inositol (mI)/ tCr ratio and −3.6%/year of NAA/mI ratio in mild cognitive impairment β-amyloid (Aβ+) patients; −0.05%/y mI/Cr and +1.2/y NAA/mI ratio in Aβ- patients, associated with MiniMental State Examination (MMSE), Apo-E, age, and sex. Low NAA/mI indicated low cognitive capacity in Aβ+ patients with high NAA/myoInositol baseline. The longitudinal changes in ratios mI/Cr and NAA/mI classified the amyloid pathology types due to dementia [105]. In Aβ+ individuals, NAA/mI ratio predicts the declining rate of cognition in future. Now, PET can image the amyloid and tau deposition to screen dementia, monitor the disease progression to evaluate molecular pathology in-vivo using MRS/PET in clinical settings. The author suggests multimodal large-scale hybrid MRS-PET clinical trials to measure metabolites and early β-amyloid and tau protein changes in AD single MRI+MRS session and show the increased mI/Cr ratio with longitudinal decline in NAA/mI ratio to track predementia AD disease progression in clinical trials.

12.6.4 RADIATION ENCEPHALOPATHY

The proton MRS measured the varied metabolite concentrations of NAA/tCr, Cho/Cho ratio, NAA, Cho, Cr in healthy control and postradiotherapy encephalopathy subjects. MRS evaluation is feasible to evaluate radiation-therapy-induced encephalopathy in nasopharyngeal carcinoma. The author suggests precise subgroup analysis including variables such as disease stage, age, and sex [106].

12.6.5 THERANOSIS OF HEPATIC ENCEPHALOPATHY

MRSI visualized the altered imaging features showing stratification, severity of the hepatic encephalopathy in non-HE cirrhosis, minimal HE, overt OHE patients by monitoring high glutamine/glutamate ratio; high choline, and high myo-inositol peaks in parietal lobe by Metafor v3.4.1 software analysis in brain regions. These MRI+MRS features showed correlation with high homogeneity and HE grade in all brain regions. However, evaluation can be biased by HE assessment cut-off, sample size, method used, geographic variation. In parallel, OHE severity and classification vary by West Haven grade [107].

12.6.6 EVALUATION OF LIVER STEATOSIS IN LIVER DONORS

A mets-analysis of hepatosteatosis by MRI + MRS in living liver donors, showed high specificity and sensitivity of hepatosteatosis to avoid liver biopsy [108]. However, the use of MRS had several drawbacks: 1. Moderate inter-study clinical or statistical heterogeneity due to diagnostic threshold variability study methodology; 2. MRI and MRS suffer from selection and recall biases with higher sensitivity and specificity or may over-estimate in retrospective studies; 3. sample size, sex ratio, mean age factors need meta-regression with these parameters and demand large data, and uniform thresholds of substantial HS limits.

12.6.7 Monitoring Hepatocellular Carcinoma and Liver Cirrhosis Development

1H MRI+ MRS predicted the probability by high Cho (Choline), Lip (lipid) contents, and Cho/Lip ratio on LCModel software to assess the magnitude of hepatocellular carcinoma (HCC) with secondary chronic hepatitis B and cirrhosis [109]. The author suggests that H-MRS visible hepatic metabolite levels may monitor both HCC and liver cirrhosis development, but more validation studies are needed.

12.6.8 Low Back Pain

In vivo MRS in low back pain LBP subjects showed: 1. Low NAA concentration in right primary motor cortex, left anterior insula, left somatosensory cortex (SSC), anterior cingulate cortex (ACC) and dorsolateral prefrontal cortex (DLPFC) regions; 2. low glutamate in ACC, 3. low myo-inositol in ACC and thalamus; 4. high choline in the right SSC; 5. high glucose in the DLPFC locations [110]. Biochemical altered LBP brain MRS profile correlatedd with possible therapy response and the physiochemical functions of brain metabolites and pain receptors. The author notices concerns such as few subjects selected, confounding factors, medication effects, unclear biochemical basis of pain specific metabolite changes; all are concerns.

12.6.9 Juvenile Myoclonic Epilepsy

1H MRS showed distinct low NAA and NAA/Cr, rise in Glx/Cr ratio in insula and striatum to indicate juvenile myoclonic epilepsy (JME) as multi-region, thalamo-frontal network epilepsy (not idiopathic general epilepsy) [111]. The frontal cortex and thalamo-cortical pathways indicated low NAA, low NAA to creatine ratio (NAA/Cr) in frontal region of JME related to memory and visual attention stroop test; low NAA/Cr ratio in thalamic linguishtic and memory regional Wisconsin card sorting test. Unaltered Glx, Glx/tCr ratio, Cho compounds and Cho/tCr ratio in frontal and thalamic regions were associated with epileptic cortical functions, neuropsychological cognitive tests as visuospatial executive functions, linguistic, memory, visual attention [112]. The MRS may show subclinical cognitive changes. The author suggests the need of validation studies due to ethical concerns.

12.6.10 Primary Focal CNS Lymphoma lesions

Hybrid approach of MRS with SPECT, PET distinguished the CNS lymphoma from HIV [113]. MR perfusion, MRI apparent diffusion coefficient ratio, regional cerebral blood volume chracteristics distinguish lymphoma from HIV-infected patients. The author suggests more clinical investigations to establish the diagnostic accuracy of hybrid MRI approaches due to ethical concerns.

12.6.11 Migraine Pathophysiology and Identification of Neuromarkers in Migraine

The 1H-MRS studies suggested many inter-ictal abnormalities in migraine patients with persistent altered mitochondrial energy loss, neuronal excitability indicated by high excitatory glumate, high inhibitory γ-aminobutyric acid (GABA) neurotransmitter peaks; low NAA levels due to mitochondrial dysfunction and abnormal energy metabolism towards excitatory stimulation or migraine attack triggers [114]. MRS can be a valuable non-invasive method to determine migraine attack, correlated severity, and monitoring medication efficacy. It has ethical concerns.

12.6.12 ADRENOLEUKODYSTROPHY

Status epilepticus or childhood cerebral X-linked adrenolukodystrophy shows abrupt pathogenic ABCD1 mutation or transient altered mental status neurodegeneration as autism with fever, diarrhea, seizures, coma, gross motor loss, fine motor loss, poor speech skills. Brain serial MRI/ MRS showed elevated lactate peak and CSF protein levels associated with diffused progressive cortex swelling, laminar necrosis, restricted diffusion indicative of mitochondrial, lysosomal, and peroxisomal disorders. Moreover, MRS showed elevated very long-chain fatty acids, lactate peaks and high CSF proteins. The acute decline in neurologic functions with elevated CSF proteins and lactate, acute decline in neurologic NAA/Cr functions by MR spectroscopy indicates MRI non-visible white matter abnormalities to predict disease progression [115].

12.6.13 HYPOXIC-ISCHEMIC ENCEPHALOPATHY

Proton MRI with MRS showed elevated Lac/NAA ratio in posterior white matter areas and deep gray nuclei of putamen and thalamus areas as indicator prognostic indicator of brain hypoxic-ischemic injury underlying neonatal hypoxic-ischemic encephalopathy (HIE) [116]. Proton MRS (1H-MRS) showed high NAA/tCr in basal ganglia/thalamus; high NAA/Cho in BG/T; high Myo-inositol/choline in cerebral cortex, as prognostic marker of therapeutic hypothermia (TH) with adverse outcomes of white matter and gray matter NAA in prediction for all HIE subjects. The author suggests the need for prospective multi-center studies by standardized protocol and analysis methods due to ethical concerns [117].

12.6.14 TRAUMATIC BRAIN INJURY

The single voxel 1H-MRS showed altered ratio of (NAA/Cr + PCr), (NAA + N-acetylaspartylglutamate NAAG)/Cr+PCr relative to creatine neuro-metabolites in complicated and uncomplicated Traumatic Brain Injury (TBI) patients. The NAA peaks and NAA values were low in complicated TBI and uncomplicated TBI groups compared to control group. Neuro-metabolite alterations indicated the onset of both complicated and uncomplicated TBI with injury severity in mTBI [118]. Furthermore, MRS is not biomarker option in management of truamatic brain injury.

Mild TBI is a risk factor for dementia. The in vivo MRS detected the metabolites in frontal lobe brain white matter as low NAA, high Glutamate, low Choline, and variable Cr as membrane high turnover changes after negative routine CCT/MRI scan of frontal lobe acute or subacute mild TBI within three months [119]. MRS correlated post-traumatic brain metabolism with cognitive dysfunction to detect mTBI patients at risk of post-traumatic neurodegeneration early. However, the author suggests that follow-up studies are needed to correlate MRS with cognitive outcome using different modified MRI sequences, high field strength, accurate voxel placements, voxel sizes, and reasonable Cramer Rao Lower bounds cut-offs.

12.6.15 EVALUATION OF PROGNOSTIC CONSCIOUSNESS RECOVERY
IN WAKEFULNESS SYNDROME INDIVIDUALS

Multi-voxel MRS of frontal cortex, temporal cortex, brain stem, fornix, internal capsule, thalamus, globus pallidus, putamen regions in patients with vegetative state/unresponsive wakefulness syndrome (VS/UWS) showed low NAA/Cr ratio in temporal cortex, calsula interna, thalamus while higher NAA/Cr ratio in these structures indicative of recovery of consciousness [120]. The low NAA/Cr ratio and NAA/(NAA+Ch+Cr) ratio in mid brain regions was associated with hypoxia in patients. The MRS accurately predicted a recovery of conciousness from hypoxix brain damage in unconscious VS/UWS patients with TBI to the level of EMCS.

12.6.16 SUBSTANCE USE DISORDERS

1H MRS is reviewed to evaluate the effect of narcotic substances methamphetamine, alcohol, 3,4-methylenedioxymethamphetamine, nicotine, marijuana, cocaine, opiates, and opioid after-use disorders. A decrease in N-acetylaspartate and choline levels in brain regions was a common feature to define the substance specific effect of above said subtance use. The author recommends more MRS clinical trials to monitor the substance abuse novel treatment due to ethical concerns. The 1H MRS showed low NAA, and low Choline levels in brain indicating altered neuro-metabolites in addicts on MDMA, cocaine, alcohol, methamphetamine, opiates/opioids, nicotine, and marijuana [121].

12.6.17 CENTRAL NERVOUS SYSTEM WITH RHEUMATIC AUTOIMMUNE DISEASES

Proton MRS measures the neuronal loss or CNS damage with autoimmune rheumatic diseases. A MRS study reviewed vasculitis, Behcet disease, Sjogren's syndrome, psoriasis, rheumatoid arthritis, juvenile idiopathic arthritis, systemic sclerosis, and systemic lupus erythematosus during November 2003 to December 2019 [122]. Low NAA/Cr ratios and high Choline/Creatine ratio in different regions of above brain diseases showed an association with varied CNS disease inflammatory activities and comorbidities with diseases. The neuro-metabolite abnormalities with non-overt CNS manifestations suggested the association of abnormal vascular reactivity, systemic inflammation with subclinical CNS manifestations.

12.6.18 DIAGNOSIS OF MESIAL TEMPORAL SCLEROSIS

Recently, MRS metabolic features of brain were reported to classify mesial lobe epilepsy, hippocampal sclerosis, mesial temporal sclerosis (MTS), mesial and temporal seizures [123]. A loss in N-acetylaspartate (NAA); NAA/(Cho+Cr) in MTS, ratio NAA/Creatine in ipsilateral hippocampus,and reduced NAA levels in extra-hippocampal regions suggested a pre-surgery epileptogenic focus.

12.6.19 THERANOSIS OF MITOCHONDRIAL DISEASE

Serial MRS visualized the response of intravenous 500 mg/kg weekly L-arginine therapy in MELAS in improvement clinically for the treatment of mitochondrial encephalopathy with lactic-acidosis and stroke-like events. L-arginine therapy normalizes brain lactate and N-acetylaspartate /Choline ratio as radiologic and clinical improvement. [124]. Consensus indicated the management of mitochondrial stroke-like episodes can be possible by MRS [125]. Mitochondrial myopathies by MRS of the brain were indicated by elevated lactate, MRS of muscle, and cardiac tissue [126].

12.7 FUTURE DIRECTIONS AND PERSPECTIVES

In future, use of clinical MRSI with CT-PET-US will be ethical, multimodal, robust, fast, equipped with new supervised segmentation, component-specific registration, disease specific spatio-spectral encoded trained pixel data sets (SPECTROMICS) by super-resolution "deep learning" towards customized precision medicine [127–131]. New perception of 3D-MRSI clinical value will certainly depend on new computer data generation and feature extraction by segmentation and multimodal registration using integrated diagnosis with high acceptance in medical practice [132, 133]. Hybrid approaches of HPLC-MR, biochemical-MR, EEG-MR, MALDI-MR, and digital histochemical-MR correlation perhaps will contribute better. It seems that quantification of metabolite absolute concentrations may indicate actual spectral metabolic screening of inborn errors of metabolism,

EEG neurophysiological properties in brain diseases and cancer comparing with test phantoms for quality assessment of clinical spectroscopy. New approaches using ^{13}C, ^{19}F, ^{7}Li, ^{11}B spectroscopic imaging may be useful. It remains to see the possibility of ethical acceptance.

12.8   CONCLUSION

With minimum computation time, deep learning approach is emerging with new hopes to utilize multimodal MRSI maps with concentrations, diffusion, perfusion, flow MRI fingerprints combined with dynamic PET, CT, SPECT, and optic molecular imaging common platform screening modality with wider ethical acceptance. The segmentation, registration, and artificial intelligence further define disease course and its clinical value jointly acceptable by scientists and clinician radiologists. Present opinion displays a common ground for the above. The possibility of ethical acceptance remains to be seen. Governments and academic clinical trials are silent on these ethical issues due to insufficient evidences.

A HELPFUL PHYSICIAN'S REPERTOIR ON AVAILABLE TOOLKITS ON DEEP LEARNING

Deep learning resources for clinical radiologists, radiomics and imaging genomics (radiogenomics), and toolkits and libraries for deep learning; deep learning in neuroimaging and neuroradiology; brain segmentation; stroke imaging; neuropsy- chiatric disorders; breast cancer; chest imaging; imaging in oncology; medical ultrasound; and more technical surveys of deep learning in medical image analysis are suggested below:

Toolkit	Ethics	Performance
AlexNet	Yes	The network that launched the current deep learning boom by winning the 2012 ILSVRC competition by a huge margin.
		Notable features include the use of RELUs, dropout regularization, splitting the computations on multiple GPUs, and
		using data augmentation during training. ZFNet [67], a relatively minor modification of AlexNet, won the 2013 ILSVRC competition.
VGG	Yes	Popularized the idea of using smaller filter kernels and therefore deeper networks (up to 19 layers for VGG19, compared
		to 7 for AlexNet and ZFNet), and training the deeper networks using pre-training on shallower versions.
GoogLeNet	Yes	Promoted the idea of stacking the layers in CNNs more creatively, as *networks in networks*, building on the idea of [70].
		Inside a relatively standard architecture (called the *stem*), GoogLeNet contains multiple *inception modules*, in which
		multiple different filter sizes are applied to the input and their results concatenated. This multi-scale processing allows
		the module to extract features at different levels of detail simultaneously. GoogLeNet also popularized the idea of not
		using fully-connected layers at the end, but rather global average pooling, significantly reducing the number of model
		parameters. It won the 2014 ILSVRC competition.
ResNet	Yes	Introduced *skip connections*, which makes it possible to train much deeper networks. A 152 layer deep ResNet won the
		2015 ILSVRC competition, and the authors also successfully trained a version with *1001* layers. Having skip

Toolkit	Ethics	Performance
		connections in addition to the standard pathway gives the network the option to simply copy the activations from layer to
		layer (more precisely, from ResNet block to ResNet block), preserving information as data goes through the layers.
		Some features are best constructed in shallow networks, while others require more depth. The skip connections facilitate
		both at the same time, increasing the network's flexibility when fed input data. As the skip connections make the
		network learn residuals, ResNets perform a kind of boosting.
Highway nets	Not yet	Another way to increase depth based on *gating units*, an idea from Long Short Term Memory (LSTM) recurrent
		networks, enabling optimization of the skip connections in the network. The gates can be trained to find useful
		combinations of the identity function (as in ResNets) and the standard nonlinearity through which to feed its input.
DenseNet	Not yet	Builds on the ideas of ResNet, but instead of adding the activations produced by one layer to later layers, they are simply
		concatenated together. The original inputs in addition to the activations from previous layers are therefore kept at each
		layer (again, more precisely, between blocks of layers), preserving some kind of global state. This encourages feature
		reuse and lowers the number of parameters for a given depth. DenseNets are therefore particularly well-suited for
		smaller data sets (outperforming others on e.g. Cifar-10 and Cifar-100).
ResNext	Not yet	Builds on ResNet and GoogLeNet by using inception modules between skip connections.
SENets		Squeeze-and-Excitation Networks, which won the ILSVRC 2017 competition, builds on ResNext but adds trainable
		parameters that the network can use to weigh each feature map, where earlier networks simply added them up. These
		SE-blocks allows the network to model the channel and spatial information separately, increasing the model capacity.
		SE-blocks can easily be added to any CNN model, with negligible increase in computational costs.
NASNet	Yes	A CNN architecture designed by a neural network, beating all the previous human-designed networks at the ILSVRC
		competition. It was created using AutoML, a Google Brain's reinforcement learning approach to architecture design
		A controller network (a recurrent neural network) proposes architectures aimed to perform at a specific level for a
		particular task, and by trial and error learns to propose better and better models. NASNet was based on Cifar-10, and has
		relatively modest computational demands, but still outperformed the previous state-of-the-art on ILSVRC data.
YOLO	Yes	Introduced a new, simplified way to do simultaneous object detection and classification in images. It uses a single CNN
		operating directly on the image and outputting bounding boxes and class probabilities. It incorporates several elements
		from the above networks, including inception modules and pretraining a smaller version of the network. It's fast enough to enable real-time processing.[b] YOLO makes it easy to trade accuracy for speed by reducing the model size.
		YOLOv3-tiny was able to process images at over 200 frames per second on a standard benchmark data set, while still

Toolkit	Ethics	Performance
		producing reasonable predictions.
GANs	Not yet	A generative adversarial network consists of two neural networks pitted against each other. The *generative network* G is
		tasked with creating samples that the *discriminative network D* is supposed to classify as coming from the generative
		network or the training data. The networks are trained simultaneously, where G aims to maximize the probability that D
		makes a mistake while D aims for high classification accuracy.
Siamese nets	Not yet	An old idea that's recently been shown to enable *one-shot learning*, i.e. learning from a single example. A
		Siamese network consists of two identical neural networks, both the architecture and the weights, attached at the end.
		They are trained together to *differentiate* pairs of inputs. Once trained, the features of the networks can be used to
		perform one-shot learning without retraining.
U-net	Yes	A very popular and successful network for segmentation in 2D images. When fed an input image, it is first downsampled
		through a "traditional" CNN, before being upsampled using transpose convolutions until it reaches its original size. In
		addition, based on the ideas of ResNet, there are skip connections that concatenates features from the downsampling to
		the upsampling paths. It is a fully-convolutional network, using the ideas first introduced.
V-net	Not yet	A three-dimensional version of U-net with volumetric convolutions and skip connections as in ResNet.

Deep learning code available at https://github.com/paras42/Hello World Deep Learning, where you'll be guided through the construction of a system that can differentiate a chest X-ray from an abdominal X-ray using the Keras/Tensor Flow frame-work through a Jupyter Notebook. Other nice tutorials are http://bit.ly/adltk tutorial, based on the Deep Learning Toolkit(DLTK) and https://github .com/usuyama/pydata-medical-image, based on the Microsoft Cognitive Toolkit(CNTK) [134].

REFERENCES

1. LeCun Y, Bengio G, Hinton B Deep learning. *Nature* 2015;521(7553):436–444.
2. Sharma R Application of segmentation in localized MR chemical shift imaging and MR spectros-copy. In Suri JS, Setarehdan SK, Singh S (eds) *Advanced Algorithmic Approaches to Medical Image Segmentation*. Advances in Computer Vision and Pattern Recognition. London: Springer 2002. doi: 10.1007/978-0-85729-333-6_7.
3. Boada FE, Christensen JD, Huang-Hellinger FR, Reese TG, Thulborn KR Quantittative in vivo tissue sodium concentration maps: The effects of biexponential relaxation. *Magn Reson Med* 1994;32(2):219.
4. Tkac I, Gruetter R Methodology of 1H NMR spectroscopy of the human brain at very high magnetic fields. *Appl Magn Reson* 2005;27.
5. Geppert C, Dreher W, Leibfritz D PRESS-based proton single-voxel spectroscopy and spectroscopic imaging with very short echo times using asymmetric RF pulses. *Magma* 2003;16(3):144–148.
6. Mountford CE, Doran SJ, Lean CL, Russell P Proton MRS can determine the pathology of human can-cers with a high level of accuracy. *Chem Rev* 2004;104(8):3677–3704.
7. Seeger U, Klose U, Mader I, Grodd W, Nagele T Parameterized evaluation of macromolecules and lipids in proton MR spectroscopy of brain diseases. *Magn Reson Med* 2003;49(1):19–28.
8. Kanowski M, Kaufmann J, Braun J, Bernarding J, Tempelmann C Quantitation of simulated short echo time 1H human brain spe- ctra by LCModel and AMARES. *Magn Reson Med* 2004;51(5):904–912.

9. Sharma R Serial amino-neurochemicals analysis in progressive lesion analysis of multiple sclerosis by magnetic resonance imaging and proton magnetic resonance spectroscopic imaging. *Magn Reson Med Sci* 2002 Nov 1;1(3):169–173. doi: 10.2463/mrms.1.169; PMID: 16082140.

10. Tzika AA, Vigneron DB, Ball WS Jr., Dunn RS, Kirks DR Localized proton MR spectroscopy of the brain in children. *J Magn Reson Imaging* 1993;3(5):719.

11. Grutter R, Rothman DL, Novotny E et al. Detection and assignment of glucose signal in 1H NMR difference spectra of human brain. *Magn Reson Med* 1992;27(1):183.

12. Zia B, Bogia DP Fast Fourier Transform and convolution in medical image reconstruction. 2020. https://www.intel.com/content/www/us/en/developer/articles/technical/fast-fourier-transform-and-convolution-in-medical-image-reconstruction.html.

13. Iqbal Z, Nguyen D, Thomas MA, Jiang S Deep learning can accelerate and quantify simulated localized correlated spectroscopy. *Sci Rep* 2021 Apr 22;11(1):8727. doi: 10.1038/s41598-021-88158-y; PMID: 33888805; PMCID: PMC8062502.

14. Yang J, Lei D, Qin K et al. Using deep learning to classify pediatric posttraumatic stress disorder at the individual level. *BMC Psychiatry* 2021 Oct 28;21(1):535. doi: 10.1186/s12888-021-03503-9; PMID: 34711200; PMCID: PMC8555083.

15. Balakrishnan R, Valdés Hernández MDC, Farrall AJ Automatic segmentation of white matter hyperintensities from brain magnetic resonance images in the era of deep learning and big data - A systematic review. *Comput Med Imaging Graph* 2021 Mar;88:101867. doi: 10.1016/j.compmedimag.2021.

16. Terpstra ML, Maspero M, Sbrizzi A, van den Berg CAT. ⊥-loss: A symmetric loss function for magnetic resonance imaging reconstruction and image registration with deep learning. *Med Image Anal* 2022 Aug;80:102509. doi: 10.1016/j.media.2022.102509.

17. Chen D, Wang Z, Guo D, Orekhov V, Qu X Review and prospect: Deep learning in nuclear magnetic resonance spectroscopy. *Chemistry* 2020 Aug 17;26(46):10391–10401.

18. Li X, Strasser B, Neuberger U et al. Deep learning super-resolution magnetic resonance spectroscopic imaging of brain metabolism and mutant isocitrate dehydrogenase glioma. *Neurooncol Adv.* 2022;4(1):vdac071. Radiologykeys: 13 Future Applications: Radiomics and Deep Learning on Breast MRI.

19. Migdadi L, Lambert J, Telfah A, Hergenröder R, Wöhler C Automated metabolic assignment: Semi-supervised learning in metabolic analysis employing two dimensional nuclear magnetic resonance (NMR). *Comput Struct Biotechnol J* 2021;19:5047–5058.

20. Sarma MK, Nagarajan R, Macey PM et al. Accelerated echo-planar J-resolved spectroscopic imaging in the human brain using compressed sensing: A pilot validation in obstructive sleep apnea. *AJNR Am J Neuroradiol* 2014;35(6 Suppl):S81–S89.

21. Marshall I, Thrippleton MJ, Bastin ME et al. Characterisation of tissue-type metabolic content in secondary progressive multiple sclerosis: A magnetic resonance spectroscopic imaging study. *J Neurol* 2018;265(8):1795–1802.

22. Filippi M, Rocca MA, Rovaris M Clinical trials and clinical practice in multiple sclerosis: Conventional and emerging magnetic resonance imaging technologies. *Curr Neurol Neurosci Rep* 2002;2(3):267–276.

23. Zuo J, Joseph GB, Li X et al. In vivo intervertebral disc characterization using magnetic resonance spectroscopy and T1ρ imaging: Association with discography and Oswestry Disability Index and Short Form-36 Health Survey. *Spine (Phila Pa 1976)* 2012;37(3):214–221.

24. Tartaglia MC, Arnold DL The role of MRS and fMRI in multiple sclerosis. *Adv Neurol* 2006;98:185–202.

25. Miller DH, Albert PS, Barkhof F et al. Guidelines for the use of magnetic resonance techniques in monitoring the treatment of multiple sclerosis. US national MS Society task force. *Ann Neurol* 1996;39(1):6–16.

26. Binesh N, Yue K, Fairbanks L, Thomas MA Reproducibility of localized 2D correlated MR spectroscopy. *Magn Reson Med* 2002;48(6):942–948.

27. Jung JA, Coakley FV, Vigneron DB et al. Prostate depiction at endorectal MR spectroscopic imaging: Investigation of a standardized evaluation system. *Radiology* 2004;233(3):701–708.

28. Boesch SM, Wolf C, Seppi K et al. Differentiation of SCA2 from MSA-C using proton magnetic resonance spectroscopic imaging. *J Magn Reson Imaging* 2007;25(3):564–569.

29. Kahleova H, Petersen KF, Shulman GI et al. Effect of a low-fat vegan diet on body weight, insulin sensitivity, postprandial metabolism, and intramyocellular and hepatocellular lipid levels in overweight adults: A randomized clinical trial. *JAMA Netw Open* 2020;3(11):e2025454.

30. Smits M MRI biomarkers in neuro-oncology. *Nat Rev Neurol* 2021;17(8):486–500.

31. Galanaud D, Haik S, Linguraru MG et al. Combined diffusion imaging and MR spectroscopy in the diagnosis of human prion diseases. *AJNR Am J Neuroradiol* 2010;31(7):1311–1318.
32. Reardon DA, Ballman KV, Buckner JC, Chang SM impact of imaging measurements on response assessment in glioblastoma clinical trials. *Neuro Oncol* 2014;16 (Suppl 7):vii24–35.
33. Auer DP In vivo imaging markers of neurodegeneration of the substantia nigra. *Exp Gerontol* 2009;44(1–2):4–9.
34. Bulik M, Kazda T, Slampa P, Jancalek R The diagnostic ability of follow-up imaging biomarkers after treatment of glioblastoma in the temozolomide era: Implications from proton MR spectroscopy and apparent diffusion coefficient mapping. *BioMed Res Int* 2015:641023.
35. Lombardo F, Frijia F, Bongioanni P. et al. Diffusion tensor MRI and MR spectroscopy in long lasting upper motor neuron involvement in amyotrophic lateral sclerosis. *Arch Ital Biol* 2009;147(3):69–82.
36. Griffiths JR, Tate AR, Howe FA, Stubbs M Group on MRS application to cancer magnetic resonance spectroscopy of cancer-practicalities of multi-centre trials and early results in non-Hodgkin's lymphoma. *Eur J Cancer* 2002;38(16):2085–2093.
37. Jacobs MA, Stearns V, Wolff AC et al. Multiparametric magnetic resonance imaging, spectroscopy and multinuclear (^{23}Na) imaging monitoring of preoperative chemotherapy for locally advanced breast cancer. *Acad Radiol* 2010;17(12):1477–1485.
38. Korteweg MA, Veldhuis WB, Visser F et al. Feasibility of 7 Tesla breast magnetic resonance imaging determination of intrinsic sensitivity and high-resolution magnetic resonance imaging, diffusion-weighted imaging, and (1)H-magnetic resonance spectroscopy of breast cancer patients receiving neoadjuvant therapy. *Invest Radiol* 2011;46(6):370–376.
39. Zuo J, Joseph GB, Li X et al. In vivo intervertebral disc characterization using magnetic resonance spectroscopy and T1ρ imaging: Association with discography and Oswestry Disability Index and Short Form-36 Health Survey. *Spine* 2012;37(3):214–221.
40. Lewis JF, McGorray SP, Pepine CJ Assessment of women with suspected myocardial ischemia: Review of findings of the Women's Ischemia Syndrome Evaluation (WISE) Study. *Curr Womens Health Rep* 2002;2(2):110–114.
41. Stivaros S, Garg S, Tziraki M et al. Randomised controlled trial of simvastatin treatment for autism in young children with neurofibromatosis type 1 (SANTA). *Mol Autism* 2018;9:12.
42. Kulyabin YY, Bogachev-Prokophiev AV, Soynov IA et al. Clinical assessment of perfusion techniques during surgical repair of coarctation of aorta with aortic arch hypoplasia in neonates: A pilot prospective randomized study. *Semin Thorac Cardiovasc Surg* 2020;32(4):860–871.
43. Magnotta VA, Heo HY, Dlouhy BJ et al. Detecting activity-evoked pH changes in human brain. *Proc Natl Acad Sci U S A* 2012;109(21):8270–8273.
44. Streeter CC, Gerbarg PL, Brown RP et al. Thalamic gamma aminobutyric acid level changes in major depressive disorder after a 12-week Iyengar yoga and coherent breathing intervention. *J Altern Complement Med* 2020;26(3):190–197.
45. Löbel U, Hwang S, Edwards A et al. Discrepant longitudinal volumetric and metabolic evolution of diffuse intrinsic pontine gliomas during treatment: Implications for current response assessment strategies. *Neuroradiology* 2016;58(10):1027–1034.
46. Goda JS, Dutta D, Raut N, et al. Can multiparametric MRI and FDG-PET predict outcome in diffuse brainstem glioma? A report from a prospective phase-II study. *Pediatr Neurosurg* 2014;49(5):274–281.
47. Vöglein J, Tüttenberg J, Weimer M et al. Treatment monitoring in gliomas: Comparison of dynamic susceptibility-weighted contrast-enhanced and spectroscopic MRI techniques for identifying treatment failure. *Invest Radiol* 2011;46(6):390–400.
48. Gonzalo N, Serruys PW, Barlis P et al. Multi-modality intra-coronary plaque characterization: A pilot study. *Int J Cardiol* 2010;138(1):32–39.
49. Chang L, Lee PL, Yiannoutsos CT et al. A multicenter in vivo proton-MRS study of HIV-associated dementia and its relationship to age. HIV MRS consortium. *Neuroimage* 2004;23(4):1336–1347.
50. Stern JM, Merritt ME, Zeltser I et al. Phase one pilot study using magnetic resonance spectroscopy to predict the histology of radiofrequency-ablated renal tissue. *Eur Urol* 2008;5(2):433–438.
51. Medical Advisory Secretariat. Functional brain imaging: An evidence-based analysis. *Ont Health Technol Assess Ser* 2006;6(22):1–79.
52. Kettelhack C, Wickede Mv, Vogl T et al. 31Phosphorus-magnetic resonance spectroscopy to assess histologic tumor response noninvasively after isolated limb perfusion for soft tissue tumorsmagnetic resonance spectroscopy to assess histologic tumor response noninvasively after isolated limb perfusion for soft tissue tumors. *Cancer* 2002;94(5):1557–1564.

53. Aragão Mde F, Otaduy MC, Melo RV et al. Multivoxel spectroscopy with short echo time: Choline/N-acetyl-aspartate ratio and the grading of cerebral astrocytomas. *Arq Neuro Psiquiatr* 2007;65(2A):286–294.

54. Kallén K, Burtscher IM, Holtås S, Ryding E, Rosén I 201Thallium SPECT and 1H-MRS compared with MRI in chemotherapy monitoring of high-grade malignant astrocytomas. *J Neurooncol* 2000;46(2):173–185.

55. Urdzik J, Bjerner T, Wanders A et al. The value of pre-operative magnetic resonance spectroscopy in the assessment of steatohepatitis in patients with colorectal liver metastasis. *J Hepatol* 2012;56(3):640–646.

56. Watanabe T, Shiino A, Akiguchi I Absolute quantification in proton magnetic resonance spectroscopy is superior to relative ratio to discriminate Alzheimer's disease from Binswanger's disease. *Dement Geriatr Cogn Disord* 2008;26(1):89–100.

57. Wallström J, Geterud K, Kohestani K et al. Prostate cancer screening with magnetic resonance imaging: Results from the second round of the Göteborg prostate cancer Screening 2 trial. *Eur Urol Oncol* 2021;5(1):54–60.

58. Schmuecking M, Boltze C, Geyer H et al. Dynamic MRI and CAD vs. choline MRS: Where is the detection level for a lesion characterisation in prostate cancer? *Int J Radiat Biol* 2009;85(9):814–824.

59. Bongiovanni A, Foca F, Oboldi D et al. 3-T magnetic resonance-guided high-intensity focused ultrasound (3 T-MR-HIFU) for the treatment of pain from bone metastases of solid tumors. *Support Care Cancer* 2022;30(7):5737–5745.

60. Kaufman MJ, Henry ME, Frederick Bd et al. Selective serotonin reuptake inhibitor discontinuation syndrome is associated with a rostral anterior cingulate choline metabolite decrease: A proton magnetic resonance spectroscopic imaging study. *Biol Psychiatry* 2003;54(5):534–539.

61. Kondo DG, Sung YH, Hellem TL et al. Open-label adjunctive creatine for female adolescents with SSRI-resistant major depressive disorder: A 31-phosphorus magnetic resonance spectroscopy study. *J Affect Disord* 2011;135(1–3):354–361.

62. Ramesh K, Eric A, Mellon EA et al. A multi-institutional pilot clinical trial of spectroscopic MRI-guided radiation dose escalation for newly diagnosed glioblastoma. *Neurooncol Adv* 2022;4(1):1–10.

63. Caivano R, Lotumolo A, Rabasco P et al. 3 Tesla magnetic resonance spectroscopy: Cerebral gliomas vs. metastatic brain tumors. Our experience and review of the literature. *Int J Neurosci*;123(8):537–543.

64. Sharma R, Narayana PA, Wolinsky JS Grey matter abnormalities in multiple sclerosis: Proton magnetic resonance spectroscopic imaging. *Mult Scler* 2001;7(4):221–226. doi: 10.1177/135245850100700402; PMID: 11548980.

65. Bedell BJ, Narayana PA, Johnston DA 3-dimensional MR image registration of the human brain. *Magn Reson Med* 1996;35(3):384.

66. Shoeibi A, Khodatars M, Jafari M et al. Applications of deep learning techniques for automated multiple sclerosis detection using magnetic resonance imaging: A review. *Comput Biol Med* 2021;136:104697.

67. Centers for Medicare & Medicaid Services (CMS) *Decision Memo for Magnetic Resonance Spectroscopy for Brain Tumors* (CAG-00141N). Baltimore, MD: CMS; January 29, 2004.

68. Ustymowicz A, Tarasow E, Zajkowska J, Walecki J, Hermanowska-Szpakowicz T Proton MR spectroscopy in neuroborreliosis: A preliminary study. *Neuroradiology* 2004;46(1):26–30.

69. Hollingworth W, Medina LS, Lenkinski RE, et al. A systematic literature review of magnetic resonance spectroscopy for the characterization of brain tumors. *AJNR Am J Neuroradiol* 2006;27(7):1404–1411.

70. Hallahan BP, Daly EM, Simmons A et al. Fragile X syndrome: A pilot proton magnetic resonance spectroscopy study in premutation carriers. *J Neurodev Disord* 2012;4(1):23. doi: 10.1186/1866-1955-4-23; PMID: 22958351; PMCID: PMC3443443.

71. Zakian K, Sircar K, Hricak H, et al. Correlation of proton MR spectroscopic imaging with Gleason score based on step-section pathologic analysis after radical prostatectomy. *Radiology* 2005;234(3):804–814.

72. Wetter A, Engl TA, Nadjmabadi D et al. Combined MRI and MR spectroscopy of the prostate before radical prostatectomy. *AJR Am J Roentgenol* 2006;187(3):724–730.

73. Wang P, Guo YM, Liu M, et al. A meta-analysis of the accuracy of prostate cancer studies which use magnetic resonance spectroscopy as a diagnostic tool. *Korean J Radiol* 2008;9(5):432–438.

74. Vedolin L, Schwartz IV, Komlos M, et al. Brain MRI in mucopolysaccharidosis: Effect of aging and correlation with biochemical findings. *Neurology* 2007;69(9):917–924.

75. Boesch SM, Wolf C, Seppi K, et al. Differentiation of SCA2 from MSA-C using proton magnetic resonance spectroscopic imaging. *J Magn Reson Imaging* 2007;25(3):564–569.

76. Dyke JP, Sanelli PC, Voss HU, et al. Monitoring the effects of BCNU chemotherapy Wafers (Gliadel) in glioblastoma multiforme with proton magnetic resonance spectroscopic imaging at 3.0 Tesla. *J Neurooncol* 2007;82(1):103–110.

77. Filippi M, Rocca MA, Arnold DL, et al. EFNS guidelines on the use of neuroimaging in the management of multiple sclerosis. *Eur J Neurol* 2006;13(4):313–325.

78. De Stefano N, Filippi M, Miller D, et al. Guidelines for using proton MR spectroscopy in multicenter clinical MS studies. *Neurology* 2007;69(20):1942–1952.

79. Keshari KR, Lotz JC, Link TM, et al. Lactic acid and proteoglycans as metabolic markers for discogenic back pain. *Spine* 2008;33(3):312–317.

80. Gornet MG, Peacock J, Claude J, et al. Magnetic resonance spectroscopy (MRS) can identify painful lumbar discs and may facilitate improved clinical outcomes of lumbar surgeries for discogenic pain. *Eur Spine J* 2019;28(4):674–687.

81. Benoist M The Michel Benoist and Robert Mulholland yearly European spine journal review: A survey of the "medical" articles in the European Spine Journal, 2018. *Eur Spine J* 2019;28(1):10–20.

82. Morrison WB, Dalinka MK, Daffner RH, et al. Expert panel on Musculoskeletal imaging. *Bone Tumors [Online Publication].* Reston, VA: American College of Radiology (ACR); 2005.

83. Shah N, Sattar A, Benanti M, Hollander S, Cheuck L Magnetic resonance spectroscopy as an imaging tool for cancer: A review of the literature. *J Am Osteopath Assoc* 2006;106(1):23–27.

84. Bartella L, Huang W, Proton (1H) MR spectroscopy of the breast. *RadioGraphics* 2007;27(Suppl 1):S241–S252.

85. Tse GM, Yeung DK, King AD, Cheung HS, Yang WT In vivo proton magnetic resonance spectroscopy of breast lesions: An update. *Breast Cancer Res Treat* 2007;104(3):249–255.

86. Bizzi A, Castelli G, Bugiani M, et al. Classification of childhood white matter disorders using proton MR spectroscopic imaging. *AJNR Am J Neuroradiol* 2008;29(7):1270–1275.

87. Umbehr M, Bachmann LM, Held U, et al. Combined magnetic resonance imaging and magnetic resonance spectroscopy imaging in the diagnosis of prostate cancer: A systematic review and meta-analysis. *Eur Urol* 2009;55(3):575–590.

88. Chuang MT, Liu YS, Tsai YS, Chen YC, Wang CK Differentiating radiation-induced necrosis from recurrent brain tumor using MR perfusion and spectroscopy: A meta-analysis. *PLOS One* 2016;11(1):e0141438. doi: 10.1371/journal.pone.0141438.

89. Weinreb JC, Blume JD, Coakley FV, et al. Prostate cancer: Sextant localization at MR imaging and MR spectroscopic imaging before prostatectomy -- Results of ACRIN prospective multi-institutional clinicopathologic study. *Radiology* 2009;251(1):122–133.

90. Lee CP, Payne GS, Oregioni A, et al. A phase I study of the nitroimidazole hypoxia marker SR4554 using 19F magnetic resonance spectroscopy. *Br J Cancer* 2009;101(11):1860–1868.

91. Beadle R, Frenneaux M Magnetic resonance spectroscopy in myocardial disease. *Expert Rev Cardiovasc Ther* 2010;8(2):269–277.

92. National Comprehensive Cancer Network (NCCN) *Central Nervous System Cancers.* NCCN Clinical Practice Guidelines in Oncology, version 1.2016. Fort Washington, PA: NCCN; 2016.

93. Sturrock A, Laule C, Decolongon J, et al. Magnetic resonance spectroscopy biomarkers in premanifest and early Huntington disease. *Neurology* 2010;75(19):1702–1710.

94. Horská A, Barker C Imaging of brain tumors: MR spectroscopy and metabolic imaging. *Neuroimaging Clin N Am* 2010;20(3):293–310.

95. Westphalen AC, Coakley FV, Roach M, McCulloch CE, Kurhanewicz J Locally recurrent prostate cancer after external beam radiation therapy: Diagnostic performance of 1.5-T endorectal MR imaging and MR spectroscopic imaging for detection. *Radiology* 2010;256(2):485–492.

96. Baltzer PA, Dietzel M Breast lesions: Diagnosis by using proton MR spectroscopy at 1.5 and 3.0 T -- Systematic review and meta-analysis. *Radiology* 2013;267(3):735–746.

97. Mowatt G, Scotland G, Boachie C, et al. The diagnostic accuracy and cost-effectiveness of magnetic resonance spectroscopy and enhanced magnetic resonance imaging techniques in aiding the localisation of prostate abnormalities for biopsy: A systematic review and economic evaluation. *Health Technol Assess* 2013;17(20):vii–xix, 1–281.

98. Gardner A, Iverson GL, Stanwell P A systematic review of proton magnetic resonance spectroscopy findings in sport-related concussion. *J Neurotrauma* 2014;31(1):1–18.

99. Harmon KG, Drezner JA, Gammons M, et al. American medical society for sports medicine position statement: Concussion in sport. *Br J Sports Med* 2013;47(1):15–26.

100. Mygland A, Ljostad U, Fingerle V, et al. European Federation of Neurological Societies. EFNS guidelines on the diagnosis and management of European Lyme neuroborreliosis. *Eur J Neurol* 2010;17(1):8–16:e1–e4.

101. Wang W, Hu Y, Lu P, et al. Evaluation of the diagnostic performance of magnetic resonance spectroscopy in brain tumors: A systematic review and meta-analysis. *PLOS One* 2014;9(11):e112577.

102. Wippold FJ II, Brown DC, Broderick DF, et al. *Expert panel on neurologic imaging. ACR Appropriateness Criteria® dementia and movement disorders [online publication].* Reston, VA: American College of Radiology (ACR); 2014.

103. Spencer AE, Uchida M, Kenworthy T, Keary CJ, Biederman J Glutamatergic dysregulation in pediatric psychiatric disorders: A systematic review of the magnetic resonance spectroscopy literature. *J Clin Psychiatry* 2014;75(11):1226–1241.

104. Wang H, Tan L, Wang HF, et al. Magnetic resonance spectroscopy in Alzheimer's disease: Systematic review and meta-analysis. *J Alzheimers Dis* 2015;46(4):1049–1070.

105. Voevodskaya O, Poulakis K, Sundgren P, et al. Swedish BioFINDER study group. Brain myoinositol as a potential marker of amyloid-related pathology: A longitudinal study. *Neurology* 2019;92(5):e395–e405.

106. Chen WS, Li JJ, Hong L, et al. Diagnostic value of magnetic resonance spectroscopy in radiation encephalopathy induced by radiotherapy for patients with nasopharyngeal carcinoma: A meta-analysis. *BioMed Res Int* 2016;2016:5126074.

107. Zeng G, Penninkilampi R, Chaganti J, et al. Meta-analysis of magnetic resonance spectroscopy in the diagnosis of hepatic encephalopathy. *Neurology* 2020 Mar 17;94(11):e1147–e1156.

108. Zheng D, Guo Z, Schroder PM, et al. Accuracy of MR imaging and MR spectroscopy for detection and quantification of hepatic steatosis in living liver donors: A meta-analysis. *Radiology* 2017;282(1):92–102.

109. Wang D, Li Y 1H magnetic resonance spectroscopy predicts hepatocellular carcinoma in a subset of patients with liver cirrhosis: A randomized trial. *Med (Baltim)* 2015;94(27):e1066.

110. Zhao X, Xu M, Jorgenson K, Kong J Neurochemical changes in patients with chronic low back pain detected by proton magnetic resonance spectroscopy: A systematic review. *NeuroImage Clin* 2016;13:33–38.

111. Zhang L, Li H, Hong P, Zou X Proton magnetic resonance spectroscopy in juvenile myoclonic epilepsy: A systematic review and meta-analysis. *Epilepsy Res* 2016;121:33–38.

112. Cevik N, Koksal A, Dogan VB, et al. Evaluation of cognitive functions of juvenile myoclonic epileptic patients by magnetic resonance spectroscopy and neuropsychiatric cognitive tests concurrently. *Neurol Sci* 2016;37(4):623–627.

113. Yang M, Sun J, Bai HX, et al. Diagnostic accuracy of SPECT, PET, and MRS for primary central nervous system lymphoma in HIV patients: A systematic review and meta-analysis. *Med (Baltim)* 2017;96(19):e6676.

114. Younis S, Hougaard A, Vestergaard MB, Larsson HBW, Ashina M Migraine and magnetic resonance spectroscopy: A systematic review. *Curr Opin Neurol* 2017;30(3):246–262.

115. Lai D, Sharma R, Wolinsky JS, Narayana PA A comparative study of correlation coefficients in spatially MRSI-observed neurochemicals from multiple sclerosis patients. *J Appl Stat* 2003;30(10):1221–1229.

116. Wu Y *Clinical Features, Diagnosis, and Treatment of Neonatal Encephalopathy. UpToDate [Online Serial].* Waltham, MA: UpToDate; reviewed May 2020.

117. Zou R, Xiong T, Zhang L, et al. Proton magnetic resonance spectroscopy biomarkers in neonates with hypoxic-ischemic encephalopathy: A systematic review and meta-analysis. *Front Neurol* 2018;9:732.

118. Veeramuthu V, Seow P, Narayanan V, et al. Neurometabolites alteration in the acute phase of mild traumatic brain injury (mTBI): An in vivo proton magnetic resonance spectroscopy (1H-MRS) study. *Acad Radiol* 2018;25(9):1167–1177.

119. Eisele A, Hill-Strathy M, Michels L, Rauen K Magnetic resonance spectroscopy following mild traumatic brain injury: A systematic review and meta-analysis on the potential to detect posttraumatic neurodegeneration 2020;20(1):2–11.

120. Kondratyeva EA, Diment SV, Kondratyev SA, et al. Magnetic resonance spectroscopy data in the prognosis of consciousness recovery in patients with vegetative state. *Zh Nevrol Psikhiatr S S Korsakova* 2019;119(10):7–14.

121. Finnell DS A clinical translation of the article titled, The utility of magnetic resonance spectroscopy for understanding substance use disorders: A systematic review of the literature. *J Am Psychiatr Nurs Assoc* 2015;21(4):276–278.

122. Frittoli RB, Pereira DR, Rittner L, Appenzeller S Proton magnetic resonance spectroscopy (1 H-MRS) in rheumatic autoimmune diseases: A systematic review. *Lupus* 2020;29(14):1873–1884.

123. Fernandez-Vega N, Ramos-Rodriguez JR, Alfaro F, Barbancho MÁ, García-Casares N Usefulness of magnetic resonance spectroscopy in mesial temporal sclerosis: A systematic review. *Neuroradiology* 2021;63(9):1395–1405.

124. Hovsepian DA, Galati A, Chong RA, et al. MELAS: Monitoring treatment with magnetic resonance spectroscopy. *Acta Neurol Scand* 2019;139(1):82–85.

125. Ng YS, Bindoff LA, Gorman GS, et al. Consensus-based statements for the management of mitochondrial stroke-like episodes. *Wellcome Open Res* 2019;4:201.

126. O'Ferrall E *Mitochondrial Myopathies: Clinical Features and Diagnosis. UpToDate [Online Serial].* Waltham, MA: UpToDate; reviewed December 2021.

127. Migdadi L, Lambert J, Telfah A, Hergenröder R, Wöhler C Automated metabolic assignment: Semi-supervised learning in metabolic analysis employing two dimensional nuclear magnetic resonance (NMR). *Comput Struct Biotechnol J* 2021 Aug 31;19:5047–5058. doi: 10.1016/j.csbj.2021.08.048; PMID: 34589182; PMCID: PMC8455648.

128. Wang Z, Li Y, Lam F High-resolution, 3D multi-TE 1 H MRSI using fast spatiospectral encoding and subspace imaging. *Magn Reson Med* 2022 Mar;87(3):1103–1118. doi: 10.1002/mrm.29015. Epub 2021 Nov 9. PMID: 34752641; PMCID: PMC9491015.

129. Li Y, Wang Z, Sun R, Lam F Separation of metabolites and macromolecules for short-TE ^{1}H-MRSI using learned component-specific representations. *IEEE Trans Med Imaging* 2021 Apr;40(4):1157–1167. doi: 10.1109/TMI.2020.3048933. Epub 2021 Apr 1. PMID: 33395390; PMCID: PMC8049099.

130. Li Y, Wang Z, Lam F. SNR SNR Enhancement for multi-TE MRSI using joint low-dimensional model and spatial constraints. *IEEE Trans Bio Med Eng* 2022 Oct;69(10):3087–3097. doi: 10.1109/TBME.2022.3161417. Epub 2022 Sep 19. PMID: 35320082; PMCID: PMC9514378.

131. Li X, Strasser B, Neuberger U et al. Deep learning super-resolution magnetic resonance spectroscopic imaging of brain metabolism and mutant isocitrate dehydrogenase glioma. *Neurooncol Adv* 2022 May 24;4(1):vdac071. doi: 10.1093/noajnl/vdac071; PMID: 35911635; PMCID: PMC9332900.

132. Li X, Strasser B, Jafari-Khouzani K et al. Super-Resolution Whole-Brain 3D MR Spectroscopic Imaging for Mapping D-2-hydroxyglutarate and Tumor Metabolism in isocitrate dehydrogenase 1-mutated Human gliomas. *Radiology* 2020 Mar;294(3):589–597. doi: 10.1148/radiol.2020191529. Epub 2020 Jan 7. PMID: 31909698; PMCID: PMC7053225.

133. Wang L, Chen G, Dai K Hydrogen proton magnetic resonance spectroscopy (MRS) in differential diagnosis of intracranial tumors: A systematic review. *Contrast Media Mol Imaging* 2022;May 18:7242192. doi: 10.1155/2022/7242192; PMID: 35655732.

134. Lundervold AS, Lundervold A An overview of deep learning in medical imaging focusing on MRI. *Z Med Phys* 2019;29(2):102–127. doi: 10.1016/j.zemedi.2018.11.002.

Index

For Product Safety Concerns and Information please contact our EU
representative GPSR@taylorandfrancis.com
Taylor & Francis Verlag GmbH, Kaufingerstraße 24, 80331 München, Germany